PRAISE FOR *HOW TO TEACH NATUR*

"From the beginning of time, we have been connected to nature, but for the first time in history, that connection threatens to be broken for most of an entire generation and perhaps generations to come. In this book and this work, I find hope. Laws and Lygren have created a powerful and practical resource for teachers to help children and adults discover, explore, love, and protect the natural world around them. Nature is magic, and everyone needs it."

—**Robert Bateman**, artist and naturalist

"Here is a natural way for parents and teachers to integrate the beauty and mystery of nature into the lives of children. The book itself is a work of art."

—**Richard Louv**, author of *Our Wild Calling* and *Last Child in the Woods*

"Charlotte Mason said that 'where science does not teach a child to wonder and admire it has perhaps no educative value.' John Muir Laws and Emilie Lygren have filled their book with the tools of nature journaling that can help us all get outdoors to wonder at and admire the nature around us. What a gift their book is!"

—**J. Carroll Smith**, EdD, Founder of the Charlotte Mason Institute, Professor of Education (Retired)

"This book beautifully demonstrates how nature journaling is a valuable, multidimensional approach to science. Through this work, students build neural pathways that enable different brain regions to communicate with each other, resulting in more flexible thinking and creative problem solving in all areas of life. This approach to nature journaling also offers students an excellent way to learn and practice a growth mindset through nature and science."

—**Jo Boaler**, Professor of Education and Equity, Stanford University; author of *Mathematical Mindsets* and *Limitless Mind*

"The writing, the illustrations, the *demonstration* of an intentional, informative, and wonderfully educational approach to nature journal writing—it's all here. From field activities with clear examples to an inquiry-based approach to learning, this book is a gem for educators, first-time journalers, and for the experienced naturalist as well. It's art, it's writing, it's education, it's science, naturally so."

—**José González**, Founder of Latino Outdoors

"In *How to Teach N*_____ _____ practices for facilitating intentional and inclusive strategies for nature journaling with students. This can spark a lifelong wonder of the natural world."

—**Arvolyn Hill**, Coordinator of Family Programs, Children's Adventure Garden at the New York Botanical Garden

"Engaging students in learning about nature is a compelling need in these turbulent times. Just as important is communicating clearly, precisely, and truthfully. Naturalists and educators John Muir Laws and Emilie Lygren help teachers achieve these goals with *How to Teach Nature Journaling*. This is a wonderful response to educators' request for meaningful ways to implement contemporary standards. The book is useful, full of activities, and manageable for classroom teachers."

—**Rodger W. Bybee**, Executive Director, BSCS Science Learning (Retired)

"*How to Teach Nature Journaling* offers step-by-step processes to reach all learners through shared experience. This can be a transformative way for students to learn to communicate and express how they view the natural world."

—**Michelle Peres**, Environmental Educator, NatureBridge

"Imagine if every teacher taught nature journaling, every student spent enough time outside to fill a journal each year, and our kids were becoming writers, artists, and scientists all at once. I'd send my kid to that school. Thanks to Emilie Lygren and Jack Laws, we now have the vision and the tools to make this our reality."

—**Craig Strang**, Associate Director, Lawrence Hall of Science

"Nature journaling is the single most valuable tool and skill a teacher can possess and use to move the Next Generation Science Standards and environmental literacy forward in their school. You need to do this!"

—**Rob Wade**, Science & Outdoor Education Coordinator, Plumas Unified School District/ Plumas County of Education

HOW TO TEACH
NATURE JOURNALING

HOW TO TEACH
NATURE JOURNALING

CURIOSITY • WONDER • ATTENTION

Written by **John Muir Laws** and **Emilie Lygren**

Illustrated by **John Muir Laws**

Foreword by **Amy Tan**

Heyday, Berkeley, California

Cover Art and Design: John Muir Laws
Back Cover Design: Marlon Rigel
Interior Design: Leigh McLellan Design
   with John Muir Laws and Emilie Lygren
Composition: Leigh McLellan Design

Published by Heyday
P.O. Box 9145, Berkeley, California 94709
(510) 549-3564
heydaybooks.com

Printed in China

10 9 8 7 6 5 4

Permissions

All reasonable attempts were made to locate the copyright holders for the materials published in this book. If you believe you may be one of them, please contact Heyday, and the publisher will include appropriate acknowledgment in subsequent editions of this book.

"New Year's Eve: The Rim" journal entry copyright © 2017 by Terry Tempest Williams. Photograph of Williams copyright © 2014 by Zoë Rodriguez Photography. Zoë & Robert Swift Rodriguez.

Photographs on pages 2, 16, 24, and 85 were taken by Éli Zaturanski Photography.

Photographs on pages x, 1, 24, 50, 158, 168, and 195 are used with permission from the BEETLES Project at the Lawrence Hall of Science, University of California, Berkeley, and are sources from video footage taken by Group 5 Media and the Lawrence Hall of Science.

Photographs on pages 13 (upper), 114 (upper), 118, and 252 were taken by Robb Hirsh.

Photograph on page 49 © fatcamera/istockphoto.com.

Photograph on page 120 © Steve Hymon/dreamtime.com.

Photograph on page 209 © by Lara G./istockphoto.com.

*We would like to dedicate this book
to our parents, Bob and Beatrice Laws,
and Rolf Lygren and Katherine Heller.
Thank you for taking us outside.*

CONTENTS

APPENDICES 257

FOREWORD by Amy Tan

BEFORE WE HAD WORDS, we had wonder. Our infant eyes took in new shapes and colors. All sounds and odors were unknown. Gradually, we observed what was familiar and what was still strange. As we grew older, our young eyes puzzled over a bug missing a leg, a blade of grass with a watery bead, a tree so tall we could not see where it ended. As first-grade scientists, we touched the backs of little frogs to make them jump. We poked ladybugs and watched them tuck themselves into their polka-dot capes. We ate mashed clover and hay to decide if we liked them as much as cows do. And we saw dead things: flies on the sill, a skunk on the road, and perhaps a pet turtle, parakeet, dog, or cat, whose sudden loss was nearly the end of our world. Through animals, we learned about goneness and grief.

As we grew older, we stopped asking as many questions, in part because we came to believe that smart kids had answers and dumb ones had questions. When we became full-fledged adults, we could use precise words to explain our experiences with nature: *Species. Raptors. Plumage. Foliage. Migration. Environment. Evolution.* They were concepts defined by details and facts. With labeled concepts, we no longer wondered as much about nature and creatures. They became as common as sunsets, and we could see them at the end of any day, should we choose, and mostly we did not. Because we were responsible adults who did productive work, we no longer had the curiosity or time to watch a troop of ants ferrying crumbs across the floor. If we needed to know something about nature—say, the reason birds do not plummet from trees while asleep—we could find the answer in a book.

• • •

When we were still waddlers, we scribbled freely on large sheets of paper. Our crayons traveled off the edges. There were no boundaries to where our eyes and fingers could roam, until we learned to make our pictures nice and neat. In kindergarten, we saw that some kids drew better than others; their drawings received praise from adults, while ours languished unnoticed. We became self-conscious out of fear of being laughed at. So it was for me. According to my father, from the time I was three, I drew pictures and made up stories to accompany them. By the age of nine, I secretly wanted to become an artist and drew pictures in the privacy of my bedroom. When I was twelve, a great-uncle who was a landscape artist looked at my sketch of a girl, and I did not cry when he corrected it by drawing black lines all over her face with a thick piece of charcoal. By age fourteen, I saw that a boy in art class could draw cartoons just like those seen in comic books. When I was seventeen, a teacher told me I had no imagination, which he said was necessary to be creative. After that, I stopped drawing. But from time to time, I told myself that one day, after I had retired from my professional career, I would return to drawing. I imagined I would have a little room

in an attic, filled with easels, paints, paper, and a peaceful view of water and trees.

When I was sixty-three, I realized that the imaginary room in the attic had been empty long enough. I would never retire as a writer, but it was not too late to learn how to draw. So I started by attending workshops in nature journaling led by a naturalist artist, author, and educator named John Muir Laws—Jack, as I later came to know him. His name was already familiar to me because I owned his guidebook on wildlife in the Sierras. I soon bought his other books: *Laws Guide to Drawing Birds* and *Laws Guide to Nature Drawing and Journaling.* I practiced techniques. I practiced drawing nearly daily, putting in my "pencil miles," as Jack called it. But it was through Jack's field trips that I learned that the more important skills I needed to develop were the ones I had left behind in childhood—to be intensely curious, to wonder aloud, to see the story in front of me and try to capture an interesting aspect of it in a drawing. There is a big difference between drawing a bird with exact details while at home and capturing its essence in the field, in the moment, when it is never still. My lively subject moved every second, and that forced me to see it afresh every second, and not based on what I already assumed it looked like. I had to see it as a child who has not seen this creature before. Wonder takes place when everything is new.

On one of those field trips, I met someone who showed me how to regain my childhood sense of wonder. Her name is Fiona Gillogly, and she was thirteen at the time. A page from her journal is within this book and shows how our noticing infinite variations in nature enlarges our view of life. On field trips, I noticed that she was excited about everything—and not just the beautiful birds we all saw. She turned over the undersides of ferns as we walked into a woodland forest. She crouched down to show me a clump of California manroot and traced how far the vines extended and coiled around other plants. When making such discoveries in the field, she crammed her pages with questions about mysteries that lead to more mysteries, unmindful that her sentences are never formed in blocks of straight lines, left to right. Her observations flow continuously, curving upward or downward, as if to avoid interrupting her train of thought. John Muir Laws has been her mentor, and although I am now sixty-seven and she is now sixteen, she has become one of mine. I am learning the joys of unbounded curiosity. Everything is again new.

I think the readers of this book—parents, teachers, and their young children or students—are much like Fiona and me. They are partners in wonderment. Curiosity provides a basket into which they can place endless questions and observations. They are in suspense about what will happen next. They take turns pointing out what they notice and what it reminds them of. No question is silly. No drawing is poor. No observation is wrong. They all exult when the next thing that happens is completely

unexpected. The clouds shift, and a phalanx of pelicans flies through the fog into sunshine. The lichen in the forest glows. A fledgling junco makes a crash landing but is unhurt. Nature journaling is happy making. What better gift of love can a parent give to a child on a daily basis? What better gift of learning can a teacher give to young students?

I think of nature drawing as a spiritual connection to nature, and nature journaling is a written testament of miracles in the wild. Each day, I wake with curiosity over what is happening in my yard. Each day holds discoveries that I write down. The sounds that juvenile birds make when they are unsuccessful in finding food. The stretchmark patterns of bark on an oak tree. The glints of gold on the bay that I mistook to be a school of anchovies. The tangled fingers of beached bull kelp drying into sculpture. Each day I can do what I loved as a child: put pencil to paper to capture life, whether a detailed rendering with colored pencils or a lively sketch in the moment noting mysteries while sitting on a log. Through the practice of nature journaling, my fear of making mistakes is gone. I have abandoned my lifelong need for perfection. I am freed from the rusty rules based on *can't* and *don't* and *won't*. My brain is more flexible. In fact, scientific research proves that active learning through nature journaling can change the brain and boost intelligence. It makes sense. If kids are free to wonder aloud without feeling dumb or tested, they remain engaged. If they are happy in what they are doing, their attention span grows. By noticing how they feel when they experience something new, they absorb ideas more quickly. By being excited with what they've created, their memory expands and becomes the wellspring for future learning. Imagine it: Whether we are six or sixty, we can forge a new brain path that goes beyond former dead ends.

In one of Jack's books, I read something profound that changed the way my brain thinks. "As you draw the bird," he writes, "try to feel the life within it." So now I look at the bird before me and imagine how it senses the world, how it feels breathing cold air, how it feels to have its feathers ruffling in the wind, how it feels to always have an eye out for possible food and possible predators. The bird sees me and is a nanosecond from flying off, but it stays. Why? By imagining the life within, the bird I am drawing is alive, no longer a shape and its parts, but a thinking, sentient being, always on the brink of doing something. By feeling the life within, I am always conscious that all creatures have personalities, and so do trees and clouds and streams. To feel the life within, I now imagine myself as the bird that is looking at me. I imagine its wariness, the many ways it has almost died in its short life. I worry over its comfort and safety, and whether I will see my little companion the next day, the next year. To feel the life within is to also feel grief in the goneness of a single creature or an entire species. Imagination is where compassion grows.

Let us join with children to imagine and wonder, to use curiosity as the guide to miracles in plain sight. Let us enter with them into wild wonder so that we become guardians together of all that is living and all that must be saved.

INTRODUCTION

WHY WE TEACH NATURE JOURNALING

EMILIE: I've kept a journal for most of my life. Over the years, I've gone for walks and recorded observations of nature or people that often turn into poems. The almost daily practice is my way of slowing down, learning, and making meaning of my life.

Curiosity is at the root of this practice and of my work as an educator, writer, and naturalist. And I've become even more curious by keeping a nature journal.

As an educator in Pescadero, California, at San Mateo Outdoor Education and Exploring New Horizons Outdoor Schools, I used nature journals in nearly every lesson with my fifth and sixth graders. If we breathlessly observed a hawk catching a snake, we journaled about what we saw, then studied up on animal behavior. We catalogued shells we found at the beach and read books about mollusks to learn more about who used to inhabit them. By using nature journaling, I got to work with students who were engaged and excited to make their own discoveries. Building my own teaching on what students recorded in their journals sent them the message that their observations, ideas, and questions mattered.

Although I'd kept written records of nature observations for much of my life, I never gave much thought to drawing until I met Jack Laws at a field station in the Sierra Nevada in 2009. Jack talked about drawing as a tool for deepening observations, finding beauty, and creating lasting memories—I used writing to reach the same goals. In our first few meetings, Jack and I spoke about our different, yet compatible, approaches to nature observation. Although Jack's mastery of drawing could have been intimidating, his enthusiasm and approachable teaching methods shoved me (as they have so many others) past my initial protest of "But I'm not an artist!" I started to sketch and draw more in my nature journal. Those early conversations evolved into a rich, rewarding collaboration focused on teaching nature journaling, one that has spanned the last 10 years.

Over the course of those 10 years, Jack and I have led dozens of teacher workshops, and I collaborated with Jack on *The Laws Guide to Nature Drawing and Journaling*. In 2015, I joined the BEETLES team at the Lawrence Hall of Science at UC Berkeley, where I develop curriculum based on research about how people best learn outdoor science. Since joining BEETLES, I have co-authored more than thirty student activities and several volumes of resources for instructors and organization leaders, all focused on teaching in the outdoors. My work with BEETLES has deepened my understanding of how to teach observation and critical thinking skills, how to use discussions and thoughtful sequencing to support the learning process, and how to encourage participation from all members of a group.

I've also studied the intersections between social emotional learning and outdoor education, which has helped me create learning experiences that develop students' self-confidence and build community. Working with educators and developing BEETLES activities have given me a deep understanding of what teachers need to create successful learning experiences for students. This understanding underpins and informs the approaches in this book.

And all the while, Jack and I have stayed in conversation about our shared approach for teaching nature journaling that forms the foundation for this book.

JACK: When I was a child, my parents inspired me to explore nature and to keep illustrated journals of my discoveries. My journals stoked my curiosity and accelerated my learning. I was and am dyslexic. In those days, I struggled academically. Nature was my refuge, and my journals were a safe place for me to wonder, think, write, and draw without fear of judgment or criticism. As my interest in natural history grew, so did my library of journals, filling shelves in my bookcase.

In 1990, I began work at the Walker Creek Outdoor School in Marin County, California. I would teach groups of fourth-through eighth-grade students about ecology and natural history through observation and drawing. I found that my students also loved to explore and journal just as I did. I could not have asked for better conditions to refine my nature drawing lessons. I learned what worked and what didn't with a new group of students every week, and I experimented and modified my lesson plans in response.

When I presented these nature drawing lessons at a conference for outdoor educators, they were enthusiastically received. The activities authentically connected students with nature, easily fit into the structure of outdoor school programs, and could be taught by educators of all levels, even those who did not consider themselves to be artists. I expanded these lessons and techniques, working with elementary, high school, and college students, and adults, while at the California Academy of Sciences. I also started the Nature Journal Club, leading monthly intergenerational workshops and field trips throughout the San Francisco Bay Area. The club is still going and growing, with branches formed throughout the US and worldwide.

I saw that these activities could be useful for everyone, not just those in the Bay Area who could attend my workshops. So, in 2010, I worked with the California Native Plant Society and Emily Breuning to write *Opening the World through Nature Journaling*, a free downloadable curriculum. Educators of all stripes—from homeschooling parents to science teachers to camp counselors—loved the student-led, inquiry-based lessons. In 2012, we further expanded and improved the lessons, publishing the second edition with writers and educators Emilie Lygren and Celeste Lopez. This book is rooted in that curriculum, but expands on it significantly to include around a dozen new activities, more strategies for supporting student engagement in the outdoors, deeper connections to inquiry-based teaching methods, discussion questions to follow every activity, advice on how to integrate journaling into longer lessons, additional teaching tips and guidance for teachers, and more.

Emilie Lygren was also instrumental in helping me with *The Laws Guide to Nature Drawing and Journaling,* whose philosophy and methods this book shares. Whereas that book supports readers in developing their own nature journaling practice, this one is meant to be a comprehensive guide to sharing nature journaling with others, especially young people, and to using the practice as an essential tool to support learning and teaching.

I love looking through other people's journals to see how they see the world and think. When I see interesting approaches or new ideas, I try to incorporate them into my own journaling—scavenging for fresh approaches keeps my process fresh and my journal ever changing. Emilie's journals and process have had a profound impact on the way I use my journal. Her writing showed me how to use words in a rich and nuanced way to document observations and record my thoughts and feelings. This has helped me record and reflect on my experience in nature and has made me a better naturalist. As an educator, Emilie has pushed me to a deeper understanding of equity in the classroom and to question my own biases.

EMILIE AND JACK: In our ongoing collaboration to bring nature journaling into children's lives, we have continued to refine our approach to teaching journaling. Our understanding of best practices in education, science communication, place-based learning, and engaging students with nature journals has grown. We teach workshops, lead field excursions, and share resources and ideas. This book is one of the fruits of that work together.

While researching this book, we interviewed working scientists about how they use journals. We studied the notebook pages of naturalists and thinkers from Leonardo da Vinci to Charles Henry Turner, Nikola Tesla to Terry Tempest Williams, looking for patterns in how they captured ideas and built meaning. Then a group of nearly one hundred educators, teachers, and homeschool parents field-tested our activities to make sure they were easy to lead and engaging for students.

We've heard educators say again and again that they see the value of nature journaling as a holistic, interdisciplinary teaching tool that is deeply engaging and meaningful. We've also heard a clear call for more support on how to teach nature journaling and integrate the practice into many different educational contexts. In teacher workshops, we're often asked to speak about how to manage groups in the outdoors, teach drawing skills, connect journaling to educational standards and frameworks, and move from stand-alone journaling activities to incorporate journaling into longer lessons and extended learning experiences. This book puts together our lessons, advice, and experience so that you can leap into journaling with students, confident that you've got a clear road map guiding the way.

Classroom teachers will find that the activities and follow-ups easily support the Common Core State Standards for English Language Arts & Literacy in History/Social Studies, Science, and Technical Subjects and the Next Generation Science Standards. Homeschool families and forest schools, who know nature journaling well, will find tools and activities to deepen their practice and help youth improve at journaling over time. Outdoor and informal educators will find learning experiences that engage students directly with their surroundings.

Along the way, we have also trusted our own sense that this book is important. It's the first comprehensive book devoted to helping educators use nature journaling to teach, but in a deeper sense, it's so important to us because it's also about how to arrive at a place of joy through teaching and learning in nature.

There is wonder, intrigue, and insight around every corner, if you know how to look. This is what sits at the core of our own practice and informs every aspect of our teaching as well. Nature journaling is one way we've found to connect to a place of joy within ourselves as we encounter the wide and wondrous world. We sincerely hope this book does the same for you and the young people in your life.

WHY NATURE JOURNALING?

Nature journaling is an extremely effective and engaging way to teach observation, curiosity, and creative thinking. Journals are the ubiquitous tool of scientists, naturalists, thinkers, poets, writers, and engineers. Using a journal is a skill that can change students' lives forever.

Children need nature. Contact with the natural world improves health and reduces stress. Nature is also a rich and meaningful place to learn. There is no computer program that can replicate the excitement of seeing a squirrel up close, the intrigue of studying leaves, or the calm of watching clouds. Our goal in creating this book is to help children and adults discover (and rediscover!) the natural world through a combination of art, writing, and science.

Journaling can be a foundational practice in classrooms, homeschool families, community organizations, and outdoor education programs. This interdisciplinary approach engages students of all ages and inspires them to be keen observers of wild places in their backyard and beyond.

TO OBSERVE AND LEARN

Journaling deepens our observations, thinking, and memory. Journal entries that include words, pictures, and numbers lead the journaler to think in different ways and make a more complete record of what they see.

Writing strengthens our thinking because we have to organize our thoughts as we put them down on the page.[1] As we describe a squirrel's behavior or the shape of an insect's wing, we articulate and clarify our ideas. Doing this also helps us form stronger memories than if we had only witnessed the event or even taken notes on a computer. (The physical action of writing cements memories better than tapping keys does.[2])

Drawing leads to close, careful observation and improved memory.[3] When we draw, we must look again and again at the least familiar parts of a subject, paying careful attention to structures and shapes. This leads us to notice biologically significant details, such as the angle of stems on a branch or the shape of a bird's beak, features we could easily miss in a written account alone.

Using numbers helps us make different kinds of observations and reveals significant patterns. Counting the spines on leaves or measuring distances between gopher holes will reveal spatial relationships and underlying processes rich for study. This process of quantifying observations will lead us to questions we wouldn't have thought to ask.

Combining writing, drawing, and numbers on a journal page creates a dynamic and rich learning experience. Focused journal entries give structure to observation and help form lasting memories. Students' observations of plants in the community garden, birds on a pond, or icicles dangling from the eaves of their classroom form the foundation of learning over the course of a lesson.

If students compare two oak trees in a journal entry, they generate focused, biologically relevant observations and questions about oak trees. After journaling, students can share observations and ideas, then learn new science concepts that will help them explain what they observed. They can turn to books and other sources of information to learn more, or design investigations based on their questions. This is active and engaged learning bound to real-life experiences.

The activities in this book support a wide variety of content learning goals. Many earth, life science, and human-impacts topics can be taught outside, beginning with direct

WHAT DO WE MEAN BY "NATURE"?

"Nature" is more than pristine wilderness. It is all around us, from ranchland to urban parks, community gardens to backyards, school playgrounds to national forests. Nature journaling is about observing whatever is in front of us. Each location is a different opportunity for learning. The leaves on an oak in the center of the city have as much to teach us as those at the base of a mountain. The approaches in this book are designed to work in a range of ecosystems and settings. Educators of any kind can use journaling to engage their students with wild places wherever they are.

observation of plants, animals, and processes in your area. This kind of learning matches what is called for by the Next Generation Science Standards, fulfills many of the expectations of the Common Core State Standards, and supports the goals of other education models, including homeschools, Montessori, and Waldorf schools.

Nature journaling encourages participation from all students. When you ask, "What are some differences between these two oak leaves? What did the leaves remind you of?" answers will come flooding back because every student had access to the same set of experiences and will have observations and ideas to contribute to the conversation.

TO BUILD TRANSFERABLE THINKING SKILLS

Journaling helps students think critically and creatively for themselves. The transferable practices and learning habits that students use in their journals will increase their success in other academic disciplines. Students can also use their journaling to become active community members and engaged citizens of the world.

> "The ultimate goal…is to help students take over the reins of their learning."
> —Zaretta Hammond[6]

Those who regularly write and sketch become better learners.[4] Drawing, writing, making observations, and analyzing those observations build brainpower. Combining pictures with text reinforces visual literacy and language acquisition.[5] The abilities to articulate ideas, connect text to pictures, and engage in critical thinking are all reinforced by a consistent journaling practice.

TO BUILD SCIENTIFIC SKILLS AND THINKING

Using the Universal Tool of Science

When we think of science tools, we imagine microscopes and test tubes. But a journal is the foundational, essential tool for naturalists and scientists. In all branches of study, in all phases of an investigation, a journal or science notebook is indispensable. Where else can you record lists, map investigations, tease out questions, meticulously record methods, gather data, and compile notes written during the day? What a scientist doesn't write down or draw won't be remembered. Much of what we know about how the natural world works, the patterns and interactions of species and landscapes, has at one time been observed and recorded in a journal.

Before technological advances and the development of data collection instruments, naturalists and engineers used paper and pencil to capture important observations in the field. Even with cameras, computers, and other tools that record data with the click of a button, many field scientists continue to use a journal.

Journals remain an essential part of a scientist's tool kit because of the clear benefits that come from drawing, writing out ideas, and putting thinking down on paper.

Engaging Students in Real Science Practices

Nature journaling is one way to make the discipline of science more accessible to all students. This authentic use of a real scientific tool helps students understand what science is and to think of themselves as scientists.

We've seen students make this connection when they journal. One early autumn day, we were working with students at a local park near Santa Cruz, California, and prompted them to look for patterns and study the landscape. They made sketches comparing different kinds of trees, wrote observations and questions, then looked through their questions to figure out which ones could be answered through making more observations in the moment and which ones would require more extended study. As teachers, we stepped back and watched as the students gathered in small groups to excitedly share what they had noticed. They talked through their questions, discussed ideas, debated possible methods of study, and occasionally bounced back to a tree or leaf to make further observations. At one point a student looked up and said, "Isn't this kind of like what researchers in college get to do, except more fun?" Another said, "This makes me want to do science when I'm older!"

These students didn't think of themselves as "into science" before this experience. Engaging authentically in journaling and practices of science made the discipline visible and accessible to them. When students understand how science is done and see it as a process for learning, not a body of facts they must memorize, they can become more interested in science classes and learning experiences. And of course, just as true inquiry is an exciting process for students, "real scientists" at universities are having fun making discoveries too!

TO CONNECT WITH NATURE

Nature is good for us. A large body of research shows that spending time in green open space has positive effects on heart disease, high blood pressure, asthma, diabetes, and obesity, as well as on many indicators of health. Exposure to nature also improves mental health, reducing stress, anxiety, and the expression of behavioral disorders in adults and children.[7] To reap these health benefits, we can aim to spend at least two hours in natural areas over the course of each week.[8]

If we offer consistent opportunities for children to engage in nature journaling, they not only meet their recommended allowance of nature time but also do so in a way that will build a lifelong affinity and connection to the outdoors.[9] Children who

STRUCTURING ACTIVITIES

If you shoo a group of students who are new to journaling outside and say, "Go journal," they are likely to get overwhelmed or discouraged. The natural world offers so many possible subjects that it can be difficult to decide what to focus on. Students will understandably fall back on using familiar and comfortable ways of recording information (writers would write, drawers would draw). Students, no matter their age, need structures to direct their attention and scaffolds to help them decide how to record information in their journal.

But too much structure is also not effective. Worksheets and "fill in the blank" exercises don't make room for students' observations and ideas—the heart of authentic journaling and learning.

In our teaching, we've found that we can best help students grow by teaching them new techniques to record information and new ways to focus their observation and thinking. When they are just starting out, we choose a subject for them. As they become more advanced, they can choose the appropriate techniques, ways to focus, and natural subjects themselves.

The activities in this book use three types of scaffolding:

1. A part of nature to study—for example, leaves, stream currents, cracks in the mud, a flock of birds, a spider. A defined focus enables students to spend their time making observations, not figuring out what to observe.

2. A focus or goal for observation and thinking—for example, making comparisons, mapping distribution, focused study of a species, timed observations.

3. Some strategies for recording information—for example, options for page layout (such as dividing the page in half for the activity Comparison [p. 45]); including labels with a drawing; types of drawings or views to show (e.g., cross sections); ways to integrate words, pictures, and numbers on the page.

ELEMENTS OF A NATURE JOURNAL

Pictures
- Icons to show weather
- Drawings at different scales

Observations, Ideas, and Thinking
- Comprehensive metadata
- Notes about colors

Numbers
- Objects are counted
- Scale is shown with relative size

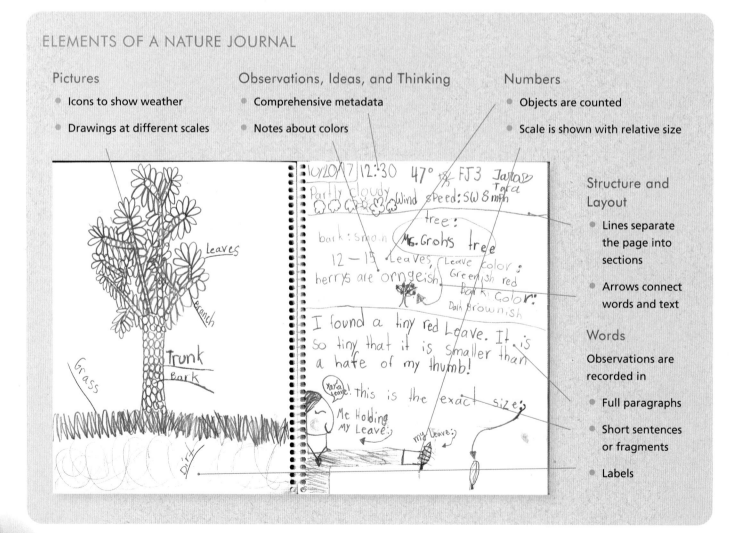

Structure and Layout
- Lines separate the page into sections
- Arrows connect words and text

Words

Observations are recorded in
- Full paragraphs
- Short sentences or fragments
- Labels

have meaningful nature experiences and exposure are more likely to continue to spend time outdoors and advocate for nature as adults.

TO PRACTICE MINDFULNESS

Beyond helping us discover joy in the practice of science and in learning, journaling offers us a way to stay present in the moment. As we journal, we slow down and focus on observation. This state of mind is similar to that at the center of many mindfulness practices, which have been shown to promote equanimity and decrease anxiety. Students with a regular nature journaling practice can build a relationship with a quiet, calm place in their mind. Journaling is an activity they can continue to use throughout their lives to slow down and be present.

Paying close attention to the world around us, instead of treating it as a mere backdrop, connects us with nature. Each subject of a journal entry becomes familiar, a touchstone of memory. The tree on the sidewalk becomes that tree you spent an hour with, not just an object you pass by on your way out the door. What we haven't yet observed is seen as an opportunity for study and intrigue. The grass along the sidewalk is a place of possible discovery, each bird out the window an invitation to slow down and look.

TO BUILD COMMUNITY

Shared experiences of journaling and being in the outdoors build connection and community. They deepen the experience of being alive, focus attention, and lead to feelings of awe and wonder. These experiences can bring groups together.[10]

Finding Connection through Observation

At a nature journaling workshop during a conference for science teachers, we started by looking at leaves. Participants picked up a leaf, shared observations out loud, then drew and described it in their journals. At the end of the 20-minute exercise, no one wanted to put their leaf back.

"I like this leaf now!"

"Yeah, it means something to me now that I've spent so much time looking at it."

"If we put the leaves back, people will just step on them when they walk by, and never look at them."

One participant tucked his leaf into his name tag, and the rest of the group followed suit. In the short time spent observing the leaves, we had become connected to them. Our attention transformed our relationship with the leaves.

This shared experience also forged connections among the workshop participants. At the end of the day, participants shared that they felt connected to the group because we had seen and observed so much together. We decided we were "leaf buddies," and for the rest of the conference, we kept our leaves tucked into our name tags. Every time we saw a fellow participant, we'd hold up our leaves and smile.

The simple process of observing and learning together built a bond between people who had spent less than a day together. Something seemingly small—a leaf—became an emblem of community. This type of experience was not hard to create. We didn't even leave the plaza in the middle of the hotel conference center. All that was required was shared time observing, wondering, and learning together. A shared practice of journaling can enrich the connections between place and people in classrooms, after-school programs, religious groups, and community organizations alike. Seeking connectedness is also a goal that can support students' science learning. As students find joy and excitement in learning together, they become excited to learn more.

> "If the point were making big discoveries or pretty pictures every time, I'd have quit long ago. The point is waking up, paying attention, noticing and exploring. The point is remembering, again and again, that the world we inhabit is infinitely rich, complex, fascinating, and beautiful. The point is learning from the world and renewing a sense of belonging within it."
>
> —Sarah Rabkin

Including Personal and Cultural Perspectives

Students come to nature journaling with different experiences and different cultural perspectives on their relationship to nature and on being in the outdoors. As teachers, we take time to learn about the experiences of different cultures in the outdoors so that we can be more responsive to our students and make the practice accessible. We also encourage students to share their personal and cultural perspectives in their journals, validating their experiences. One simple way to invite students to connect their new nature journaling experience to their lives is the phrase "It reminds me of." When students record their perspectives alongside their scientific observations, it will enrich their experience and lead to deeper insights and stronger memories. We teach students to distinguish between their personal perspectives and their factual observations, and we welcome both.

When students share personal perspectives with one another, they learn from others' knowledge and experiences. Students who regularly share their experiences and listen to and respect those of others have deeper conversations and learn to work well together. This builds community and student participation.

TO EXPERIENCE AWE AND WONDER

We may live surrounded by a beautiful community, or travel to the farthest corners of the earth, but if we are not present and fully awake in these experiences, we will miss most of what they have to offer. If we know how to pay attention, we can find wonder, novelty, insight, and new aspects of even the most commonplace events that fill our days.

Love itself is an act of sustained, compassionate attention. Think of deep relationships with your loved ones, the love of a special place that you have known over time, or the richness of your favorite pieces of music when you focus on the structure and layers of sound. All these relationships grow and develop through attention. Learning to pay attention allows you to open yourself to love in all aspects of your life—in your family, your community, nature, and the world.

In *The Sense of Wonder*, Rachel Carson wrote: "One way to open your eyes to unnoticed beauty is to ask yourself, 'What if I had never seen this before? What if I knew I would never see it again?'"[11]

Imagine how carefully we might attend to every second if we were seeing a friend for what we knew would be the last time. Our attention to every nuance would be heightened. We would be focused on taking in the moment as fully as possible.

As mundane as everyday objects and occurrences may seem, each moment has never existed in history and never will again. We might have seen robins before, but we have never seen *this* robin, on *this* day. It is easy to slip back to inattention and to forget this impermanence. Each moment, each observation, is a singular gift that will never come again. The world is infinitely complex, mysterious, deep, and unknown. William Stafford said, "If we only knew how to listen better, even the grasses by the roadside might have something to tell us about how to live our lives."[12]

There is wisdom and wonder in even the seemingly ordinary, and attention is one tool we can use to come into contact with this richness. Attention is not an innate characteristic but a skill we can develop with intentional practice, and teaching nature journaling is a way to deepen both the teacher's and students' ability to pay attention. It takes work, but every drop of that sweat grows new connections in our brain.

Deep attention, curiosity, and creativity enrich everyone's lives. We have a responsibility to share these fruits with others: with children, with our peers, and with our elders. Developing youths' capacity for paying attention can deepen their resilience, thinking power, and joy in being alive. A nature journaling practice is a gift that can stay with students their whole lives.

YOU CAN DO THIS, AND IT IS WORTH IT

Nature journaling is a set of skills that can be taught and learned. You do not have to be an artist to teach nature journaling, but you and your students will need a growth mindset.

One of the biggest obstacles to journaling with students is a belief that the practice is only for "gifted" artists or that you must be an artist yourself to teach journaling activities. Both are myths. You and your students can start wherever you are and learn the skills as you go.

Keeping a nature journal is a part of the science curriculum in many European countries. Students maintain regular logs of observations of the comings and goings of natural phenomena—for example, weather; bird migrations; and when plants bud, bloom, take seed, or drop leaves. As a result of regular practice, many adults continue to keep such journals. Because simple sketching is a part of these journals, adults emerge from such a program unintimidated by drawing or journaling and are more likely to consider themselves artists. Were all of these adults somehow born with a "gift" for drawing? Not in the slightest. They honed their skill over time through constant practice.

If you feel intimidated about starting a journaling program, perhaps because you haven't drawn since the third grade, have messy handwriting, or do not have a science background, know that this door will open for you too. Over time, you will grow your skills as a journaler and a teacher of the practice, and this book will guide you along the way. When you take risks and commit pencil to paper, you are modeling the same behavior you want your children or students to take.

THE GROWTH MINDSET AND NEUROPLASTICITY

"I just can't draw." "She's a math genius." "I don't have musical talent." "He's a born artist." "There's no way for me to change; that's just who I am!"

Does this sound familiar? Many of us might have had similar thoughts before, and for many years what we understood about neuroscience supported these fixed views of our abilities. Before the second half of the twentieth century, it was thought that the brain developed early in life and then remained static in physical shape and mental capacity.[13] This "fixed mindset" view says, You're stuck with what you've got. You are a drawer, or you're not.

Understanding the Growth Mindset

We now have a better understanding of how brains work, and it is good news. Our brains grow and change throughout our lifetime, and we can drive this process by learning.[14] New neural pathways form when we learn new skills or ideas, rewiring and rebuilding the brain.[15] As the psychologist Carol Dweck writes, "In a growth mindset, people believe that their most basic abilities can be developed through dedication and hard work—brains and talent are just the starting point. This view creates a love of learning and a resilience that is essential for great accomplishment."[16]

Understanding this concept can change the way we view ourselves as learners, and is often a key piece in encouraging students to pick up their pencils and start journaling.

We need to coach students to believe in their ability to gain new skills through hard work. Students who know they can grow will have the perspective and perseverance to keep going even when what they're doing is hard. A student might begin by thinking, *Well, I'm just not a drawer like Thomas, so there's no point in really trying*, but as teachers we can help them come to think, *Drawing is challenging for me, but if I work at it and get support, I can probably get better at it.*

This does not mean that learning new things is easy. But "productive struggle"—when we are working just outside our comfort zone or just above our current skill level and understanding—builds gray matter in the brain.[17] When we feel challenged or stuck as learners, it is often a sign that we are engaged in that productive struggle, building competency and new understandings. As students begin to journal, try to reframe mistakes and challenges as moments of growth. This kind of understanding can be key to success with journaling.

Encouraging a Growth Mindset

Journaling skills—drawing, writing, and math—are capacities that can be worked at, not gifts allotted to a chosen few. When even skeptical students journal once a week for six months (or several times over the course of a shorter program), they will be able to look back at their first pages and see a marked difference. This is evidence of their ability to grow and a record of the effort it took to get them there. It is also exciting. When students journal and see their progress, it can kick off a self-sustaining cycle of improvement as well as investment in the practice.

If your students are struggling to see their potential for growth, ask them to share stories of resilience and of challenges they have overcome. Share similar stories about people in your local community, or role models your students respect. Frame these people not as heroes with extreme and unusual talent but as people who worked hard, asked for help, and approached challenges with intention. Model a growth mindset in your own learning

BUILDING SKILLS AND THE MYTH OF LEARNING STYLES

The concept of learning styles or multiple intelligences (the idea that each person is a visual, verbal, kinesthetic, or auditory learner) gained popularity in the 1970s. Many educators began teaching lessons in different "styles" in an attempt to meet students' needs. Schools were founded with tracks for different learning styles. This was an admirable response given the information available at the time, but the concept of learning styles has little support from a neurological standpoint.[18] It also has not held true in the face of research on learning.[19] Although different people may have preferences for how they like to learn or some innate skill in certain types of learning and thinking, everyone can learn through many different modalities. Approaches to learning that are difficult at first can develop with practice. It is still valuable to include a range of teaching methods (discussion, drawing, writing) in learning experiences because doing so leads to varied, dynamic lessons and gives students the opportunity to develop new skills. The danger comes when we "track" students by identifying a student's learning style or giving them a quiz that does the same.

This is a disservice in both the short and the long run. A student who is told they are a kinesthetic learner, for example, may never attempt to improve their writing skills because they do not believe they can. They might also tune out or resist instruction that isn't given in their "style," missing out on learning opportunities, and further cementing their idea that they are only capable of one type of learning.

process, being honest with students when you are struggling to learn something new. Designating a celebration time during which students share and honor their "growth moments" or "little wins" is another way to highlight students' process. These regular reflections can reveal changes and patterns that might go unrecognized over a longer period of time.

The concept of the growth mindset is not the latest incarnation of the self-esteem movement. It is not about giving everyone a medal and just saying "good effort" whether or not someone is learning. Instead it's about acknowledging that effort and encouraging deeper engagement: "Great effort! I know you are working hard, and I know you can keep getting better. Let's talk about what you've tried, what went well, and what you can try next." This message acknowledges the work the student has done already and helps them look toward the future. The teacher expresses confidence in the student's ability, then affirms what the student did, then offers next steps. This is how growth happens.

Statements That Reinforce a Growth Mindset

If you catch yourself saying "I can't draw," just add the word "yet" at the end of the sentence.

The point isn't to get it all right away. The point is to grow your understanding step-by-step. What can you try next?

Learning new things is challenging. If you feel challenged, remember that when you learn new journaling techniques and approaches, you grow your brain.

This is a challenge, and I believe you can do it. What did you learn from this journal entry?

What journaling strategies did you use that were successful? How might you use them again?

What approaches and new ideas can you learn from looking at other people's journals?

What parts of this journal entry were the most difficult or challenging for you? How did you handle that?

What new strategies did you use today? Did you use techniques or make kinds of observations that are different than what you usually do?

Statements That Reinforce a Fixed Mindset

Not everybody is good at [drawing, writing, math]. Just do your best.

You are such a [math, drawing, writing, artistic] person!

That's OK, maybe [drawing, writing, math] is not one of your strengths.

Don't worry, you'll get it if you keep trying. [If students are using the wrong strategies, their efforts might not work, and they might feel particularly inept if their efforts are fruitless. They need new strategies along with encouragement.]

Wherever you are, you have what it takes to start. As you teach others, your practice will grow along with them. You and your students can do this. With practice, you will develop quickly. Stick with it and look for your improvement. If you stop, start again, and then again. Be playful. Be brave. Take risks. Innovate your own way and share it. Have fun and celebrate the rich connections and knowledge that grow from your hard work. Your experience of being alive, and your students', will be deeper for it.

Additional Resource

Mindset: The New Psychology of Success, by Carol Dweck

MANAGING THE OUTDOOR CLASSROOM

Keep your students' individual needs and relationship to the outdoors in mind at all times. Use routines to create a baseline of safety and comfort that enables students to focus on journaling and learning.

TAKING STUDENTS OUTSIDE

Field journaling activities take concentration and focus. We want to set students up for success by taking care of their basic needs and journaling at a time when they'll be engaged.

Meeting Students' Physical Needs: A Plug for Maslow

Journaling (and outdoor learning in general) is not about "toughing it out" in the elements to build character. Students' basic needs are important, and if those needs are not taken care of, students' attention will be on them rather than on journaling. To tend to students' physical needs outside, we

- Pay attention to weather conditions and adjust our plans accordingly. If it is excessively hot, cold, wet, or windy, students will be less capable of focusing on journaling activities. Children are more susceptible than adults to the effects of temperature. If we are hot, they may be verging on heatstroke. If we are cold, they may be in the early stages of hypothermia.

- Allow students to wear clothing that is comfortable for them, and make sure they know about outings ahead of time. If getting their clothes or shoes dirty on outings is a serious concern for students, we collect low-cost boots and clothes from thrift stores that they can borrow for field trips to areas with muddy or wet conditions.

Students' comfort and safety are the priority. If teachers tend to emotional and physical needs, students can focus on being present, having fun, and learning.

- Wear a hat and face the sun while we are giving directions, so that students don't have to look into the sun.

- Carry low-cost materials to mitigate the effects of weather, such as squares of plastic from thick garbage bags or a similar material for students to sit on. These are particularly useful in wet conditions or if students are worried about getting dirty or sitting on the ground. (In colder conditions, we use squares cut from a cheap foam pad (e.g., an Ensolite pad) to insulate students from the ground. Without it, they will lose heat quickly, and sitting down long enough to journal will not be possible.)

- Encourage students to bring hats, or bring a class set of inexpensive foam-rimmed visors or hats to help shield them from the sun so that they can see their journal page.

- Carry a small spray bottle to mist students' faces and necks during hotter seasons. This works wonders for their endurance in hot weather. (Spray bottles are also great for misting spider webs, making them more visible to the eye.)

Supporting Students Emotionally

Tending to students' emotional needs helps them learn outdoors. Each of our students will have a different relationship to nature, informed by their prior experiences and cultural perspectives. We try not to assume any level of comfort or discomfort in our students. We meet them with curiosity and ask about their previous experiences, feelings about being outside, and what we can do as a leader to make them comfortable. As we choose outdoor locations and prepare for outings, we then take the individual experiences of our students into account.

Before you take your class outside, tell them how you'll take care of their physical needs, and be clear about what students should expect in the outdoor experience—how long they will be there and whether there will be access to a bathroom (and, if there will not be a bathroom accessible, telling them how to go outside).

When students are new to outdoor learning, it is better to visit one place several times rather than switching locations every outing. Returning to a place allows students to get familiar and comfortable, leading them to focus more and more on learning instead of navigating a new location.

These basic practices go a long way. Tending to students' needs and establishing routines help them feel safe and know what to expect. As a result, they don't have to worry about whether they will be OK or what the experience will be like, which enables them to focus on journaling and their learning experience.

Setting Expectations

It's important on the first outing as a group to establish the outdoors as a learning space. If we only play games the first time we walk to the local park, students will assume that's what they will do the next time. Instead, we communicate that the rules of the classroom or the agreements of the program still apply, and we state any additional expectations for the new context. When we are firm yet kind in setting boundaries and establishing structure, we support students' sense of safety and act as reliable leaders.

Yet we don't structure every moment of this first excursion. There is a lot going on outdoors. Students will be naturally drawn toward different things, and this is OK. If every moment of students' time outside is directed by the teacher, they may become distracted from the learning experience because they never get the chance to explore and check out what is interesting to them.

The routine *I Notice, I Wonder, It Reminds Me Of* (otherwise known as *INIWIRMO*; see p. 36) is an ideal tone-setting activity and a way to establish the outdoors as a learning space right away. It's the first thing we do with any group we work with in the outdoors. It is simple and gives students tools for exploring nature that they can continue to employ throughout the rest of the program. It also gives students a framework for gathering information that can apply directly to their journaling.

INIWIRMO begins with structure and focused learning, then offers students the autonomy to explore whatever they find interesting, using the framework of the routine to guide them. Autonomy is a key part of intrinsic motivation and leads to increased buy-in to learning experiences.

Using Routines and Rituals

Just as we use *INIWIRMO* to set outdoor learning in motion, regular routines and rituals can support students' engagement and focus while nature journaling.

You can give students roles and responsibilities that help you manage the outdoor classroom and empower them as active participants. For example, one student can be in charge of handing out journals, another of making sure the group drinks water, another of noting the time the group begins and ends journaling, another of recording the moment with a photograph, and so on. You can permanently assign these roles to individual students, or the jobs can be rotated.

Having a role or job can also engage extra-energetic students. The responsibility focuses their energy in a positive way, gives them a role they are appreciated for, and honors the assets they bring to a group experience.

To help students drop in to the journaling process, we sometimes develop a short ritual for opening and closing a journaling or outdoor experience. An opening ritual can be as simple as taking a few breaths together, making a group coyote howl, or walking between two trees as a symbol of a "gateway" to exploration. As a part of an opening ritual, we can also acknowledge the indigenous people native to the land we are on. (For example, if we're at Point Reyes National Seashore, we might say, "Before we begin, we would like to acknowledge that we are on ancestral land of the Coast Miwok, who have an enduring presence here.")

Closing rituals might include a circle of gratitude sharing, or a moment for students to share something with the group that they want to remember about their experience. Rituals like these set an attentive tone and provide a routine that marks the mindset students get into when they journal.

Deciding When to Journal

Many student management issues can be avoided by planning when journaling experiences will take place and making adjustments in the moment when necessary. When we journal with students, we

- Choose a time of day when we know students are likely to be able to focus. (For example, if our students are consistently wired right before lunch, we'll plan to journal at another time.)

- Avoid doing several journaling activities in a row. Journaling is hard work, and we want to recognize that students may burn out if we push them too hard.

- Intersperse focused journal activities with walking, discussion, classroom work, or other nature exploration.

- Don't force it. If a group has a lot of energy when we've planned to do a journaling activity, we come up with a quick way for students to move around instead—take a walk up a hill, pretend to be squirrels, offer free time, or play a short, high-energy game. Afterward, students will be ready to slow down and focus again.

FREE PLAY IN NATURE

If young people experience free play in nature, they are more likely to value and want to protect nature when they are older.[21] Don't be afraid to give students free time in the outdoors now and then, setting them loose to explore without an agenda. If you do this after you've taught observation skills, rarely will students complain of being bored. Instead, they will each find a way to be in nature that's exciting for them, and the autonomy will actually make them more likely to practice and retain the skills they find important.

Scouting before Activities

We've had the uncomfortable experience of planning to journal about fungi, only to realize as we walked outside with students that there weren't any mushrooms where we expected them to be! We try to regularly visit the outdoor areas where we plan to take our students journaling. Even a short weekly excursion can help us pay deeper attention and notice phenomena for students to study. With consistent observation, we see the seasons change and notice a sudden influx of ants or intriguing snow formations. Knowing what is around at any given time gives us flexibility and insight into what to focus students on in their journaling.

Working toward specific learning goals (such as a lesson on geology) requires more advance planning and familiarity with the local natural phenomena than does a one-off field activity. When we're planning longer-term learning with nature journaling, we shift back and forth between looking at journaling activities and the natural phenomena in our chosen place, considering the kinds of observations the combinations we choose will lead students to make.

Bringing the Outdoors Indoors

It's not always possible to journal outside. If weather conditions prevent us from going outside, we bring small natural objects such as leaves, acorns, or pinecones inside. Growing classroom plants is a great opportunity to observe change over time. During rainy, snowy, or otherwise harsh weather conditions, observing classroom pets or specimens is a great way to keep students journaling. Students can also make quick trips outside to gather materials or make observations, then return indoors to journal. Over time, you can build a classroom nature museum, filled with shed snake skins, pinecones, abandoned wasp nests, bones, and other treasures. This cabinet of curiosities adds life to the classroom and is readily available any time you need natural objects on a rainy day. (Nevertheless, the best journaling opportunities are still out in the field where found objects are in context with the rest of nature.)

Nature is everywhere. You do not need to travel to a national park to do nature journaling. Journal in the vacant lot around the corner, at the unmowed edge of a ball field, in a neighborhood park, at a neighboring farm, in a garden, or in the plantings in front of your home or school. Both the vacant lot and the mountain meadow will reveal patterns and surprises that could fill a lifetime of study and wonder. Look for areas with a diversity of species and those that have been left untended to let nature sort what grows where.

Building Community Partnerships

Reach out to local parks, community gardens, or other organizations and see whether they are open to forming a partnership with your class. Some teachers have arranged exchanges whereby students are allowed to access an outdoor area for free in exchange for creating a field guide to the area or educating local community members on what they find there. If a local community group is interested in hosting a nature journaling workshop, you can encourage students to teach the basic observation skills they've learned or showcase their journal entries. This can be a powerful opportunity for students to share what they have learned through journaling.

Indigenous communities are important thought partners in nature study, and hold deep relationships with and understandings of the land. Nonindigenous people must seek to engage with native knowledge and culture with respect. An authentic and direct way to do this is to develop relationships with local indigenous communities and to invite members of those communities to cocreate programming and influence your teaching practices (if they want to). This takes time, humility, and learning to listen.

Additional Resources

- An excellent tone-setting activity is the discussion routine *Walk and Talk* from the BEETLES Project (beetlesproject.org/resources/for-field-instructors/walk-and-talk/).

- See the BEETLES Project guide, *Engaging and Managing Students in Outdoor Science* (beetlesproject.org/engaging-and-managing-students/), for more research and practice in creating engaging learning experiences for students in the outdoors.

- *Culturally Responsive Teaching and the Brain,* by Zaretta Hammond, is a great book that includes guidance on building supportive learning partnerships with students.

HOW TO LEAD JOURNALING ACTIVITIES

These key tips and strategies set students up for success when you introduce each activity, keep them engaged while they journal, and help them make meaning afterward.

INTRODUCING ACTIVITIES

The strategies described in the next sections set students up for success as we introduce activities. It's helpful to review the instructions for each activity before we are in the field giving them. Understanding the instructions helps us speak with clarity and be responsive to students' questions.

Explain Why You Are Doing the Activity

We tell students the purpose behind our choice of activity. For example, we might say, "This journaling outing will be an opportunity to hone our observation skills" or "This activity will deepen our understanding of animal behavior." Pointing out specific skills that students will develop helps them recognize the activity as an opportunity to grow and increases intrinsic motivation. We also may explain how the journaling activity fits within a wider learning experience and connects to other topics in class. This helps students view journaling as a critical part of the learning process and as related to the rest of class.

Set Boundaries and Identify Hazards

We set boundaries for students' explorations based on the location, students' comfort with the place, and our own familiarity with the group. In an area with high visibility, such as an open meadow, we allow students to wander farther; in a dense forest or area with varied terrain, we create boundaries that will keep students in sight. As we become more familiar with a group of students and they become more familiar with an outdoor location, we can consider expanding these boundaries if it feels safe to do so. Clear boundaries communicate expectations to students and can make them feel safe. Our boundaries can also keep them away from any sensitive or hazardous natural features; any talk about boundaries should include information on hazards such as nettles, poison oak or ivy, cliffs, bodies of water, or unstable ground.

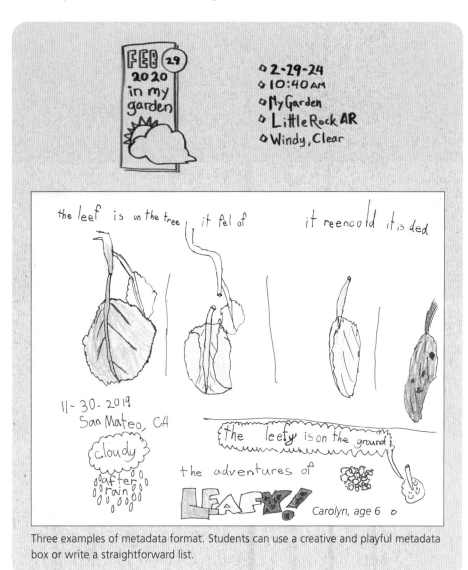

Three examples of metadata format. Students can use a creative and playful metadata box or write a straightforward list.

Record Metadata

Metadata is the "data behind the data" of every journal entry. It's a good habit to begin any journaling activity by asking students to record in the corner of the page the date, location, and weather. With metadata, a journal entry becomes historically relevant and scientifically useful. This information gives important context to the journal entry and will help students mark the passage of time and the growth of their abilities. It also helps focus students on the big-picture context of that journal entry and is a nonthreatening way to break in the new blank page.

Use Words, Pictures, and Numbers

We suggest saying, "You must include words, pictures, and numbers on your journal page, but use more of whichever is most comfortable for you." This makes your expectations clear and offers a bit of a push to use approaches for recording information that are more challenging.

De-Emphasize Pretty Pictures

Many students (kids and adults) are nervous about starting journaling because they feel they must produce a piece of Art. But the goal is observation, not perfect art. When we introduce journaling activities to students, especially students new to journaling, we always say, "It's not about making a pretty picture; it's about recording accurate observations." This reminder can help soothe worries that students might have about drawing and reminds them of the goal of the experience. Partway through an activity, we remind students, "If you're starting to worry about making a pretty picture, turn it into a diagram. Add labels everywhere and focus on showing your observations."

Demonstrate the Activity on a Whiteboard

As we give students the instructions for the journaling activity, we do a quick demonstration on a portable whiteboard or pad of paper, adding quick sketches and using horizontal lines to represent words. This shows how students can lay out their journal pages. Nearly every activity in this book includes an example of what this demonstration could look like. (For a few activities that don't involve journaling or don't have a structured prompt, there is no whiteboard demonstration in the instructions.)

Seeing what the activity can look like on paper helps all students visualize your expectations. This is especially helpful for students who are language learners. For students who are new to journaling, this also offers a place to start, reducing the anxiety of facing a blank page.

Prime Students with Big Ideas

As we noted earlier, the key elements of guided nature journaling are picking a small piece of nature, focusing observation and thinking, and using a variety of techniques to record data. But if you're teaching nature journaling as part of a broader curriculum, you can also begin with a "big idea" to further focus students' observations and questioning. If we prime students to think about structure and function while creating a journal entry about a plant, they'll focus on very different details than if we'd asked them to think about the plant's interactions with other organisms. Priming can be as simple as suggesting that students think about a key concept from your teaching as they journal, or asking them to write "Cause and effect—why is it like that; what might have caused that?" at the bottom of the page before starting a journaling activity. A written reminder gently encourages observations or questions that relate to the concept as students work.

Preempt "Done"

You've likely experienced it: 5 minutes into an activity, a student gets up from their seat, approaches you, and says, "I'm done." We preempt "done" as we introduce the activity, telling students,

"There will come a point during your journaling where you feel you are 'done.' This is just a feeling. Notice it and keep working. Often you may notice the most interesting or surprising things when you keep going! At the end, we'll have the chance to share the most interesting things we all noticed AFTER we thought we were done."

MANAGING STUDENTS DURING AN ACTIVITY

Classroom management is still important when your classroom has no walls! We use these strategies to keep students engaged and to track their progress.

Model Engaging in the Task

When we send students out to journal, we briefly do the journaling activities ourselves (even just for a couple of minutes), unless students really need our attention and support in that moment. Children watch what we do all the time. Our participation shows students that journaling is not just busywork we are asking them to complete. If they see parents, teachers, and chaperones eagerly grabbing their journals to take notes about a cactus or a sparrow, they will do the same.

Check In with Individuals

While students journal, we circulate and talk to students about their observations and journaling. We try to check in with each student one-on-one during this time, asking how the activity is going and briefly engaging them in conversation about their observations.

This is a golden opportunity to notice where students are with different journaling skills and to offer in-the-moment feedback and support. When we talk to students about their journals, we notice what they have done on the page and show them how it relates to the expectations we set out at the beginning of the activity. For example: "I see you've added hairs to the stem of the flower and used words to describe the hairs. That shows more information than just a drawing or writing alone." This reflection can offer a boost of confidence for students who are nervous about journaling, and it reinforces the expectations of the activity.

We also offer "feed-forward"—ideas for how students can push their journaling skills. For example, we might say, "I notice you've shown the stump from one perspective in your drawing. Is there some other point of view that you think it would be helpful to include?" or "I'm noticing you've used one label. Are there other ways you could include words on the page?" In contrast to a grade or evaluation after the fact (feed*back*), these feed-forward ideas can be applied to students' journaling right away.

It takes practice to give meaningful feedback on journal entries. Our natural instincts are to comment on the prettiness or realism of the picture, but we must instead focus on reinforcing the expectations we set out. See the section "Appropriate, Timely Feedback" (p. 228) for more detail on giving feedback.

Redirect Distracted Students

As we're circulating, we keep a lookout for students who are distracted or off task. Without reprimanding them, we'll gently redirect their attention by asking questions about their journal page or the part of nature they're focused on. For example: "What else can you notice about your plant?" Or we try a more focused question that will lead them to make more observations, such as, "Is it that color green everywhere?" or "What does it look like on the other side?"

If, despite our best efforts, a student insists they are "done" before the time has elapsed, we'll ask them, "What have you observed so far?" Then we listen to what they say. They'll often share observations they haven't recorded in their notes. We'll respond, "Great! Write that down or show it in a drawing!"

Offer Reminders

As students are journaling, we'll make a few announcements out loud to the whole group to remind them of expectations. For example: "If you've been focused only on your drawing so far, be sure to include some writing and labels. If you've been mostly writing, be sure to use some drawing." Or "Remember to ask yourself: What do I notice? What does it remind me of?"

Manage Timing

Most of the activities in this book can be completed in 30 minutes to 1 hour, including instructions, journaling time, discussion, and wrap-up. When we send them out, we tell students how much time they will have to journal, and we let students know we will call them back or blow a whistle when it is time to regroup. Many students become so engrossed in journaling that time slips by quickly, and we pay attention to their body language as a cue to when to call them back in. If we told students they would have 10 minutes but see that they are focused on their work, we'll let the clock run. They are doing the important work of making their own observations.

By contrast, if many students are becoming restless and distracted, our 10 minutes could end up being only 7. This is particularly important in challenging environmental conditions, such as strong wind, a cold snap, or hot weather.

We'll also offer time checks to let students know when half the time has elapsed and when they have a few minutes left and should start wrapping up. You can also say to the group, "Raise your hand if you would like more than two minutes to finish." This lets students know that time is almost up and nudges them to finish their work.

This flexible approach is not appropriate in every situation, however. Students with autism and those who have experienced trauma may prefer specific and consistent expectations. Talk to these students first if you change the timing of an activity.

DISCUSSION: MEANING MAKING AFTER ACTIVITIES

Each activity in this book includes a discussion section. We see discussion as a critical part of the learning process, not an afterthought to be included only if there is extra time. We recommend planning for at least 10–15 minutes of your time to be spent in conversation with students after their observations, though you could spend more time if it works for your students, or revisit discussion questions as you build on students' journaling through follow-up activities and extended learning. When students discuss which details helped them match a leaf to a diagram in *To Each Its Own* (p. 42), they are thinking about how to create a data-rich, content-specific journal entry. When they talk about the differences they noticed between plant species in *Collection or Field Guide* (p. 57), they benefit from the observations the other students made. As students try to explain the patterns they saw after "mapping" (p. 107) the locations of spider webs, they engage with science content and process their ideas. Discussion *is* the learning process.

We always give students time to "turn and talk" with a partner about each question, before launching into a brief whole-group discussion. Pair talk ensures that every student gets the chance to speak and process their thoughts, and it increases the likelihood that more students will share in the large group. We follow pair talk with a whole-group discussion because this allows students to hear and build on one another's observations, ideas, and perspectives, and gives us as instructors a sense of where students are in their understanding.

In the suggested discussion questions at the end of each activity, we've included general questions that guide students to reflect on the journaling process itself (e.g., "How did shifting your perspective affect your observations?") and to gather their thoughts about the observations they made (e.g., "What did you notice about the leaves you compared?"). We begin our discussion with these questions to support students' growth with journaling and to reinforce the observations students made while journaling.

Each discussion section also includes sets of questions around "big ideas" such as cause and effect or structure and function (e.g., "What are some possible explanations for the dandelion distribution you observed?"). These questions launch students into the next phase of the learning experience as they use their observations to build toward an understanding of larger, more abstract concepts. Although these questions are based on the Science and Engineering Practices and/or Crosscutting Concepts from the Next Generation Science Standards (NGSS; see the section "Journaling and the NGSS" on p. 238 for more on how you can use NGSS with nature journaling), they work in any learning context to deepen students' understanding of natural processes.

Do not try to ask all the questions in the discussion sections. Beyond the general questions, limit the discussion to questions from one Crosscutting Concept category to keep the discussion short and focused.

GETTING MORE FROM A JOURNALING ACTIVITY

After students have completed a page of observations in their journal, you can guide them to think more deeply about what they saw and to

NGSS CROSSCUTTING CONCEPTS IN NATURE JOURNALING

The NGSS Crosscutting Concepts are "big ideas" that are useful across different disciplines in science.

Patterns. Nature is full of patterns, and scientists often use patterns to group or categorize species, or as clues to underlying processes or forces at work. What patterns do you notice? How can you describe the pattern? Are there exceptions to the pattern? What might be causing the pattern? What does the pattern remind you of?

Cause and Effect. Events have causes. We can think of nature as a world full of mysteries, or "effects," and try to explain what might have caused what we see. What happened here? What caused this to happen? How might you explain that? What are other possible explanations? Why is it like that? What would happen if...? How does this affect other things?

Scale, Proportion, and Quantity. Different things are relevant at different scales, and the size of an object determines aspects of its function. When we observe the world at different scales, we notice different properties and functions. How might you think differently if you zoomed in or out? How can you measure change at this scale? How does size or number affect how this works?

Structure and Function. An object's shape and structure relate directly to how it functions. We can observe structures and features, and attempt to explain how they work. What is it like? How does it work? How is this similar to or different from others you have seen? What are some possible explanations for how the differences in structure impact how they work? How might this structure function to help the organism survive in its environment?

Systems and System Models. Thinking of a natural feature as a system helps us think about the impacts and interactions that lead to changes we can observe. What are the parts of this system? What are the boundaries? How does the system work? How do the parts of the system interact? What would happen if X were removed? What would happen if Y were added? Can you find feedback mechanisms in the system?

Energy and Matter. Energy flows and matter cycles through natural systems. Students can look for evidence of these processes when they make careful observations of a natural area. Where does the matter in this system come from? How does it change within the system? Where does it go? Where does the energy in this system come from? Where does it go? What does the energy do in this system? How does energy flow and matter cycle in this system? What evidence can you find of this movement?

Stability and Change. Change is a constant in natural systems; looking at what causes change is one way of deepening understanding. What are the forces of change in this system? How fast does it change? Is change steady and slow or in periodic bursts? What causes change in this system? What is stable in this system? What isn't? What changes quickly in this system? What would happen if X were different? What was it before? What will it be next?

get more out of the experience. This may be done in addition to facilitating a discussion or as a wrap-up in itself. Ask the students a series of guided questions and give them time to modify their page as you do. This activity add-on takes only a little time, but it makes a big difference in what the students get on the page. Although not required, the use of a small set of colored pencils adds a lot to this exercise.

Questions and Prompts

- Look over the observations you just made. Are there any things you observed but did not record? Take a moment to add those in.

- Are there any other questions you could add? Or can you add to one of your questions? Sometimes the question behind the question is really interesting.

- What big idea is this page about? Add a title in bold or block letters on the top or down the side of the page. Extra points if you can come up with a pun!

- Are there different topics of information on the page? Try to make your page easier to scan by adding subheadings or color coding, or by putting lines or boxes around related information.

- Would it help to draw arrows to connect related ideas, observations, and questions?

- Look over all of your observations and questions. If you saw or asked something that is particularly interesting, give it a little emphasis by drawing a big block letter, exclamation point, question mark, star, heart, or happy face next to it.

- Last, check to make sure you added the metadata: date, location, weather, and other pertinent information.

NATURE JOURNALING IN DIFFERENT CONTEXTS

Nature journaling is compatible with and already a part of many widely used educational systems.

CHOOSING ACTIVITIES TO MEET LEARNING GOALS

The activities in this book are flexible and can support many different learning goals and educational philosophies. As we mentioned earlier, the type of learning students do through the journaling activities in this book supports Montessori and Waldorf educations and fulfills the goals of the Common Core State Standards and the Next Generation Science Standards. These activities are also in line with homeschool frameworks such as the Charlotte Mason method. Inquiry-based, student-led learning driven by curiosity and wonder can be part of every school environment.

This book is not a stand-alone curriculum, and we don't attempt to cover how every activity can be applied to each grade level and every subject. But know that doing field activities with students can be integrated easily into how you already teach and can support your goals for student learning. You know your students, school, and context best. The journaling approaches and activities in this book can be used to support most learning frameworks. The table on the facing page offers examples of how teachers in different disciplines and educational contexts can use nature journaling activities.

FINDING PHENOMENA

One of the wonderful freedoms of using nature journaling as a teaching tool is that each combination of journaling activity and natural phenomena is a unique experience and can lead students to different types of observations. Making a collection of leaves leads to very different questions than making a collection of things that are red. When our learning goals for students are flexible, we head out in search of what is interesting and exciting, and pick an activity based on that. The curriculum is whatever is going on and whatever students are excited about. Using journaling activities to meet specific learning goals or standards requires more advance planning.

To choose journaling activities, ask yourself these questions:

- What are my learning goals?
- What activities will lead students to make observations or think about ideas relevant to those learning goals?
- What natural phenomena are in my area, or will be when I want to do this activity?
- What combination of phenomena and journaling activity will build toward my learning goals?

Wherever you are, you do not need to look far to discover something intriguing that can become the subject of students' journal entries. You might notice a busy network of ant trails that seems perfect for *Mapping*, a group of lizards that students want to focus on for *Species Account*, or a rush of birdsong that students could capture through *Forest Karaoke*. In appendix B, we've offered a guide summarizing each activity, possible phenomena, and how each activity can support different learning goals.

ORDER OF ACTIVITIES: DECIDING WHAT TO DO FIRST, NEXT, AND BEYOND

The activities in this book are organized by the kind of learning or type of skills you will learn doing an activity. Choose from among the thirty-one activities to find the ones best for your students and your teaching focus. We do recommend that when you're just starting with students, you begin with the foundational skills of journaling and nature observation, as taught through *I Notice, I Wonder, It Reminds Me Of*. And we suggest following that with a journaling activity from our "Introductory Journaling Activities" section. That sets students up with basic skills for future journaling experiences. Once students have this foundation, choose activities that make sense for your group and location. The activities in this book don't need to be done in the order listed in the table of contents.

You can build students' journaling skills over time by interspersing skill-specific journaling activities (from the chapters emphasizing drawing, writing, or numbers) with activities focused on natural history and science. This combination of activities gives students new journaling approaches, then opportunity to practice them in context. We'd advise against doing several skill-focused activities in a row at the beginning of students' journey with journaling. All students, of every skill level, can learn about nature from the moment they pick up their journals.

Create a Theme by Repeating Phenomena

Students can do several different activities focused on the same phenomenon. For example, students could complete *Species Account* for a wildflower, then use *Mapping* to show its distribution, *Inside Out* to home in on the structural details, or *Timeline* to show the flower from bud to seed. Each would lead students to make different types of observations.

STANDARDS OR FRAMEWORK	HOW TO USE THIS BOOK TO SUPPORT A FRAMEWORK OR STANDARDS
Charlotte Mason Homeschooling Method	Time in nature was a key element in Charlotte Mason's vision for education, and she introduced the practice in her schools of having children keep "nature notebooks." The activities in this book continue this practice of nurturing children's appreciation of the diversity and beauty of nature. Our field activities integrate art, science, and reflection, all core tenets in Mason's vision of education. In Mason's day, writing in the nature notebook typically took place after the nature walk, and painting was done using specimens collected during the walk. Although this remains the common practice, some contemporary Charlotte Mason educators bring their notebooks with them to the field, and the activities in this book support that.
Montessori	We're inspired by the hands-on learning approach of Maria Montessori, and nature journaling is already a part of many Montessori programs. The longer blocks of time used in Montessori teaching sessions (often 3 or 4 hours) lend themselves to going outdoors and having the time to explore, discover, and journal. Because Montessori supports children's autonomy by setting up activities where they are free to explore and find subjects that interest them, the individual exploration that occurs in journaling fits in well. Mixed-age groups in Montessori classes can be used to build journaling and mentoring partnerships, with more experienced students supporting younger ones as they grow together and share their skills.
Waldorf Education	From early childhood through adulthood, a personal relationship with nature holds a central place in Waldorf schools. Nature journaling supports this goal. Waldorf education's focus on the arts, including sketching, drawing, painting, and others, can also be reinforced through nature journaling. Journaling can be woven into the main lesson blocks in Waldorf schools, such as *Botany* and *Man and Animal*; integrated into art lessons; and used to enhance drawing and observation skills during a variety of other lessons (such as, but not limited to, biology, physics, and chemistry).
Common Core State Standards	Collaboration, communication, thinking, and engaging with information are at the center of the Common Core. The approach to journaling in this book directly relates to many aspects of these standards. As students use words, pictures, and numbers to describe their observations, they build foundational literacy skills that support their development as writers and communicators. Speaking and listening skills are also a critical part of the Common Core. Discussions as a part of journaling and inquiry give students practice listening to one another's ideas and evidence, critiquing reasoning, and building communication skills. Journaling and nature observation are also an opportunity to learn important math skills in context—counting, measuring, and estimating at lower grades and using averages, calculations, and statistics in middle and high school.
Next Generation Science Standards (NGSS)	The journaling activities in this book offer students authentic opportunities to participate in the kind of deep and rigorous learning called for by the NGSS. As students journal about, discuss, and make explanations for natural phenomena, they engage in Science and Engineering Practices; the discussion questions connect directly to Crosscutting Concepts and lead students to use them as thinking tools; and journaling, discussion, and follow-up activities build understanding of Disciplinary Core Ideas. The section "The Next Generation Science Standards" on page 238 includes specific information about connections between journaling, Science and Engineering Practices, and Crosscutting Concepts.
Non-NGSS State Science Standards	Journaling can also play a key role for states with science standards not based on the NGSS. The activities and approaches in this book can help students develop their understanding of science content related to life and earth sciences. Journaling is also an authentic way for students to engage in the inquiry and process skills that are a part of many science standards.

Repeating Activities

Similarly, we can repeat the same activity focused on different facets of nature. *Timeline* could focus on wildflowers, mushrooms, or rocks in a meandering stream, yielding a range of different observations.

Repeating an activity strengthens students' journaling skills because they are more likely to integrate the approach of the activity into their journaling tool kit. Once students identify a technique (such as *Comparison* or *Zoom In, Zoom Out*) as a tool for making observations, they can apply it during future journaling outings.

Building on Activities

Some of the richest rewards of nature journaling can emerge when the practice is a springboard into further inquiry and learning. Although every journaling activity in this book can stand alone, each can also function as one phase in a longer lesson that meets specific learning goals. We can guide students to build on the knowledge they gained while journaling through continued research and follow-up activities and, in doing so, deepen their understanding of key concepts and ideas.

In the section "From Activities to Longer Lessons," we've offered guidance for incorporating journaling activities into extended learning experiences that meet specific teaching goals and standards.

ADJUSTING ACTIVITIES FOR AGE AND EXPERIENCE

All activities in this book are appropriate for fourth grade and up (unless otherwise noted at the beginning of the activity). There are no adjustments needed to use journaling activities with older audiences. Older students or more experienced journalers will engage with the same activities at a deeper level.

One of the most inspiring things about teaching nature journaling is watching even the youngest students create knowledge from their observations. We do suggest scaffolding and adjusting journaling activities to set up younger students (in this context, third grade and below) for success. Open-ended instructions for students to find their own journaling subjects (e.g., "Go find two trees and compare their leaves") can be specified for younger children to make it clear what part of nature students should focus on (e.g., "Compare the leaves of these two trees"). Are your children or students too young to write, or emergent writers? Let them dictate their observations and questions to you or to another student. Transcribe what they say word for word. This catches the freshness and beauty in the way children describe the world and demonstrates your respect for what they notice and wonder.

In most journaling activities in this book, the teacher gives instructions, and this is followed by extended time for student journaling and discussion. With younger students, we break instructions up into subtasks and complete each step along with the group. The table on the facing page shows some examples.

Journaling Activity, Instructions as Written

"Using words and pictures together, make a labeled sketch of a leaf in your journal. [Instructor models making a quick sketch and labels on a whiteboard.] Try to include a few different views of your leaf and add written observations and questions." [Instructor adds a side view to their whiteboard demo, then adds horizontal lines to indicate written observations and questions.]

[Students spend ~10 minutes sketching leaves, adding written notes, labels, and different views of their leaves.]

Journaling Activity, Scaffolded Instructions

"We are going to journal about leaves. Start by sketching your leaf. Draw the outside shape and any shapes or holes on the leaf." [Students and teacher spend a few minutes sketching leaves.]

"Now let's include a side view. Take a moment and draw your leaf from the side on your paper." [Students and teacher spend a few minutes sketching the side view.]

"We want to use words and pictures together on our journal pages. Let's add labels. What are things we might label?" [Students respond: stem, holes, colors.]

"What are questions we have about our leaves?" [Students respond.]

"I'm going to write 'stem' and add an arrow to show where it is. Then I'm going to add labels for holes that I see on my leaf. [Teacher adds labels on sketch.] I'm also going to write my questions. Now you add some labels for stems, holes, or other interesting features, and write questions." [Students write.]

Discussion Question, Instructions as Written

"Look back at your journal entry. Discuss with a partner: What were some differences and similarities between the two kinds of leaves you studied? What are some possible explanations for the differences you noticed? How might the different shapes or structures of leaves help the tree survive?" [Students discuss.]

Discussion Question, Scaffolded Instructions

"Look at your journal entry. With a partner, talk about how the two leaves you drew are different from each other. They might be different shapes or colors." [Students discuss.]

"Keep looking at your journal entry, and now discuss similarities between your leaves. Maybe the stems or veins are the same size, or they feel similar to the touch." [Students discuss.]

"Pick one difference between the two leaves, like how one is fuzzy on the outside and the other isn't. Talk to your partner about why they might be different. What would be helpful for a tree about having a fuzzy leaf?" [Students discuss.]

Observation Activity, Instructions as Written

"Observations are what we notice through our senses. They're not our opinions like 'I notice the leaf is gross,' or identifications like 'I notice it's a leaf.' Observations might include 'I notice the top is smooth and the bottom is rough. There's a hole in the middle. It's a grayish-brownish color.' Take two minutes to make observations out loud about your leaf with your partner." [Students take turns making observations about leaves with a partner.]

Observation Activity, Scaffolded Instructions

"Observations are what we notice through our senses, like 'I notice there is a hole in the center of the leaf' or 'I notice it is brown and green.' Let's practice together, starting with colors. What colors do you notice? Greens, reds, grays? Share with a partner." [Students notice colors and talk with their partner.]

"Now let's focus on textures. Use your sense of touch to notice different textures on the leaf. Is it rough? Smooth? Bumpy? Smooth in one place, but rough in another? Share with a partner." [Students notice textures and talk with their partner.]

NATURE
JOURNALING
ACTIVITIES

NATURE JOURNALING ACTIVITY CHAPTERS

Each chapter comprises a group of activities that focus on developing a specific set of skills or engaging with nature in a similar way. In addition to activities, each chapter includes the following:

- Information and background for the instructor. If you're not sure how to teach drawing or are intimidated by the thought of bringing numbers into the game, don't worry. We've included techniques to share these different skill sets with students.

- Example journal pages from working scientists and naturalists. These "mentor texts" showcase a wide variety of approaches to journaling, illustrate how journaling supports learning, and highlight key practices and techniques.

The next sections outline the content of each chapter.

GETTING STARTED: INTRODUCTORY JOURNALING TECHNIQUES AND ACTIVITIES 31

Activities in this chapter build foundational journaling skills and techniques. We recommend beginning with some of these foundational activities to set students up for success in subsequent journaling experiences.

OBSERVATION AND NATURAL HISTORY 51

Activities in this chapter pull students' attention to different parts of nature. Ideal for teaching life science or building students' sense of place, these activities focus on observing organisms and attuning to details of the landscape.

INQUIRY, INVESTIGATION, AND SCIENTIFIC THINKING 85

This chapter offers a deeper dive into scientific thinking. Activities introduce new inquiry skills, offer questioning scaffolds, reveal natural patterns and processes, and guide students to create visual explanations.

WORDS: ARTICULATED THOUGHT AND STORYTELLING 127

This chapter focuses on how to put words on a journal page. The activities guide students toward scientific and creative writing as well as storytelling, and they build visual and written communication skills.

PICTURES: DRAWING AND VISUAL THINKING 159

Activities in this chapter advance students' drawing skills and offer scaffolds for creating dynamic visual layouts. Drawing and diagramming exercises will provide students with approaches for quickly and efficiently sketching in the field.

NUMBERS: QUANTIFICATION AND MATHEMATICAL THINKING 195

Activities in this chapter offer structures for gathering and representing quantitative data in journaling. Some activities focus on how to find the numbers behind any observation; others offer structures for gathering quantitative data on natural processes.

ACTIVITY FORMAT

The activities in this book have a consistent style that makes them easy to scan and present to your students. Each journaling activity includes the following sections and information.

Summary

The summary at the beginning of each activity includes information about the kind of learning the activity will support and a brief synopsis of what the students will do during the activity. This is a good place to scan when you are deciding which activity to use.

Sidebars (Time, Materials, Teaching Notes)

The sidebar provides an "at a glance" summary of the timing, setting, and materials for the activity as well as notes about leading the activity, management, or context specific to the activity.

Natural Phenomena

This section offers suggestions for possible phenomena. Check here for ideas about the parts of nature that students could focus on during the activity.

For many activities, there are nearly infinite possibilities for phenomena to focus on. For example, students could engage in *Comparison* using almost anything, from seed types to bird beaks, or could do *Mapping* focused on a creek bed, plants on a schoolyard, or location of gopher holes. Activities such as *Species Account* and *Timed Observations* focus on observing one kind of organism.

Procedure Summary

This is a summary of what students will do during their independent journaling time. After you've given the directions for each activity to students, write these bullet points out on your whiteboard (or write them on a piece of paper ahead of time) and set it out somewhere so that students can be reminded of what they are to do.

SPECIES ACCOUNT

Students choose one species that they can readily observe, and document as many details as they can about it through direct observation.

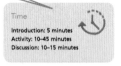

Time
Introduction: 5 minutes
Activity: 10–45 minutes
Discussion: 10–15 minutes

Materials
☑ Journals and pe...

optional
☑ Binoculars
☑ Example species account field notes from local scientists

Teaching Notes
Scientists in many disciplines of life science make focused species accounts. These thorough records describe species, including markings of individual organisms, where they were found, what was nearby, and interesting behaviors. This is a simple and powerful approach to learning in nature, one that students can continue to use in their journaling. This is also a great jumping-off point for studying the species in more depth, giving students the background to dive into relevant research or to think about the species' interactions with its environment.

Much of what we know about nature started with direct observation and experimentation. There are many species that have been deeply studied, but there is always more to learn and discover. Each observation, if recorded and shared, becomes part of a growing understanding of the world. Species accounts are a common approach to cataloging organisms and building a database of information. In a species account, the observer attempts to learn as much as they can about the type of organism, using words, pictures, and numbers to record details about structures, behaviors, and location in and interaction with the surrounding environment. Once students learn and develop an approach for doing a focused species study, they can apply the skills anywhere they go.

NATURAL PHENOMENA

Any plant or animal that can be observed for a sustained period can be used for a species account. If you think an animal might scamper away, use the *Animal Encounters* protocol instead. Students don't necessarily need to focus on the same organism, unless you want the whole group to build a base of observations to use to reach specific learning goals. Find an area with enough plants, animals, or fungi that individual students could choose their own subject to observe. Plants are very cooperative and will not walk away. Animals are fun because they exhibit behavior that can also be recorded. Encourage students to choose animals that will not crawl or fly away halfway through the observation period. Catching small insects, macroinvertebrates, or other critters in clear plastic cups is a way to deal with this issue. Captive animals are often easy to observe, but may exhibit behavioral and structural differences compared to wild animals.

PROCEDURE SUMMARY

1. Record as many observations and questions about this species as you can, using words, pictures, and numbers.

2. Include information about how the organism looks, its behaviors and feeding habits, where it was found, and the like.

3. Focus on specific observations, not explanations.

DEMONSTRATION

When the whiteboard icon appears in the procedure description: As you suggest things to include in a species account, create a sample page that reflects those suggestions. Do not worry about making a pretty picture. Your bunny can be a circle with two lines for ears. Demonstrate making more than one sketch, to show different

72

Demonstration

As you give the step-by-step instructions, use a small, portable whiteboard to demonstrate what you want students to do. Your example need not look exactly like ours or be a pretty picture. Just give an overall structure to the page, use simple line diagrams to represent drawings, and draw horizontal lines to simulate writing. The whiteboard icon in the procedure indicates when to do the demonstration.

Procedure Step-by-Step

The procedure is directions to give to students to do each activity. The numbered steps summarize the instructions. Lettered steps are directions to give students as you might say them.

---

views; changing scale; using words, pictures, and numbers; and including metadata.

PROCEDURE STEP-BY-STEP

1. **Tell students that they will get the chance to learn as much as they can about a specific species by studying it.**

   a. "We are about to practice our observation skills by using them to learn as much about [ants, this type of tree, these worms, a species of students' choosing, etc.] as possible."

   b. "This is your chance to do an in-depth study and become more familiar with this part of nature. We can learn a lot when we focus in on one species and give it directed attention."

2. **Explain that the goal is to describe the species in as much detail as possible using words, pictures, and numbers; and students can use "I notice, I wonder, it reminds me of" to help them focus on specific observations.**

   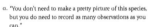

   a. "You don't need to make a pretty picture of this species, but you do need to record as many observations as you can."

   b. "Use words, pictures, and numbers together to describe what you see, relying more on whichever approach is most comfortable."

   c. "You can use the frame 'I notice, I wonder, It reminds me of' to help guide your observations and what to write down."

3. **Encourage students to be specific with their observations and language.**

   a. "When you make observations, be as specific as possible. Don't just say 'The leaf is green'; rather, say 'It is deep blue-green at the base, shading to yellow-green within two millimeters of the edges.' It's important to come up with as accurate a description of this organism as possible."

4. **Tell students to focus on making observations (e.g., "There are yellow leaves at the ends of the branches"), not assumptions or explanations ("The leaves at the ends of the branches are dead").**

   a. "Sometimes when I record observations, I make assumptions or explanations for what I see. I might say, 'Older leaves on the tree are yellow,'

but I don't really know that those leaves are older. I saw the yellow leaves and without realizing it assumed that they were older. This is an explanation for why the leaves are yellow. My assumption that the yellow leaves are older may be wrong."

   b. "When you record your observations, try to avoid assumptions or explanations. Just describe what you see, not what you think is going on."

   c. "I could say instead, 'Larger leaves that are farther from the branch tips are yellow.'"

5. **Ask students for suggestions of observations they could include that go beyond just describing how the organism looks.**

   a. "We can do more than just record observations to show what the organism looks like."

DISCUSSION

Lead a discussion using the general discussion questions and questions from one of the Crosscutting Concept categories. Intersperse pair talk with group discussion.

General Discussion

   a. "Let's share some of our observations. Look at your notes and find an example where you provided rich and specific details. Let's share some of these with the group."

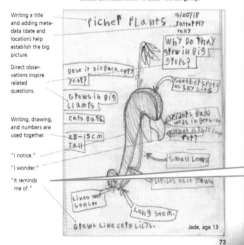

Writing a title and adding metadata (date and location) help establish the big picture.

Direct observations inspire related questions.

Writing, drawing, and numbers are used together.

"I notice."

"I wonder."

"It reminds me of."

Jade, age 13

73

Discussion

Sets of questions help you facilitate a discussion with your students after they do an activity. It is in these discussions that they will draw greater meaning from the activities. General discussion questions wrap up a specific activity. Crosscutting Concept questions and Science and Engineering Practice questions help students see a bigger picture, build toward conceptual understanding, and see greater application for their newly developing skills.

As you look through the question categories to decide which ones will help students reach your learning goals, imagine how your students might respond. We recommend discussing the general questions, and questions from one of the other categories. It is better to have a deeper discussion about fewer ideas than try to cover everything and not allow students to build on their thinking and questions. Give students time to talk about each question in pairs before discussing it with the whole group.

Student Work

For many activities, we have included an inset journal page from a student. The callouts around the illustration point to aspects of the page that demonstrate some of the targeted journaling skills. Pointing out these kinds of details in your own students' journals motivates them to continue to use these strategies throughout the journal.

GETTING STARTED

Introductory Journaling

Techniques and Activities

GETTING STARTED: INTRODUCTORY JOURNALING TECHNIQUES AND ACTIVITIES

FOUNDATIONAL TECHNIQUES

Activities in this chapter will build foundational journaling skills and get students excited to keep at it. We recommend beginning with *I Notice, I Wonder, It Reminds Me Of*. It is always the first activity we do when we take a group of students outside for the first time because it sets a tone of inquiry, curiosity, and excitement, and teaches them observation tools that can translate directly to the journal page.

To motivate students to keep journaling, their first few experiences should be confidence building, not overwhelming. *My Secret Plant* and *To Each Its Own* help build the skills of capturing observations and thinking in words, pictures, and numbers on the journal page. The game-like quality of these activities takes students' attention off the fact that they are journaling. Then, at the end of the "game," they have produced a journal page full of words, pictures, and numbers that can serve as a model for all of their future journal entries. This first experience of success reinforces a growth mindset and shows students that they can do it. *Comparison* and *Zoom In, Zoom Out* are similarly accessible, approachable activities. They also introduce techniques students can include in future journal entries.

ACTIVITIES IN THIS CHAPTER

USE WORDS, PICTURES, AND NUMBERS

Using words, pictures, and numbers leads to deep observation and different types of thinking. We coach students to use all three to stimulate creative thought and capture varied information.

Eriko Kobayashi

"I have been sketching for 25 years as a nature artist. I used to not write that much in my sketchbooks. When I restarted journaling again last year after my daughter turned 3, I realized that writing is as important as drawing. I have many old sketchbooks with less writing; however, they tell me only indistinct memories. The drawing and writing combination for recording information definitely leads me to observe nature more and more deeply."

Translation

1. In the morning, we found lady spiders busily working to fix up their webs, which were broken by the powerful typhoon No. 21.

2. Legs look shorter here because it bends them.

3. I realized one of the lady spiders was plump.

4. But the one right next to her was skinny.

5. Only 40 cm difference between each of their webs. Does the quantity of food make the difference?

6. They often eat Aobahagoromo—*Geisha distinctissima.* [Arrow points to a small diagram of an insect remnant.]

Multiple Points of View

Showing the spider from different angles leads to more observations and a more thorough record of its structures.

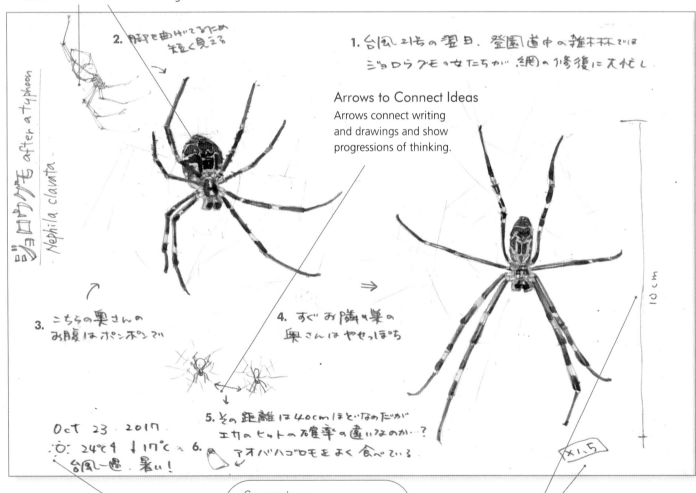

Arrows to Connect Ideas

Arrows connect writing and drawings and show progressions of thinking.

Icons to Show Information

Developing a set of icons to describe different weather conditions can be a fun way to start playing with symbols.

Comparison

Details and differences are often easier to notice and describe when you compare similar or related objects.

Size and Scale

Draw small organisms larger than they appear to show interesting details. Record the amount of the enlargement.

MIX IT UP

There is more than one way to write, draw, or use numbers to explore nature. We can scaffold different approaches for recording information to build students' tool kits.

BUILDING THE TOOL KIT

Different modes of recording data are appropriate for different parts of a subject. As we offer different tools and approaches, we give students more flexibility in their journaling, and support their thinking. When we take out a hammer, we see the nails; a screwdriver, we notice the screws. It is the same with words or pictures. When we ask students to show a side view in a drawing or we show them how to vary writing modalities, students gain tools they can use in any future journaling entry.

Emilie Lygren

"Journaling is a tool that helps me follow threads of curiosity, moving from question to observation, mystery to discovery. My nature journal is my favorite method for learning wherever I go. What's more, journaling leaves me with a durable sense of awe. It is as much a practice in resilience as it is a learning tool. When I slow down and pay attention, I'm pulled into awareness of and deeper relationship with the world."

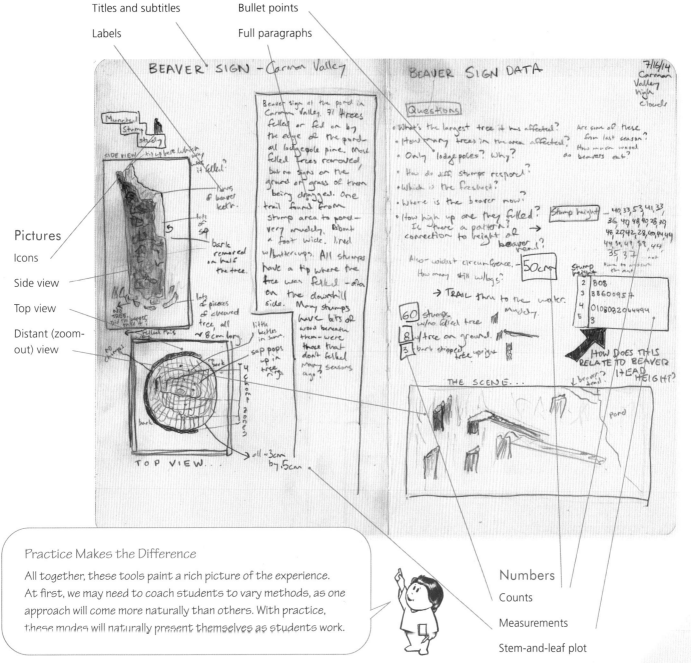

Words
- Titles and subtitles
- Labels
- Bullet points
- Full paragraphs

Pictures
- Icons
- Side view
- Top view
- Distant (zoom-out) view

Numbers
- Counts
- Measurements
- Stem-and-leaf plot

Practice Makes the Difference

All together, these tools paint a rich picture of the experience. At first, we may need to coach students to vary methods, as one approach will come more naturally than others. With practice, these modes will naturally present themselves as students work.

35

I NOTICE, I WONDER, IT REMINDS ME OF

Learning how to observe, ask questions, and make connections in nature is learning how to learn. This powerful routine can be the foundation of exploration, discovery, and wonder in students' nature study and in any other academic discipline.

Time

Activity: 20–30 minutes
Discussion: 15 minutes

Materials

☐ Journals and pencils

optional
☐ Hand lenses

Teaching Notes

Take the time to introduce this routine to students as written. Just saying the words is not enough. The structure of this activity is designed to offer guided practice so that students can successfully learn and apply these new skills. The portion of this activity where students take a few minutes to study whatever is interesting to them is essential. Applying the routine in a new context helps students internalize these prompts as skills. The autonomy to decide what to explore gives students agency, engages them in the task, and makes this way of thinking "their own."

This "core" routine is the first thing we do with a group of students in the outdoors. It sets up a culture of curiosity and exploration, and gives students a way to engage with their surroundings. It also forms the foundation of students' journaling.

You and your students can learn observation and fundamental inquiry skills through this simple routine. "I notice" focuses our attention and helps us articulate and remember our observations. "I wonder" sparks inquiry and invites us to question deeply and broadly. "It reminds me of" leads us to connect what we observe to what we already know, which builds stronger memories. The ability to make useful connections between seemingly unrelated things is an important aspect of creativity, and practicing "It reminds me of" builds this skill. Together, these prompts can change the way you and your students experience the world, offering a routine and practice for learning about anything.

NATURAL PHENOMENA

For the first part of *I Notice, I Wonder, It Reminds Me Of* (INIWIRMO), focus students' attention on a small part of nature they can hold in their hand, such as a leaf, pinecone, acorn, or seed pod. A larger organism that a whole group can observe together, such as a tree, will also work. Then introduce each of the three prompts as described in the "Procedure Step-by-Step" section, giving students time to practice responding to each one before moving on. When students get time to practice using the prompts on their own, any natural area that is safe for them to explore will work.

PROCEDURE SUMMARY

1. Make observations (I notice…).
2. Ask questions (I wonder…).
3. Make connections (It reminds me of…).
4. Use all three observation tools to learn about anything interesting to you.

DEMONSTRATION

When the whiteboard icon appears in the procedure description: Write each of the prompts (i.e., the words "I notice…," "I wonder…," and "It reminds me of…") on a whiteboard as you introduce them.

PROCEDURE STEP-BY-STEP

1. **Tell students that you will share some tools and skills that will help them be better observers.**

Note: If the group has a lot of energy and seems to be losing patience with being still, introduce "I notice," then hike a short distance or give students a moment to run around, then add "I wonder." Then hike or walk a little further and add "It reminds me of" with another found object. Then give students time to practice all three prompts focused on whatever is interesting to them.

2. **Explain the first prompt: "I notice," and offer examples**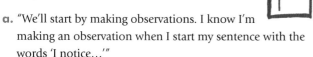

   a. "We'll start by making observations. I know I'm making an observation when I start my sentence with the words 'I notice...'"

   b. "An observation is something perceived through the senses like sight, smell, touch, hearing, or taste. It is not an opinion about or an identification of an object."

   c. "An example of an observation could be, 'I notice the veins stick out of the surface of the leaf' or 'I notice the side of the leaf feels sharp.'"

3. **Tell students that they'll take the next minute to observe their leaves (or other object), saying what they notice out loud with a partner.**

   a. "Take the next minute to come up with as many 'I notices' as possible, saying all of them out loud to the person next to you. You all will be surprised by how much you see and remember."

   b. "Anything you see—like shapes, structures, behavior, or colors—or perceive through your other senses is fair game."

   c. "If you get stuck, you can just be silent for a moment, try changing your perspective, or try using a different sense to observe. You can also listen to the observations that others share, and build on or confirm what you hear. Give it a try."

4. **Give students ~1 minute to say observations out loud with their partners, then ask a couple of students to share any interesting observations with the group.**

5. **Explain the second prompt: "I wonder."**

   a. "Now we're going to focus on asking questions, using the frame 'I wonder.'"

   b. "Start asking questions ('I wonders') out loud. Do not be afraid of asking questions. The point is not to answer them now but just to get them out there. Any question is OK. Saying the question aloud will help you remember it later."

   c. "If no questions come to you, try saying 'I wonder...' and see what fills the silence afterward. A question may come."

   d. "You can also go back to making observations and then ask questions about your observations."

e. "If you make this a regular practice, questions will flow more easily. You can make yourself a more curious person! Take the next minute to come up with questions, saying them out loud."

6. **Give students ~1 minute to ask questions out loud with their partners, then ask a couple of students to share any interesting questions with the group.**

7. **Introduce the third prompt: "It reminds me of."**

   a. "Now ask yourself what you are reminded of. Try to come up with as many connections as you can."

   b. "Is this like something you studied before, observed somewhere else, read about, or saw on a nature special?"

   c. "Have you seen this kind of object before? Does the shape or structure look like or remind you of something else?"

   d. "Your connections could also be to an experience you were reminded of. Do you have any other memories that surface when you look at this?"

   e. "Give yourself permission to be creative and playful. Any connection that comes to mind is important to share."

8. **Tell students to take ~1 minute to share "It reminds me ofs" with a partner, then give a few students the opportunity to share some of their interesting connections with the group.**

   a. "Connecting this new observation to what you already know will help you remember what you are seeing. This can also help you ask interesting questions."

   b. "Take a minute to make connections and say them out loud."

9. **Ask students to reflect on how much they were able to learn in just a few minutes of focused attention.**

   a. "Take a look at your leaf [rock, stick, or other part of nature]. You were able to learn a lot about that leaf in a very short time, all by making your own observations and doing your own thinking."

   b. "Look around you at the other natural objects and things nearby. There are mysteries and cool things everywhere in nature."

   c. "When you have tools for making observations and connections and for asking questions, you can learn and discover wherever you go."

10. **Tell students that they'll have a few minutes to move around and use their observation skills to learn about anything that is interesting to them.**

    a. "You're going to have some time to use your new skills to study and learn about anything that's interesting to you in this area."

    b. "You can be by yourself or in a group of two or three."

c. "No matter how many others you are with, say your observations, questions, and connections out loud."

d. "You might choose to move around and study different things, or you might find something really interesting and exciting and stay with it for a while."

11. **Offer any relevant safety precautions and set boundaries based on the needs of your students and the area they're exploring, then send them off.**

12. **As students work, take time to circulate, troubleshoot, and engage students in conversation about what they're noticing.**

a. (Optional) If a group of students finds something especially cool or interesting, call the whole class over, then ask students to share their observations, questions, and connections out loud.

DISCUSSION

Facilitate a discussion with the general discussion questions.

General Discussion

a. "What are some things you learned through your observations, thinking, and questioning?"

b. "What skills do you feel like you just got better at?"

c. "Are there any other places where you could use your observation skills to learn?"

FOLLOW-UP ACTIVITIES

A Routine to Use Anytime

INIWIRMO can become a regular part of instruction. Use it as the entry point for students' learning whenever they encounter an unfamiliar phenomenon or whenever you are beginning a lesson or unit of curriculum.

Once a group of students knows this routine, it can be applied on a moment's notice. When a bird lands nearby, you can tell students "Quick! While the bird is here, let's see what you can learn about it using your observation skills. Go!" By the time the bird flies away, your group will have made rich and detailed observations. You can ask, "What were some of the most interesting observations you made or heard one of your classmates say? What were some questions that came up? What things did this remind you of?" Responses will come flooding back. These responses can be recorded in journals or included in future meaning-making discussions that take place as a class.

Thinking on the Page

This routine can become a foundation of students' journaling. Students can record written or sketched "I notice, I wonder, It reminds me ofs" in any future journaling activity. If students are unsure where to begin, or get stuck, they can always fall back to noticing, wondering, and making connections.

Working in Other Disciplines

I Notice, I Wonder, It Reminds Me Of can enrich any subject, not just nature observation. The prompts in this activity are sentence frames you can use in active reading strategies or to help students analyze art. Similar frameworks are sometimes used in theater, where actors develop characters by writing down what they notice, wonder, and are reminded of while reading a script. Using these three sentence starters can be a way to begin to write poems or other creative works. Many mindfulness practices center on the act of noticing, separating the observations from interpretations of them. This use of the routine can become a part of students' social emotional learning, helping them slow down and develop self-regulation. Share these connections with students and ask them if they can think of any other uses of the routine.

MY SECRET PLANT

Students record observations of a plant using words, pictures, and numbers, then challenge a partner to find the plant using their notes.

This activity helps students develop essential journaling skills: focusing on details; selecting relevant and useful observations to record; using words, pictures, and numbers; and communicating ideas on a data-rich page. Knowing that the journal entry will have an audience helps students focus on the task.

NATURAL PHENOMENA

Find an outdoor area with a diversity of plants and enough room for students to spread out. This could be a natural area; vacant lot; or the "weeds" growing around a lawn, an unmanaged margin of a sports field, or a playground.

PROCEDURE SUMMARY

1. Use words, pictures, and numbers to describe a plant.

2. Record unique features, such as growth forms, holes, and colors

3. When I call "time," give your journal to a partner and challenge them to find your plant by using your notes.

DEMONSTRATION

When the whiteboard icon appears in the procedure description: Draw a plant and add written notes. As students suggest elements that would help their partner find their plant (color notes, scale, location map, details about insect bites or other unique markings), add the ideas to the demonstration.

PROCEDURE STEP-BY-STEP

1. **Tell students that they will create a treasure hunt for a classmate in their journals, using words, pictures, and numbers to describe a plant.**

   a. "In a moment you are going to create a treasure hunt for one of your classmates. We will spread out in this area, and each of you will choose one plant, then describe it as accurately as you can using words, pictures, and numbers."

   b. "You'll need to include enough information so that another person will be able to pick out your individual plant (not just the species or type of plant) from the others around it. You win if your partner can identify your plant using your journal entry."

2. **Explain that students' journal entries should include words, pictures, and numbers and that they can rely more on whichever mode is most comfortable for them.**

    a. "This is not about making a pretty picture. It's about recording accurate information."

    b. "You will need to use words, pictures, and numbers to record your observations."

    c. "If you're more comfortable writing, you may write more. If you're more comfortable drawing, you may draw more. If you like using numbers, you can do more of that. But you must use all approaches to show what you see."

3. **Ask students what clues they could include to help their partner find the plant, then record their suggestions in an example demonstration on the whiteboard.**

    a. "What are some of the clues you could include in your notes that would help your partner find your plant?"

    b. If students don't mention the following ideas, then share them yourself: notes on color and size of the whole plant and individual parts; numbers of things such as leaves, seed pods, or branches; notes on what is near the plant; a small map showing where it is; or locations and numbers of unusual details such as bug bites or holes.

4. **Define boundaries for the activity. If you've done a lot of outdoor exploration with your students, you may simply instruct them to stay within earshot. Give clear and specific boundaries for larger groups or in a context where students should stay close.**

5. **Explain that when you call "time," students will use their partner's journal to find their partner's plant.**

    a. "You will have only thirteen minutes to record your notes. At the end of that time, I will call for you to come in. When you hear the call, come back here to find a partner, but first, look at your surroundings and pick out a couple of landmarks to make sure you can find your own plant again!"

    b. (Optional, if it will work for your group of students) "You can keep working for a short while if you aren't ready."

c. "Then you will bring your partner to the area near your plant and give them your journal notes. Give some boundaries for where they should search for your plant. Make this area bigger for larger or more obvious plants, and smaller for smaller plants. You want it to be challenging, but still possible to find your plant."

    d. "If your partner has trouble finding your plant, narrow down the area for them or show them the angle you were observing from. If they still have trouble, you can make the search area even smaller."

6. **Remind students of expectations, ask whether there are any questions, and tell the group to begin.**

    a. "We will be working separately and quietly. Do not distract other students. If there is someone whom you really want to be your partner, make sure you are not working near them, because you do not want to see what plant they are working on."

    b. "Any questions? Go for it!"

7. **Circulate and troubleshoot as students are getting started with their work, helping students who are struggling to choose a plant or focus on the activity.**

8. **When the time is two-thirds done, give reminders about using words, pictures, and numbers and adding metadata.**

    a. "Be sure you are using pictures, numbers, and words in your diagram. If you've only used one approach for

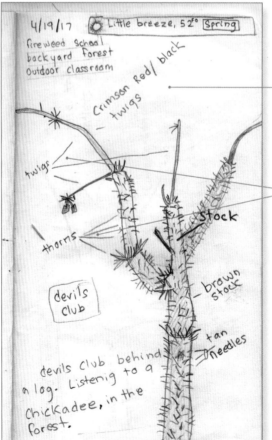

Encourage students to make rich color descriptions. *Crimson* is a wonderful elaboration of *red*. The descriptive palette of red, orange, yellow, green, and blue does not fully describe the variations in color.

Label lines integrate writing and drawing. This helps students make specific notes about unusual details. It also takes much of the Art pressure off drawing. The drawing does not need to stand alone. What details are best shown with drawings? What details are best shown with words?

Cyrus, age 11

recording information, be sure to incorporate other approaches into your note taking."

b. "If you have not already done so, add the date and location [metadata] to your work."

9. **When time is up or students seem ready, give the signal to find the plants with a partner, then help any pairs that are having trouble finding their plants.**

Note: If some students do not come in immediately, this is a great sign. They are so into it that they are forgoing the game to record more details about their plant!

DISCUSSION

Lead a discussion using the general discussion questions and questions from one of the Science and Engineering Practices or Crosscutting Concept categories. Intersperse pair talk with group discussion.

General Discussion

a. "How many of you were able to find your partner's plant? Did your partner find yours?"

b. "What were some of the details that really helped you find your partner's plant?"

c. "When your partner was using your notes to try to find your plant, were there any details you wished you had included?"

d. "What kinds of details might be helpful to record in future journal entries?"

e. "Were there any interesting approaches you saw in your partner's entry that you might want to use in the future?"

Obtaining, Evaluating, and Communicating Information

Tell students to put out their journals, then circulate, noticing patterns or interesting approaches their classmates used to record information.

a. "Let's take a look at our journal entries. Everyone has a different way of recording information, and that's OK, but we can learn new ideas from looking at one another's work."

"Open your journal to the page from this activity and put it on the ground in front of you."

"If you don't want anyone to know which page is yours, you can put it down and walk quickly to another place in the circle. In a moment, when everyone's journal is on the ground, no one will remember which one is yours."

"Walk around and look for patterns, similarities, and differences in the journal entries. How did different people lay out the page? How did people use words and pictures together to show observations? Discuss what you notice with the people around you."

Patterns

a. "Compare your notes with other students in your group. Can you see any patterns in the plants—such as forms of growth, branching, flower shape, leaf color, or leaf shape—across the species and individual plants you studied?"

b. "What common characteristics—similarities in growth pattern, leaf shape, leaf color, and so on—would you use to categorize these plants into groups?"

Cause and Effect

a. "Look at some of the unique features you recorded on your plants, like holes, spots, tears, shapes, or unexpected patterns of growth. What might have caused some of these features to occur?"

b. "What things in the surrounding environment might impact or affect this plant? Can you see any evidence that these could have influenced your plant?"

c. "How might this plant affect its surroundings? Can you find any evidence of this?"

d. "Were there any features, like certain leaf shapes or types of holes, common to many plants in the area? How might you explain them?"

Structure and Function

a. "Pick an interesting structure, or part, of the plant that you drew in your journal. Then discuss: How might its shape, texture, or other feature help it function in this environment?"

b. "With a partner, compare the leaves, stems, fruits, seeds, and other features of your two plants. Discuss: How are they different? How might they function differently?"

TO EACH ITS OWN

Students draw and describe one item from a set of similar objects (such as leaves or shells), then play a matching game in which they pair the objects to the notes their classmates made.

Time

Introduction: 5 minutes
Activity: 10–30 minutes
Discussion: 10–15 minutes

Materials

☐ Journals and pencils

☐ Tape

optional

☐ Hand lenses or magnifying glasses

Teaching Notes

This activity is an ideal introduction to journaling. The goal is for students to practice recording data accurately, and the scaffolding in the instructions is critical to their success. If you just say, "Go out and draw a leaf," students who do not like drawing will disengage, and even your "drawers" might just make a symbolic drawing without focusing on details.

This introductory activity gives students practice creating accurate, data-rich journal entries through the lens of a game. It is a great confidence builder for students who might be insecure about writing or drawing—they are (pleasantly) surprised when they see a classmate successfully use their notes to match their leaf or rock to their journal entry. This approach reframes drawing and journaling as tools used to communicate information rather than as means of producing something that "looks good." The discussion afterward helps students notice different approaches to recording observations and information. In the process, students learn new strategies for journaling and practice producing scientific text.

NATURAL PHENOMENA

Assemble (or tell students to find) a set of small and similar objects, such as fallen leaves, shells, or acorns. Make sure that students have objects of the same species. In rain or extreme weather, bring a collection of objects inside. If your students are younger, or newer to journaling, pick an object that is relatively easy to draw. Simpler leaves are an often successful choice because they are more two-dimensional than a rock or shell.

PROCEDURE SUMMARY

1. Use writing, drawing, and numbers to describe your object.
2. Record unique features, such as growth forms, holes, shapes, and colors.
3. Others in your group will try to match your object to your entry, so make your page accurate and detailed.

DEMONSTRATION

When the whiteboard icon appears in the procedure description: Model tracing a leaf (or other natural object). Add details using written notes and numbers, including suggestions students bring up.

PROCEDURE STEP-BY-STEP

1. Tell students to select a natural object of the same type as the objects in the set you have assembled, or from their surroundings.

   a. "This is a maple leaf. [We will use the example of a leaf for this procedure description.] You have one minute to look around and find one that you like, one that is interesting to you. When you have picked a leaf, return to this spot."

2. Explain that students will make a diagram of their natural object in their journal, then play a matching game.

   a. "We're going to play a matching game. Here are the rules: When I say 'go,' you will only have twelve minutes to make a diagram of your leaf in your journal, in as much detail as possible."

   b. "When we are done, we will all try to match each leaf to the journal page describing it, so do not crumple or throw away your leaf."

3. Explain that students' journal entries should include words, pictures, and numbers and that they can rely more on whichever mode is most comfortable for them.

   a. "This is not about making a pretty picture. It's about recording accurate information."

   b. "You will need to use words, pictures, and numbers to record your observations."

   c. "If you're more comfortable writing, you may write more. If you're more comfortable drawing, you may draw more. If you like using numbers, you can do more of that. But you must use all approaches to show what you see."

4. Offer drawing tricks that will help students start making a diagram of their object.

   a. "I've got a trick to help you draw any leaf. Place your leaf on your journal page, lightly but firmly hold it in place with one finger, and lightly trace it without pressing in on the leaf. The key here is to draw lightly. The tracing gives a rough outline of the object that you can then refine and detail."

   b. (For a three-dimensional object [acorn, shell]) "I've got a trick to get you started. Lightly trace around the object to block in the basic shape. You can also slightly angle your pencil inward so that the rough tracing is not too large. The key here is to draw lightly. The tracing gives a rough outline of the object that you can then refine and detail."

5. Ask students to list the kinds of details it will be helpful to include.

   a. "Your notes need to be accurate and detailed so that others can match your diagram with the leaf. What kinds of clues or details can you include to help us tell one leaf from another?"

6. Listen to students' responses, asking for clarification and emphasizing key ideas. Add examples of their suggestions to the whiteboard.

7. If students don't bring up any of the following strategies, share them yourself: measuring the leaf to exact size and writing "actual size" next to it; labeling colors or adding color; recording numbers of things, such as leaf points or holes; drawing from more than one perspective; including unique characteristics, such as bug bites or broken pieces.

8. Tell students to begin making their diagrams, then offer reminders, or support students who are struggling as the group works.

   a. "Keep adding details, even if you think you're done. There's always more to observe. If you finish a first level of detail quickly, go back in at a higher level of resolution. You can enlarge interesting parts in an inset."

9. When the time is two-thirds done, give reminders about using words, pictures, and numbers and adding metadata.

   a. "Be sure you are using words, pictures, and numbers in your diagram. If you've only used one mode of recording information, be sure to incorporate other modes into your note taking."

   b. "If you have not already done so, add the date and location [metadata] to your work."

   c. "Keep your leaf; you will need it intact for the next part of the activity."

10. Call students back together, then facilitate the matching game: Groups of about fifteen students clear leaves from a small patch of ground or rock, place their journals in a circle around the bare area, and put the leaves they drew in the center. Next, students move to a position in front of someone else's journal. Then each student studies the journal entries before them, picks up a leaf when they think they know the entry it matches, then puts it down on that journal.

   a. "Please move any fallen leaves from the center of the circle, then put the leaves you drew down there."

   b. "Put your journals in a ring around the leaves, opened to the page you were just working on, then move to stand at a different part of the circle so that you're not in front of your own journal."

   c. "Study the journal in front of you, then move around the circle to study some other journals. Notice details that the observer recorded. Then look at the details of some of the leaves in the center."

d. "When you think you know which leaf matches one of the journal entries, take it from the center and place it on the journal."

11. **If necessary for your group, facilitate this process with students. For example:**

a. "Does everyone agree that the leaves are in the right place? If you don't agree with the placement of a leaf, pick it up and move it to another journal, and say why you think it belongs there instead."

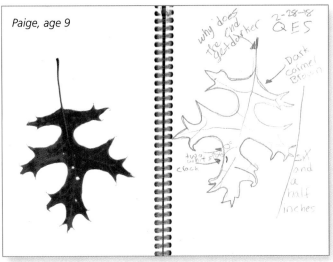

Paige, age 9

Lightly tracing the leaf or making a series of small dots along the edge helps quickly and accurately match the shape of the leaf. It is not cheating. It is a practical shortcut.

12. (Optional) Tape the leaves inside students' notebooks.

DISCUSSION

Lead a discussion using the general discussion questions and questions from one of the Science and Engineering Practices or Crosscutting Concept categories. Intersperse pair talk with group discussion.

General Discussion

Once all the leaves have been matched, ask students to turn and talk about the following questions with a partner, then share with the group. Highlight any important ideas that connect the group's journaling abilities as a whole.

a. "What details were most helpful for identifying a leaf? Why?"

b. "What kinds of details were easier to show through writing? What details were easier to show through drawing? How did you use numbers to show your observations?"

c. "Can you notice any patterns in how we recorded information? Are there any strategies in others' journal entries that you could incorporate in your future journaling?"

Obtaining, Evaluating, and Communicating Information

Put out students' journals, then tell students to circulate, noticing patterns or interesting approaches their group used to record information.

a. "Let's take a look at our journal entries. Everyone has a different way of recording information, and that's OK, but we can learn new ideas from looking at one another's work."

b. "Open your journal to the page for this activity and put it on the ground in front of you. If you don't want anyone to know it is your page, you can put it down and walk quickly to the other side of the circle. In a moment, when everyone's journal is on the ground, no one will remember which one is yours."

c. "Move around and look for patterns, similarities, and differences in the journal entries. How did different people lay out the page?"

d. "How did different people use words, pictures, and numbers together to show observations? Discuss what you notice with the people around you."

Patterns

a. "What similarities in growth and rot patterns do you see when you look at these leaf observations together?"

b. "How could we group these leaves based on shared characteristics?"

c. "Do you think you could find two leaves that are exactly the same? Why or why not? What kinds of things cause variations in leaves? What are some of the differences you'd expect to see between different leaves?"

Cause and Effect

a. "What might have caused the patterns we observed? For example, why might all the leaves have circle-shaped holes?"

b. "Look at the unique features of your leaf, such as holes, tears, or discolorations. What are some possible explanations for the causes of these features?"

Structure and Function

a. "Compare these leaves to the leaves of another kind of tree. How are their structures different? How are they similar? How might they function differently? Connect your explanation to what you can observe in the surrounding environment."

COMPARISON

Students observe two similar species or objects (such as two types of trees, mushrooms, or flowers) and sketch them side by side, noting differences between them.

Comparing two similar subjects makes their unique details stand out. Comparison is a basic tool used in scientific investigations to reveal patterns or collect data. When students make comparisons in nature, they will generate numerous relevant observations. This baseline of observations can support students to start thinking about how different structures work or how varied conditions might have affected the subjects they observed. Students can continue to use comparison as a tool in their journals.

NATURAL PHENOMENA

Find a location where students can observe two species or objects that are generally similar, such as two kinds of willows; gopher and mole mounds; two kinds of eggs, seeds, or fruit; elk and deer; pill bugs and sow bugs; two kinds of shells; or poison oak and blackberry. Students can also compare two organisms of the same species or two areas subject to different conditions—for example, a puddle in sun vs. one in shade, a plant grown inside vs. one grown outside, an exposed vs. sheltered location, or wet vs. dry. If you would like to use the activity to build students' understanding of specific science concepts, prompt the whole group to journal about the same thing so that they have common experiences and observations to work from.

PROCEDURE SUMMARY

1. Compare the two subjects using words, pictures, and numbers to show similarities and differences.

2. Record any questions that come up.

DEMONSTRATION

When the whiteboard icon appears in the procedure description: Draw a T chart to organize the data and object titles. Add descriptive elements and drawings in a parallel structure, emphasizing that you are showing the difference between the elements on each side.

PROCEDURE STEP-BY-STEP

1. **Explain that students will compare two different species or objects, using words, pictures, and numbers to record what they see.**

   a. "You will be making comparisons between these two [trees, plant species, insect exoskeletons, rock types, leaf shapes, etc.]."

   b. "Your goal is to notice similarities and differences, then record them using words, pictures, and numbers."

Time

Introduction: 5 minutes
Activity: 10–40 minutes
Discussion: 10–15 minutes

Materials

☐ Journals and pencils

optional
☐ Rulers

☐ Hand lenses

Teaching Notes

Choose two subjects that are relatively similar to each other. This is key to the success of this activity. Compare an orange to an elephant, and the differences are so vast that it becomes overwhelming to describe them. Compare an orange to a lemon, though, and all sorts of relevant details jump out—the texture of the skin, the shape of the fruit, the attachment point of the stem, and the like.

2. Give some suggestions for how students could record their observations and structure their journal entry.

a. "Think about how to structure your page to show the two subjects."

b. "You could divide your paper with a line down the middle and put one object on either side. Or, if you have another idea of how to organize your page, go for it! If you note a detail on one side of the line, show how it is different on the other side."

3. Tell students that they must use words, pictures, and numbers in their journal entries, but can use more of whichever form of note taking they are most comfortable with.

a. "You must use a combination of words, pictures, and numbers to record your observations. But you can use more of whichever approach is comfortable for you."

b. "Some similarities or differences may be easier to show in writing, and others may be easier to show in drawing. Be thoughtful about which approach you use to show different kinds of details."

4. Share some specific strategies for diagramming the subject based on its size, and remind students of any drawing strategies or approaches they already know.

a. (If students are comparing two very large things, such as two tree species) "You don't have to draw the whole tree. You could show one leaf and a sample of the bark, then do a very small drawing of the overall shape of the trees next to each other to show their differences in stature."

b. (If students are comparing something smaller than their journal) "Make your drawing life size or larger than life, and be sure to note how much you have enlarged the drawing. You might choose to do an inset and just magnify part of the object."

c. "Remember, the goal isn't making a pretty picture, it's recording accurate observations. Use any of the drawing or diagramming strategies you already know to record information efficiently."

5. Ask students whether they have any questions about what they'll be doing, set boundaries, then send them out to journal.

6. As students work, take time to keep track of time, circulate and troubleshoot, and engage students in dialogue about their choices and strategies for recording information.

7. (A few minutes before you will call students back) Say: "Take about two minutes to wrap up and add any final details to your journal entry."

DISCUSSION

Lead a discussion using questions from one of the Crosscutting Concept categories. Intersperse pair talk with group discussion.

Patterns

a. "What are some patterns you noticed as you made your comparisons? What features were similar in each subject? What was different?"

b. "Why might those features be similar or different?"

Cause and Effect

a. "Look back at your journal page and focus on the differences between the two subjects or areas of study. What are some possible causes for the differences you observed?"

b. "Why do you think there was more or less _____ in each of the areas you observed?"

c. "What might have caused any of the patterns you observed?"

Structure and Function

a. "What were some of the differences that you observed? How were the structures [such as leaves, bark, or branches] different from one another? What differences were there in overall structure and growth pattern?"

b. "Pick one type of structure [such as leaf, bark, or branches] to focus on. Turn and talk to someone near you about how the two structures are different from one another. Then discuss: How might they function differently?"

Side-by-side comparisons help students see subtle differences they would otherwise overlook.

This color-coded Venn diagram shows high-level thinking and graphic visualization skills. Looking at student journals enables you to see how and what students think!

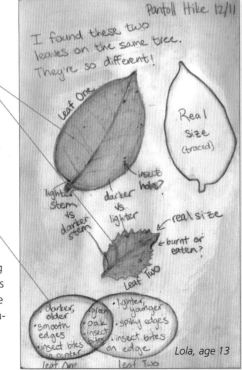

Lola, age 13

ZOOM IN, ZOOM OUT

Students investigate and draw an object in three scales (magnified, life size, and distant) and reflect on the kinds of observations made at each level.

Getting close to a subject reveals different details than those you see from a distance. Close up, micro textures and fine details stand out. Standing back from a subject in nature, you begin to notice where it is in the environment, its relationship to other things around it, and its overall shape. This activity gives students a framework to intentionally record observations at close, medium, and long range. In the process, they will make varied observations about the subject and pick up the valuable tool of changing perspective, which they can apply in future journaling entries.

NATURAL PHENOMENA

Any natural object can be used for this activity, including trees or other plants, rocks, fungi, or any animal that does not move too much. *Zoom In, Zoom Out* is best completed in the field, where students' "zoomed-out" view can include the environment and context around their subject. Pick subject matter that supports other lessons you are doing with your class. If your goal in doing this activity is to support a lesson about a certain topic (e.g., plant structures or a fungus's relationship to its environment), make sure the whole class does the activity focused on that part of nature. Although it is ideal to do this activity with an object in its natural environment, if students are working with an object that has been removed from its surroundings (e.g., a seashell in a classroom), they can skip the zoomed-out view and focus more on their life-size diagrams and enlargements.

PROCEDURE SUMMARY

1. Record observations at three scales: close up, life size, and far away.

2. Use circles to show magnified views for the "zoom in." In the zoomed-out view, you can include details such as growth forms, where the subject is, or a small map.

3. Use writing, drawing, and numbers to record your observations.

DEMONSTRATION

When the whiteboard icon appears in the procedure description: Make a quick example sketch of a plant (or other subject) at life size; circle a feature you want to enlarge, and draw the enlargement within a larger circle. Then add a more distant view of the plant (perhaps showing the entire plant if it is large, e.g., a tree) or where the plant is in the environment, in a side-view diagram, map, or landscape sketch. Add sets of lines to suggest written notes.

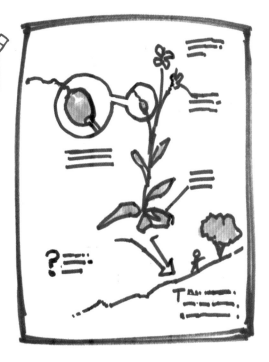

PROCEDURE STEP-BY-STEP

1. **Tell students that they are to observe a subject from three points of view: close up, life size, and more distant.**

   a. "We will be looking at [leaves, cottonwood trees, these fungi, river stones, etc.] from three levels of focus: life size, a magnified close-up view, and a more distant view."

   b. "You will record your observations at these different scales all on the same page. There's a lot you can observe by changing your level of focus."

2. **Give suggestions for the types of details students could show and focus on at each perspective.**

   a. (Life size) "In the middle of your page, draw a view of your subject that is exactly life size. If the object is larger than your page, only draw part of it. Add written notes about what you notice at this distance."

   b. (Zoom in) "Then choose some part of the subject that you find interesting, and 'zoom in' to observe it in detail. To show this view, draw a circle around that part of your drawing. At the side of the paper, draw a larger circle and draw a magnified view of that same area showing details that are too small to be shown in the life-size picture. Include written notes and questions."

   c. (Zoom out) "Step back and make a final sketch, this time zoomed out to take in the whole subject and some of its environment. You could show a side view of the subject, or a small map of its overall shape and where it is in the environment. Again, add written notes about what you see at this scale."

3. **Remind students to use words, pictures, and numbers to record their observations at every scale.**

   a. "At every point of view, be sure to use a combination of words, pictures, and numbers to show what you observe, but you can use more of whichever approach is most comfortable for you."

   b. "Some observations might be easier to show in drawing, others in writing. Think about which mode is best for the observations you're recording."

4. **Set a time limit and the boundaries of the study area, then send students out to journal.**

   a. (About halfway through) "We've got about half our time left. If you've only worked on one perspective so far, shift to another one."

DISCUSSION

Lead a discussion using the general discussion questions and questions from one of the Crosscutting Concept categories. Intersperse pair talk with group discussion.

Zoomed-in details and zoomed-out big-picture diagrams can be used in any study, from botanical drawings to animal behavior investigations.

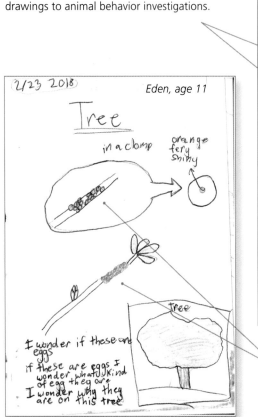

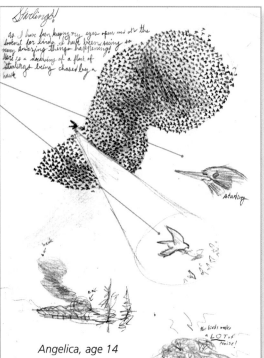

Students can include drawings of the same subject at different scales as a strategy for capturing a range of details and observations.

General Discussion

a. "What kinds of details do you notice when up close? Far away?"

b. "How does shifting your perspective on something change the way you think about it?"

c. "What structures might we be able to see if we had better magnification?"

d. "How could you use this 'zoom-in and zoom-out' approach in other journal entries?"

Structure and Function

a. "What are some of the structures (or parts) you observed at the zoomed-in scale? At the life-size scale? At the zoomed-out scale?"

b. "When scientists are learning about a part of nature, they think about how a structure might work, or function, to help the plant, animal, or other living thing survive. Pick a structure, such as a leaf or the husk of a seed, and make some possible explanations of how it might function. Connect this explanation to the surroundings, thinking about how they influence your journal subject."

Scale, Proportion, and Quantity

a. "Are there structures or patterns of shapes you're seeing in the magnified view that are similar to structures you see in the distant view? How are they alike and different?"

b. "How might the size of a structure affect how it functions? For example, if one of the micro structures you observed were much larger, how might its function change?"

Cause and Effect

a. "Look at some of the unique features of the subject you focused on in your journal. What did you notice?"

b. "What might have caused some of these features to occur?"

c. "What things in the surrounding environment might impact or influence the subject of your journal entry? Can you see any evidence of this?"

d. "Is there any evidence you see of organisms that might have interacted with the subject of your journal entry?"

e. "Can you come up with any explanations based on evidence about how the environment influences and interacts with your subject?"

FOLLOW-UP ACTIVITIES

Showing Scale in Magnified Views

Teach students how to show the scale of magnified views (three methods). The first two are useful if you plan to scan and resize the notes. The third does not require any measuring device.

1. Measure the actual size of the enlarged object or area and write the length next to the enlargement.

2. Measure the actual size of the enlarged object or area and add a scale bar next to the enlargement.

3. Compare the size of the real object with the enlarged drawing. Use the real object as a measurement unit and count how many could be placed end to end to match the length of the enlargement. This number is the magnification. If you could line up three-and-a-half seeds across your enlargement, you have magnified three-and-a-half times (or 3.5x).

Shrinking Adventure

Build a shrinking machine that can shrink the student down to the size of an ant, then have them complete the activity as written. Be careful not to step on any of them.

Reflecting on Perspective

Prompt students to look for places in previous journal entries where they'd already unintentionally included zoomed-in and zoomed-out views, or places where they would have liked to use this strategy.

OBSERVATION
AND
NATURAL HISTORY

OBSERVATION AND NATURAL HISTORY

FOCUSED OBSERVATION

Each place holds unique mysteries and little wonders. The activities in this chapter guide students to slow down and deepen their relationship with place through studying species, narrowing their focus, and focusing on patterns and processes.

As students look through different lenses, from studying what falls within a circle of string to observing flowers as they bloom, they will tune in to the processes at work around them and notice biologically significant details that can deepen their understanding of how the natural world works.

ACTIVITIES IN THIS CHAPTER

LOOK BIG, LOOK SMALL

There are big and small wonders everywhere in nature. Slowing down and shifting our perspective lead us to notice more. We can offer structures to students to focus their attention in different places and deepen their relationship with place.

Marley Alexander Peifer

"Journaling is my fundamental practice. It synthesizes my interests and accelerates my learning."

Be Patient, and Look Again

Most people trust their first impressions. But when we sit quietly and continue to look, we will often find that there is more going on than we thought. We can invite students to watch patiently and document what they see, reflecting changes in their thinking on the page.

See the Forest and the Trees

Recording close-up details and distant views of different organisms all on the same page is a great way for students to develop a sense of place. The process gets them looking everywhere.

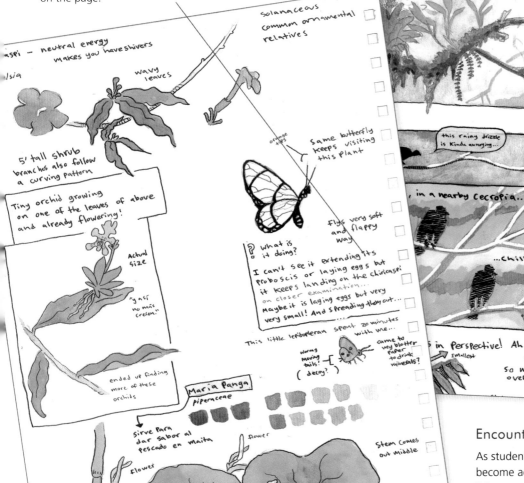

Encountering Wildlife

As students sit quietly journaling, animals will become accustomed to their presence and emerge. Whenever an animal is in close view, we can encourage students to observe and document as much as they can, recording observations and making sketches. Then, when the animals move along, students will have a rich record of the encounter.

MAKE COMPARISONS AND STUDY CATEGORIES

Looking for similarities and differences within a category focuses students on structure, variation, and relationships. This deepens their understanding of nature as a whole.

> **Using Abbreviations or Codes for Long-Term Projects**
>
> On these pages, Paola uses standardized abbreviations when she measures snakes. She uses the same codes in all her field notes. This makes it easier to make comparisons between different specimens and saves time in the field.

Paola Carrasco

"I am a professor at the National University of Cordoba and researcher with the National Council of Scientific and Technical Research, Argentina. I use a personal standardized list of abbreviations, drawings, exclamations or question marks, arrows, and written observations. I use these approaches to maximize time, to highlight information, and to remember things later. Using drawings is very useful to remember certain observations in a less subjective way. I complement these field notes with a series of photos."

Focused Species Accounts

A focused study of one organism or type of organism is a valuable approach to nature study. Slowing down to spend time with one individual leads to deeper observations and insights.

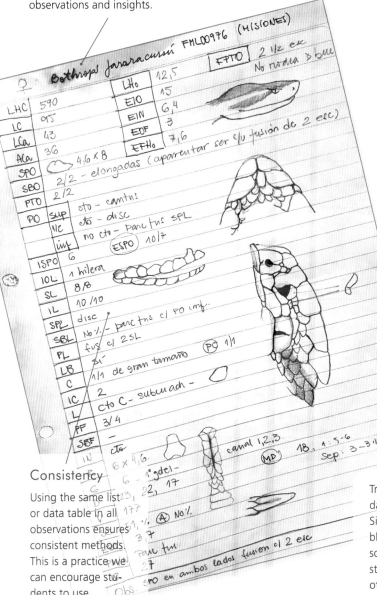

Consistency

Using the same list or data table in all observations ensures consistent methods. This is a practice we can encourage students to use.

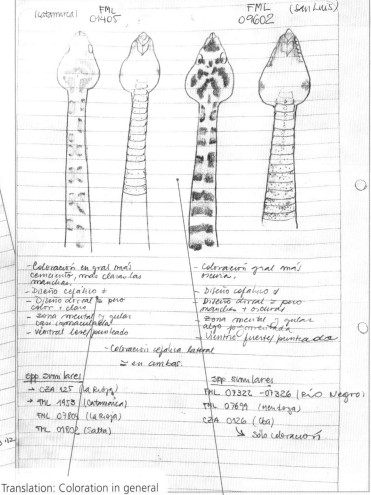

Translation: Coloration in general darker. Different cephalic pattern. Similar dorsal pattern but darker blotches. Mental and gular area somewhat pigmented. Venter strongly dotted. Similar specimens of the species. Lateral cephalic coloration similar in both.

Collection

A group of drawings or "collection" focused on similar organisms makes the similarities and differences stand out.

GET LOST IN WONDER

Nature is infinite and wonderful. We can shift students' attention to different aspects of the natural world and encourage them to find something that delights them and get lost in it.

Akshay Mahajan

"I work as a hardware engineer most of the week, which means my days are usually packed with attempting to make sense of the unknown unknowns, churning plausible solutions and then trying to diagram my thoughts in front of my team. Nature journaling for me is no different! The unique structure and mechanism of nature start a series of cognitive contraptions, which then become fun to connect when I dump it all out on my journal. There is just so much going on in nature that I am unconsciously stimulated to capture as much information as I can through the means of text, arrows, diagrams, color keys, memory associations, or even just mere cartoons—I love cartoons! And not everything on the page has to match with what Google says. For me, simple guesswork of why things are the way they are is just as fascinating as the real thing!"

Interesting Icons

Using icons to identify questions, observations, or feelings (such as surprise) is fun and helps make the journal page organized and easy to scan. It can also help students reflect on their experience.

Focusing on Process

Observing patterns and processes is a foundational practice of natural history. When students study processes and try to explain what they see, they learn about that aspect of nature and build learning skills at the same time.

What We See along the Way

As we guide students to focus in on one facet of the natural world, they'll notice other nearby phenomena. We can encourage students to include these surprises on their journal page instead of seeing them as distractions, leading to a richer record of a moment in time.

Focused Attention

When we prompt students to focus on one part of nature, a world opens up. The activities in this chapter will shift students' attention and help them build a relationship with nature as a whole as they spend quality time with the parts.

COLLECTION OR FIELD GUIDE

Students make a field guide or "collection" of things within a focused category, such as leaf types, rocks in a stream, things that are red, or tracks.

Children are natural collectors. Kids (and adults) collect cards, stamps, coins, bird eggs, or any number of other objects. Focused observation of a category leads to a deeper understanding of it. A stamp collector scans every envelope that enters the house and notices unusual stamps and postmarks that most people would miss. Students can harness that type of focus to deepen nature observations by making a field guide. Any topic will lead students to explore a world that they might otherwise overlook and to develop understanding of a category of things. If a student makes a collection of fall fruit in their journal, they begin to see fruit on every bush and vine. If students make a field guide to "fuzziness," they will begin to see objects, relationships, and patterns among fuzzy things that would otherwise go undetected. Exploring with narrow focus will spark questions about similarities and differences among the objects, such as "What might be common functions of fuzziness?" paving the way for deeper learning about science concepts.

NATURAL PHENOMENA

Students can make a collection in any outdoor space, even a seemingly bare schoolyard. Most published field guides are identification manuals: guides to birds, mammals, tracks, plants, or the biota of entire regions (e.g., the Sierra Nevada or Southwest deserts). These same topics make great subjects for your students' field guides. Or you can offer more creative topics, such as things caught in spider webs, patterns made by melting snow, icicle shapes, insects visiting a creosote bush, or things that are striped.

Field Guide Categories

- Classic field guide categories: trees, leaves, rocks, landscape features, bird feathers, macroinvertebrates, or any other facet of the natural world

- Field guide of phenomena or evidence of an effect: things impacted or shaped by water, signs of fall, things affected by wind, things that snow does, shapes of icicles, evidence of drought, things that are broken, things that are soft, things with a strong odor, things attracted to porch lights at night, signs of the season

- Field guide to a pattern: 120° angles, spirals, branching patterns (not just in tree branches), things found under rocks

- Field guide to a system or a "tiny world": things on a lawn, under rocks or a log, on a rotting log, in puddles, or on windowsills

PROCEDURE SUMMARY

1. Make a field guide to (your chosen subject).

2. Include three to five things in this category in your field guide.

3. Arrange the page so that you show a drawing with words next to it.

4. Record observations with words, pictures, and numbers, paying attention to similarities and differences.

Time

Introduction: 5 minutes
Activity: 30–45 minutes
Discussion: 10–20 minutes

Materials

- ☐ Journals and pencils

- ☐ Examples of field guides, including small regional guides

- ☐ Examples of collections

optional
- ☐ Rulers

- ☐ Hand lenses

Teaching Notes

A broad topic can be overwhelming. In a diverse meadow, yellow flowers might be a more manageable topic than wildflowers. A narrow focus also makes it easier to notice patterns and generate relevant questions (e.g., "What pollinators are attracted to yellow flowers?")

If you would like to use students' field guide observations to deepen their understanding of specific concepts, make sure they create a field guide focused on a common subject.

DEMONSTRATION

When the whiteboard icon appears in the procedure description: Choose a subject for your demonstration field guide. Create an informative title across the top of the page. Make a series of quick sketches of similar objects and indicate the addition of written notes with sets of lines. Give some verbal suggestions of different ways that students could structure their journal page.

PROCEDURE STEP-BY-STEP

Depending on your goals and your students, you can introduce this activity as making a field guide or making a collection.

Field Guide Introduction

1. **Show students examples of field guides and explain that they are tools useful for identifying and learning about plants, animals, and other parts of nature in a specific area.**

   a. "This is a field guide. It is a tool used to identify and learn about kinds of plants and animals that live in a specific area. For example, this book is a field guide to [wildflowers of the Sierra Nevada]; this pamphlet is a field guide to [trees along a nature trail in southern Florida]."

2. **Pass out a field guide to a group of four students, asking them to flip through the pages, noticing what kind of information is included and how it is arranged on the pages.**

   a. "Look through these books in small groups. What information is shown? How is it arranged to make it easy to understand?" (Typically, a field guide will have pictures showing different stages or forms for each subject, written information describing key points, and maps. Subjects are arranged in an order that helps compare similar species.)

Collection Introduction

1. **Discuss the kinds of objects that students have collected.**

   a. "Have any of you made a collection of some kind of object? What sorts of things do you collect?"

2. **Discuss the impact of making a collection on attention and discovery, highlighting how being a collector of something attunes you to details others might miss.**

   a. "What have you have noticed about the things you collect that other people might miss?"

   b. "How does making a collection impact your ability to notice and learn things about what you collect?"

   c. "When you collect something, you notice details about the types of things you collect that others don't."

Field Guide/Collection Procedure

Depending on your goals and your students, you can continue the introduction of the activity by giving students the same field guide subject or letting them choose their own topic. If all students will focus on the same subject, give them the "teacher-driven" instructions. If your students will be choosing their own subjects, give them the "student-driven" instructions.

1. **Teacher driven: Tell students that they will make a field guide to, or collection of (seed pods, fruits, animal tracks, holes in leaves, clouds, etc.), recording subjects in that category in their journal.**

   a. "You are going to make a field guide/collection to [seed pods/fruit, animal tracks, leaf buds, etc.]. You will find and record as many subjects as you can in this area."

2. **Student driven: Explain that students will make a field guide to, or collection of, any subject they want, encouraging them to narrow their focus to a specific topic and giving them a moment to brainstorm ideas with a partner.**

   a. "You're going to make a field guide/collection, and you will pick the area and the subject. This can be as scientific or as playful as you wish."

   b. "Generally it is easier to focus on a specific topic. So instead of doing a guide to all the plants and animals, you could narrow it to something like spider webs, insect damage on leaves, or tracks in the mud (if there are a lot of tracks in the mud)."

   c. "Take a moment, find two partners, and come up with a list of topics you could do in this area."

3. **Tell students to use words, pictures, and numbers to describe each subject of their field guide/collection, highlighting differences and similarities.**

a. "Try to include three to five subjects in your field guide/collection, and use words, pictures, and numbers to describe each subject."

b. "Show what makes each subject different from the others. Your goal is not to make pretty pictures. Your goal is to make accurate observations."

4. **Ask students to share ideas about how to structure their field guides or collections, based on their initial observations of field guides and their own ideas.**

a. "Describe the way the information was laid out on the pages of the field guides you looked at earlier."

b. "What kind of information was included? What did the pages look like?"

c. "Do you have any other creative ideas for how you could structure the pages of your field guide/collection?"

5. **(Optional) Give additional instructions that will focus students' observations to meet a learning goal, such as structure and function or interactions between organisms. For example:**

a. "As you make your field guide/collection of macroinvertebrates, pay specific attention to their structures or body shapes and how they are similar or different."

b. "As you make your field guide/collection of things found under this log, try to notice and record any evidence of how they're interacting with each other or their surroundings."

6. **Remind students of boundaries, ask if there are any questions, and tell them to begin, suggesting that they observe a few possible subjects first.**

a. "Please begin. You will have twenty-three minutes to make your field guide/collection."

b. "You may want to observe a few possible subjects first to help you think ahead about how to organize the information on your page."

7. **As students work, take time to circulate, check in, ask them about their observations, support students who are struggling, and give feedback on observations and strategies for recording information.**

a. (After about half the time has elapsed) "You have about ten minutes left to finish your field guide/collection. If you've only recorded one subject so far, you should move on so you can include at least three subjects in your field guide/collection."

DISCUSSION

Lead a discussion using the general discussion questions and questions from one of the Crosscutting Concept categories. Intersperse pair talk with group discussion.

General Discussion

a. "Did you notice anything interesting as you made your field guide, or did any cool questions arise? Share some with a partner."

Thoughts about similarities and differences in the collection

Drawing feathers life size and to scale makes them easy to compare.

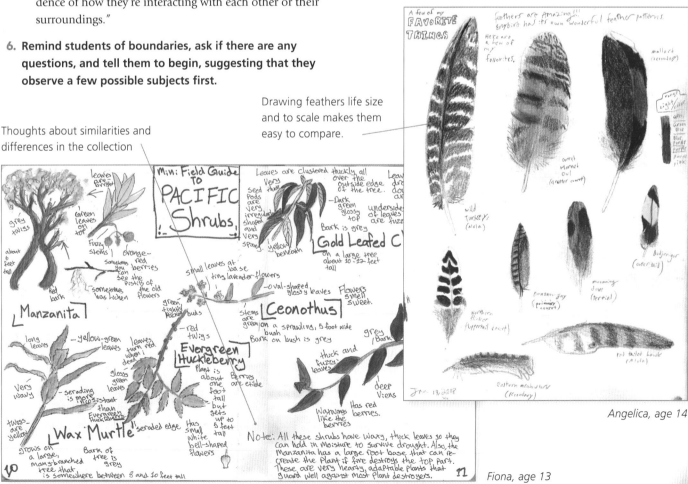

Angelica, age 14

Fiona, age 13

b. "What were some of the similarities and differences between the subjects you recorded in your field guide?"

c. "Check out some of the field guides that other students made. What are some differences and similarities between your subjects or how you chose to record information?"

Patterns

a. "Are there any features or structures that are shared by several or all of your field guide subjects? If so, describe them."

b. "What are some general statements we can make about icicles [or name the field guide category] based on what we observed?"

c. "What are some possible explanations for the similarities and differences we saw?"

Cause and Effect

a. "What are some features that all of the [water catchers, icicles, broken things, etc.] share? How might we describe this category of things in general? What were some of the differences between the subjects of your field guide?"

b. "This discussion might sound like: 'Well, all of the icicles weren't clear all the way through. They all were cylindrical in shape, but their outsides were rough. Some were very long and thin, while others were very thick at the base.'"

c. "What are some possible explanations for why these features occur?"

d. (If you made a field guide to evidence of an effect, such as evidence of drought) "What can we say about how drought is impacting this place?"

Systems and System Models

a. "What were the different things you found? Review a list with a partner."

b. "Did you see any interactions or evidence of interactions between different organisms, or between the organisms and the environment?"

c. "What are some other possible interactions you think might happen?"

d. "Take a moment to draw lines between the plants and animals that you think may interact with each other, and label the line with what you think the interaction might be."

e. "What are the nonliving factors that affect the organisms you saw?"

f. "How might the organisms or the interactions between them be impacted if some of these environmental conditions were to change?"

A SENSE OF PLACE: PHENOLOGY COLLECTION

Making a field guide to signs that represent a specific place at a specific time inspires rich observation and interesting discussions. This observation process gives students a sense of place. What objects would represent this habitat, historic location, stretch of coast, or natural area: characteristic trees, landscape views, rocks, animal behaviors, historical features? What would indicate this moment, this day, this season: morning dew, cloud shapes, seasonal plants (in bud, flower, or fruit), leaves turning color, or migratory birds? Encourage students also to think about sounds, smells, or feelings that could be described with words. Anything that catches a child's attention could be included as it is here now, but this prompt often helps students think more deeply about place. Make similar collections or field guides at different times of the year or different locations.

Structure and Function

a. "Are there any features or structures that are shared by several or all of your objects, such as leaves or bark? Pick one of these common structures and discuss how that structure is different or similar in each of your field guide subjects." "This kind of discussion might sound like: 'Well, the leaf on the oak tree in my field guide was very small and kind of tough, and a little curled. But the leaf on the maple tree was very wide, and way more flimsy."

b. "Now, discuss: How might those different structures function differently?"

c. "What other differences do you see from one object to another? How might they lead to different functions?"

FOLLOW-UP ACTIVITY

Make a Class Field Guide

Pick a local habitat or park and make a field guide as a class project. All the little decisions, from what area the guide should cover to what sorts of details should be included for each species (e.g., drawings, range map, description, seasonal changes or stages, behavior), make for interesting class discussions. Generally a more restricted guide, with a smaller area and narrower scope of contents, is easier to make. A guide to the water birds found in a neighboring marsh is more manageable than the plants, animals, and fungi of Great Smoky Mountain National Park. Divide the work so that all students are involved in development and production of the guide. Make copies for every student and extras for parents or a school fundraiser.

TIMELINE

Students explore plant development through sorting and drawing stages of budding, flowering, and fruiting. Then they discuss possible functions of plant parts.

This activity focuses students on life cycles of plants. As students search for flowers in different stages of growth and attempt to put them in order, they will recognize the major structures and notice how they change. Students will not only learn about that individual plant but also make meaning about the function of different plant parts and how this corresponds to the plant's life cycle.

NATURAL PHENOMENON

This activity should be conducted in an outdoor space where students can observe plants and flowers in different stages of growth, such as a school garden or natural area. Look for species that show stages from bud to flower to fruit or seed at one time. Students could also make a timeline of fungi if you find mushrooms in different stages of development, or of leaves in different phases of decomposition.

PROCEDURE SUMMARY

1. Find a flower in the peak of its bloom and describe it in the center of the page with words and pictures.

2. Find the oldest and youngest versions of this flower and draw them on the left and right sides of your page.

3. Find as many intermediate stages as you can and draw them, arranging them sequentially on the page.

4. Write down any questions that come to you, and include written notes.

DEMONSTRATION

When the whiteboard icon appears in the procedure description: Demonstrate finding and adding stages of flower development, sorting them, and adding notes or questions. Start with one flower in full bloom and show how you can work backward or forward as you add stages. Add stages out of sequence, showing how to leave space for other elements later.

Time

Introduction: 10 minutes
Activity: 20–50 minutes
Discussion: 10–20 minutes

Materials

☐ Journals and pencils

Teaching Notes

This activity won't work in every season. Plan to do it at a time of year when you know there will be enough plants in different stages of growth for students to observe, and scout a location ahead of time.

Certain flowers with complex structures or curling petals can be challenging to draw. If your group is newer to drawing, encourage students to pick a flower with a simple structure to draw, or offer students strategies for drawing complex flowers efficiently.

See instructional videos on johnmuirlaws.com and *The Laws Guide to Nature Drawing and Journaling* for more details about specific strategies for drawing different types of flowers

PROCEDURE STEP-BY-STEP

1. **Explain that students will record observations of a flower at its peak bloom, then will add drawings of older flowers to the right of the original diagram and younger flowers to the left.**

   a. "Find a flower that's at the peak of its bloom. Make a careful drawing or diagram of it in the middle of your page and include written observations and numbers. Leave room to either side for making other notes."

   b. "Then try to find stages of growth that are more developed (older) than your first drawing. Draw and describe these with notes to the right of the central drawing. Try to sort these drawings on your page with the oldest ones farther to the right. Challenge yourself—what is the oldest stage you can find?"

   c. "Next, look for stages that are younger than the central flower. Draw and describe these on the left side of the page. Try to sort these so that the youngest stages are farther to the left."

2. **Tell students to use words, pictures, and numbers to record their observations, and encourage them to write down any questions they have as they're working.**

   a. "Write, draw, and use numbers to describe what you see."

   b. "If any questions occur to you while you are working, write them down in your journal so that you don't forget them, then focus back on the project."

3. **Set boundaries, send students out to journal, then circulate, troubleshoot, and talk to students about their observations.**

4. **(When half the allotted journaling time has elapsed) Say:**

   a. "We're about halfway done. If you've only been working on finding stages older than the original flower, move on to looking for and recording some younger ones."

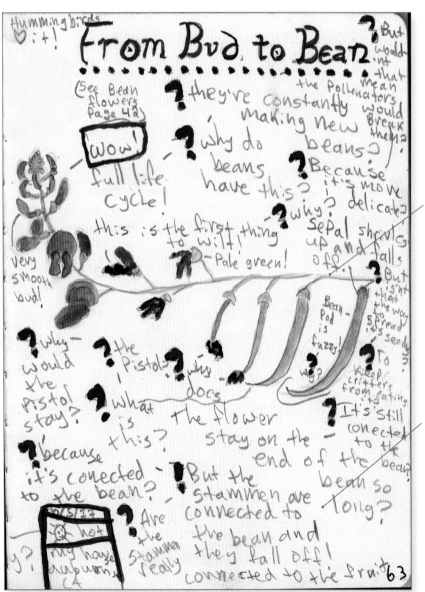

Here, all the stages of the life cycle appear on a single plant. On other species, you will have to look at several plants to find all the stages.

Curiosity chain: Study this series of questions and observations. Notes about the persistence of the flower and stamen on developing fruit lead to a discovery about the closer relationship between the fruit and the pistil.

Fiona, age 13

DISCUSSION

Lead a discussion using the general discussion questions and questions from one of the Crosscutting Concept categories. Intersperse pair talk with group discussion.

General Discussion

a. "In groups of four, share your notes and observations. Find out: Did everyone make the same decisions about what was young or old? If you do not all agree, share your evidence for why you each arranged them the way you did."

b. "It's OK to respectfully disagree with each other. Be willing to change your mind in the face of new evidence. After you are done with your discussion, number your observations to sort them from youngest to oldest."

c. "Look at the different stages you drew, and discuss: What might the function of the flower be at this stage? Then group the stages together based on a common function by labeling each group with a name that describes what you think it might do."

Patterns

a. "Please lay your journals on the ground [picnic table, rock…], then look through the observation notes made by your group. What similar patterns do you see among the different stages of flowers in your journal entries?"

b. "Are there any flower structures that you see at every stage? What's different among them?"

c. "Now look at the ways the plant structures change over time. Do you see any patterns in what changes, and when? Which structures are the first to change?"

d. "How do you think the growth patterns of other flowers might compare?"

Cause and Effect

a. "What changes did you see between different stages of flowers? Which of those changes could have been caused by environmental factors, such as temperature changes or impacts from organisms? Which ones do you think might have been caused by genetic factors?"

b. "Why are these flowers changing? How might this help the plant survive?"

Stability and Change

a. "Take a look at your drawings. What parts of the flowers changed? What stayed the same?"

b. "Did you see any evidence of a change in the rate of change—in other words, were there bursts of activity and pauses? Why? What might have caused those changes?"

c. "What was the youngest stage we found? What do you think came before that? What was the oldest stage we found? What do you think would come after that?"

Structure and Function

a. "With a partner, describe in detail some of the structures you observed."

b. "Pick one structure to focus on, and try to trace how it changed from young flower to old flower."

c. "Pick one structure to focus on, and come up with some possible explanations for its function, thinking about how its function might have changed over time. For example, if a leaf petal became withered at a certain point, why might that be the case? How is its function changing?"

d. "Did you see any evidence of a change in function of the flower or of any individual structures in the flower? How might the functions of this plant change over time? What about individual structures? How did their function change?"

STRING SAFARI

Students discover a world of wonders within the boundaries of a loop of string. Using maps, drawings, and diagrams, they describe their discoveries in the pages of their journals.

Time

Introduction: 5 minutes
Activity: 20–45 minutes
Discussion: 10–15 minutes
Extension: 20–30 minutes

Materials

- ☐ Journals and pencils
- ☐ Pieces of string, 1.5–3 m (5–10 ft.), or hula hoops, one per student

optional

- ☐ Printout of Dürer's "The Great Piece of Turf"
- ☐ Hand lenses

Teaching Notes

To give students a sustained experience of connection with a patch of land, integrate this activity with *Sit Spot* and offer repeated opportunities for students to return to their "string world." They will notice small shifts over time and engage in an authentic study of phenology (seasonal changes). This deepens students' connection with nature and offers a set of observations and data that can be used to better understand the place as a whole.

In 1503, the artist Albrecht Dürer painted "The Great Piece of Turf," a portrayal of a section of weeds and grasses from a German meadow. Each blade is rendered in crisp detail. The painting reveals the complexity and beauty of roadside weeds that are easily overlooked and shows how a small patch of ground can be a rich area for study. In this activity, the string loop focuses students on a particular area and creates a special world in which they find surprising wonders and treasures. Just as field scientists' focused study is a foundation for their thinking, these discoveries can become the driving force of students' learning about life science concepts.

NATURAL PHENOMENA

This activity can be conducted in any outdoor area with varied ground cover. This could be the unmanaged border of a sports field, a vacant lot, or other natural area. Look for spots with different species of weeds and other small plants, and unusual objects such as pinecones, lichens, mushrooms, or rocks. The loop of string does not to be large. In fact, giving students a very large area to observe might be overwhelming. We suggest cutting the string into 1.5–3 m (5–10 ft.) sections.

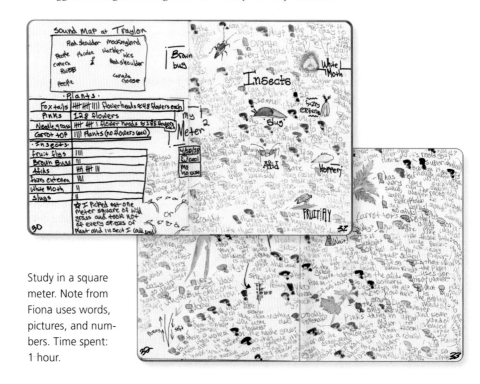

Study in a square meter. Note from Fiona uses words, pictures, and numbers. Time spent: 1 hour.

PROCEDURE SUMMARY

1. Put your loop of string on the ground.

2. Use writing and drawing to record observations about as many subjects as you can within the loop.

3. Use numbers to record amounts of interesting subjects.

DEMONSTRATION

When the whiteboard icon appears in the procedure description: Make quick sketches of plants and animals found within a demonstration loop of string as you describe the activity to students. Simulate adding written notes with sets of horizontal lines. If you plan to do the *Making a System Model* extension activity, leave blocks of white paper on the right and left sides of the paper for the input and output arrows, and instruct students to do the same. You may want to have the students draw light circles in these areas to remind them to leave these sections blank.

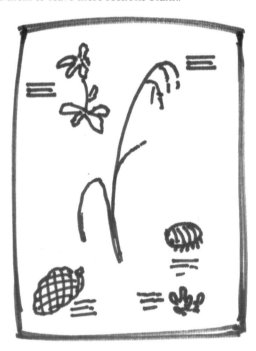

PROCEDURE STEP-BY-STEP

1. **(Optional) Show students Dürer's "The Great Piece of Turf," asking them to share observations about it out loud with a partner.**

2. **Explain that students will get to explore their own "tiny world" by using a loop of string to mark off an area to study.**

   a. "You are about to discover a world hidden in plain sight. When you train yourself to look closely, you can find amazing things anywhere."

   b. "In a moment you will each get a piece of string. We will then spread out and each find a patch of ground that contains interesting things."

   c. "You will use your piece of string to make a circle on the ground, and that will be your 'tiny world' to explore and study."

3. **Explain that students will observe objects within the string circle, then use words, pictures, and numbers to record each subject they find.**

   a. "Look for plants, animals, natural objects, evidence of animals, or any other interesting treasures you can discover in your loop of string. You will need to observe closely to find as much as you can within the boundary."

   b. "Use words, pictures, and numbers to describe what you find."

   c. "What are some things you expect to find that you could include in your journal entry?" (Students might say: soil, dead leaves, insects, etc.)

4. **Explain that students don't need to make detailed drawings of every subject, and can use words or a small map to show where the subjects appear in the circle.**

   a. "Trying to make a detailed drawing of everything in the loop of string exactly where it appears could take a long time, especially if there are leaves or plants on top of each other."

   b. "You could draw a small circle in the center of your page, make individual drawings of each thing you find in the loop of string around the circle, and then use writing and arrows to describe where they are found."

   c. "You could even make a small overhead map to show where each object you draw is found within the loop of string."

   d. "Does anyone else have any ideas about creative or interesting ways to record what we find?"

5. **Suggest counting and observing differences and similarities among objects students find a lot of, such as pinecones or leaves.**

   a. "If you find a lot of some kind of object, like a lot of pinecones or rocks, you could count them or observe similarities and differences."

6. **Tell students that it's OK if they get engrossed for a little while in observing a certain object or organism within their loop of string.**

   a. "You might find one object or organism that is very intriguing to you, and if you get lost in observing it and recording observations about it, that's OK."

7. **If you will be doing the *Making a System Model* extension activity, instruct students to leave a blank spot to the right and left of the page.**

8. **Tell students to keep looking for things even if they feel "done," and to see what interesting things they can notice after they feel as though they are done.**

a. "At some point before the time is up, you will probably get the sense that you are done. Remember, this is just a feeling, and has nothing to do with how much or how deeply you can observe."

b. "If you feel done, notice the feeling and then see how much more you can discover after that."

9. **Set boundaries, give a time limit, ask if there are any questions, then send students out to journal. Be a role model by working in your own journal.**

10. **When half the allotted journaling time has elapsed, circulate, troubleshoot, refocus, encourage, and talk to students about their observations, using questions like these:**

a. "What else do you notice?" (Student makes observations.) "Good—put that down in your journal."

b. "What are other ways you could describe what you find in this space or add more details to your notes?" Possibilities include maps, cross sections, close-ups, alternate drawing angles, more written notes (paragraphs, bullet points, or labels).

DISCUSSION

Lead a discussion using the general discussion questions and questions from one of the Crosscutting Concept categories. Intersperse pair talk with group discussion.

General Discussion

a. "In groups of four, share your notes and observations, and discuss: How did limiting your observations to the area inside the string affect the way you observed? What was it like studying a 'tiny world'?"

b. "What was the most interesting observation you made? What questions do you have about what you found in your loop of string?"

c. "During your study, did you come up with any new ways of taking notes or recording information you could use again in the future?"

d. "If you could do this same activity again in any location on earth, where would it be, and why?"

Stability and Change

a. "Did you see any evidence of things that are changing in your observation area? How can you tell?"

b. "If you came back in one week, do you think you would notice any changes in the area you studied? What might those be?"

c. "If you set up a camera to record your area over one year, what changes would you expect to see?"

d. "What thing in the area you observed would change the most? What would change the least?"

Patterns

a. "Take a look at the notes or observation areas of a couple of people nearby. What are similarities and differences that you notice in the types of things that are in your circles?"

b. "What about the location of things in the circle? For example, are there leaves all over in every person's circle, or are there other patterns of things that were found?"

Cause and Effect

a. "Take a look at your observation area and journal page. Did you notice any mysteries to unravel? A mystery might be a hole in a leaf, the location of a pinecone, or a piece of grass that has been bent."

b. "When scientists notice something, they often think about how it came to be, or what caused it. See if you can come up with some possible explanations for the mysteries in your circle."

Energy and Matter

a. "Did you see any evidence of decomposition [or, if students don't know that term, "things breaking down"] in the area you studied? What did it look like? Say as much as you can about what this process looks like."

b. "Did you see any evidence of last year's plants? What do they look like now? Where are the plants from two years ago?"

STRING SAFARI EXTENSION: MAKING A SYSTEM MODEL

The area within the string can be a jumping-off point for beginning to think about ecosystem modeling. Systems thinking can help students explain their observations or make nuanced predictions about what might happen in a given area if conditions change. All system models have boundaries that define what is inside and outside the system. Even in a seemingly clear system such as a pond, exactly where do you draw the boundary? Do you draw the line at the water's edge, the zone of wet soil, the extent of pond-related vegetation, or the watershed boundary? These boundaries must be decided, but they are somewhat arbitrary. Practicing making a system with the string as a boundary will prepare students to apply systems thinking in other contexts.

After defining the boundary, students think about interactions between plants, animals, and microbes within the system; consider outside forces entering the system (inputs); then discuss things leaving the system (outputs).

PROCEDURE SUMMARY

1. Convert the objects in the diagram to a system model: use words, pictures, and arrows to identify boundaries, system components, and the inputs and outputs of the system.
2. Use words, pictures, and numbers to identify and show interactions between system components, inputs, and outputs.

DEMONSTRATION

When the whiteboard icon appears in the procedure description: Draw arrows between system elements that may interact. Add labels to the arrows, briefly describing the interaction.

Add an input arrow and list things that enter or influence the system. Add an output arrow and list things that leave or are produced by the system.

Time

30–40 minutes

PROCEDURE STEP BY STEP

1. **Briefly define what a system is, offering some examples that students will be familiar with (e.g., the respiratory system, a bicycle).**

   a. "A system is a group of related parts that work together. For example, your body is a system of organs that function together to keep you alive. The parts of a bicycle are a system."

   b. "The parts of a system interact and affect each other. For example, a bicycle pedal is connected to the gears with a chain, which turns the wheels when you push on the pedal."

   c. "A system can be any size, from the digestive system contained in one small organism, to a small pond, to entire forests, to climate systems that encompass the entire earth."

   d. "You can think of parts of nature, such as a tree, as systems too; parts of a tree interact with and impact each other."

2. **Tell students that they will practice using the idea of systems by looking at their loop of string as a boundary.**

   a. "To practice thinking about things as a system, we are making these loops of string the boundaries of our system."

   b. "Draw a dotted line around the things on your page to represent the boundary of this system."

3. **Guide students through the process of identifying parts of their system, thinking about how the parts interact with each other, and using arrows to represent these interactions.**

   a. "Discuss with a partner: What are the parts of the system in your loop of string?"

   b. "Can you think of any ways that the parts within the system might interact with each other?"

   c. "Take a moment to draw arrows between elements in the system that you think interact with each other. Label the arrows with a brief description of the interaction it represents."

4. **Guide students through the process of thinking about the inputs to the system by asking them about forces outside the loop of string that interact with the system, and how those forces affect the system.**

   a. "What are some forces that may affect this system, but are not inside this loop of string? Consider things you cannot see, and things that may have happened in the past but that you do not see going on right now."

   b. "These are called inputs. Draw a big arrow pointing into the system and label it with the inputs you named."

   c. "How do these inputs affect the parts of the system—the things you found, labeled, and described there?"

5. **Guide students through the process of thinking about outputs of the system by asking them what leaves the loop of string, and how that affects nearby areas.**

   a. "What are some of the things that leave the loop of string? Does the system create or produce anything? Again, think about things you both can and can't see."

   b. "These are called outputs. Write down some of the outputs you named, then draw an arrow around them, pointing out of the system."

   c. "How do these outputs interact with or affect nearby areas?"

6. **Ask students to make some predictions about what might happen if the quantity of an input, such as water, were to change.**

   a. "Now that we've labeled some interactions among members of the system, and inputs and outputs in it, we can use our model to think about what could happen to the parts of the system if the conditions of this area were to change in the future."

   b. "Look at the set of inputs you listed. What might happen if one of these inputs, such as water, increased or decreased in quantity?"

   c. "Use the interactions you labeled to trace how that would impact the things inside the system."

   d. "How would those changes, in turn, affect the outputs of the system?"

7. **Explain that any part of the system with lots of arrows pointing to it interacts with many other parts of the system, and that changes in that one part of the system can greatly affect everything else.**

   a. "Was there any part of your system that had a lot of arrows pointing to it? This means that a lot of things within the system interact with it."

   b. "If it were to disappear or decrease in quantity, that could significantly affect the other parts of the system."

8. **Point out that the thinking students are doing is similar to the thinking that scientists do when they study ecosystems and how they might be affected by changes in the environment.**

   a. "Scientists often use systems thinking and have these kinds of conversations when they are looking at ecosystems and trying to make predictions about what might happen in the future."

9. **Explain that the boundaries of a system affect what we learn from it, then ask students to think about what might be different if the boundary were much smaller or much larger.**

   a. "For this system model, our boundaries were a loop of string."

   b. "When you're making a system, you can choose to put the boundary in many different places. In the case of a pond, you could draw a boundary where the water ends, or around the plants next to the pond, or where you stop seeing the influence of water."

   c. "Where you draw the boundaries of the system affects the kind of explanations you make about what is happening within the system."

   d. "If you were instead to define the boundary of the system as one plant, how would the inputs and outputs change? What about if you defined the system as this whole [forest, meadow, schoolyard] instead of the loop of string?"

FOLLOW-UP ACTIVITIES

Repeating the Activity in a Different Environment

Students could repeat this activity in a different ecosystem or biome, then think about the differences between the two areas.

"The Magic Circle": Team Observation

A desert ecologist once drew a 100-foot circle around a Joshua tree, then directed his whole group of students to observe, categorize, and draw everything they saw within it. This larger riff on the circle of string can build a group's understanding of a specific area and environment, and could be a fun way to frame the activity *Team Observation*.

ANIMAL ENCOUNTERS

When your students encounter wildlife, the opportunity for observation may be brief. This activity will help your students take advantage of the situation, offering a structure for deliberate observation and getting the observations to paper as fast as possible.

When you find an animal in the field, there is no way to know how long you will be able to observe it. It helps to have an plan to allow you and your students to get the most out of what you see. This activity is a template for action: Begin by having students verbalize their observations as they get out journals, then briefly prompt students to use words, pictures, and numbers to record what they observe. If the animal sticks around and students are still engaged, offer different suggestions to focus their journaling without interrupting the group. When the animal leaves, debrief by sharing questions. The level of detail that students remember about the animal encounter will be much higher than if they had not used their journals or verbalized their observations. Having an approach for animal encounters in your "back pocket" allows you to be flexible and to respond to the environment. You can also weave this activity into a learning experience, use it as an opportunity to jump into further research about the animal, or make connections to science concepts.

NATURAL PHENOMENA

You can use this protocol with any animal encounter in the field. This approach works for mammals, birds, insects, reptiles, and amphibians. The longer, closer, and less obstructed the view the better, but go with what you get. Even a quick glimpse of a deer in passing is rich with information if you intentionally remember what you see.

PROCEDURE SUMMARY

1. Say observations out loud as long as you can see the animal.

2. Use words, pictures, and numbers to describe the appearance and behavior of the animal, using words to capture what is difficult to draw quickly.

3. Because the animal will move, start several drawings and work on one whenever an animal assumes that pose.

Note: There is no whiteboard demonstration for this activity because students should be looking at the animal, not you. Direct their attention toward the animal and tell them to verbalize their observations while they get out their journals and pencils. Then give verbal instructions as quickly as possible, instructing students to keep their eyes trained on the animal while they listen to you.

PROCEDURE STEP-BY-STEP

1. **Make sure everyone can see the animal.**

   a. "James found an animal! Show us where it is. Raise your hand if you don't see it yet. Who can give clear directions to describe how to find it?"

2. **Get the group started verbalizing observations (in a soft conversational voice) as they get out their journals, encouraging them to listen to one another's ideas.**

   a. "To help us remember the details we see and to make sure we observe accurately, I want everyone to start describing the details of how the animal looks and what it is doing, out loud in a soft conversational voice. [This is generally not loud enough to scare animals away.]"

Time

Introduction: 1 minute
Activity: Variable depending on the cooperativeness of the animal
Discussion: 10–15 minutes

Materials

☐ Journals and pencils

optional
☐ Binoculars

Teaching Notes

Verbalizing observations is an important aspect of nature study and a critical part of this activity. If students describe what they see out loud, they will remember with greater detail and clarity what they have observed. The first moments of observation are the most critical, as the animal you are watching may disappear the next moment. Don't wait until you have passed all the journals out to start observing. Tell students to start a stream-of-consciousness list of observations, describing how the animal looks, what it is doing, and anything else they notice. After a minute of this intense directed observation, students will be ready to start transferring observations to paper.

b. "Listen to the observations of people standing near you, and add to their observations. Are you seeing the same things?"

c. "Continue saying your observations out loud while I pass out your journals and pencils [or while you get your journals out of your bags or backpacks]."

3. **Tell students to use words, pictures, and numbers to describe the appearance and behavior of the animal, thinking carefully about which note-taking approach is best for the information they're trying to capture.**

a. "Let's use this opportunity to get as much information down on the page as we can. Use words, pictures, and numbers to record your observations."

b. "Some observations are easier to show with written notes than with drawings. Use written notes along with your sketches. These can be in paragraph form to describe behaviors or structures that are difficult to draw, a list of observations down the side, and labels that connect to the drawings with little lines."

4. **Offer a quick strategy for dealing with the animal's movements: Students start several drawings of different poses, shifting the drawing they are working on when the animal moves.**

a. "This animal is going to move, but don't let that stop you from drawing it."

b. "On your journal page, start three drawings to describe this animal. Make two drawings that show different views

or poses. These could be side and back views, or two different positions the animal returns to often (like head up and head down). The third view will be a close-up of some detail that is interesting to you."

c. "Bounce around from one drawing to another as the animal moves. When you get a good look at the detailed pose you are interested in, work on that drawing. When you get your side view, work on that drawing. You do not need to finish all three drawings. One will probably get further developed than the others, and that's OK."

5. **Encourage students to keep working in their journals as long as the animal is there.**

6. **As time passes, drop in different prompts to focus students' attention, offering the prompts out loud to the group as suggestions (but not requiring that students stop their journaling to listen to you). For example:**

a. "If you would like, you can begin to find the numbers hidden in your observations. Count, measure, and time things. For example, you might time the number of seconds the animal spends doing different behaviors, estimate the distance between the animal and the forest, count the number of stripes on the animal, and so on."

b. "Take a moment to focus on asking and recording questions. If you have not already, make a question mark icon and list as many questions as you can come up with below it. Then go back to observing and add in questions as they come to you."

c. (If you are watching a single individual) "How does it look? Begin to focus on recording its body shape, and markings. What is it doing? What behaviors and

Draw different views and postures of the same animal. Combine enlargements, details, and fast posture sketches.

Reinforce key observations that you have drawn with written notes. Add "I wonders" and "It reminds me ofs."

Angelica, age 14

Bree, age 10

movements do you see? Focus on context—where is it in relation to cover or other parts of the environment?"

d. (If you are watching a group of animals) "Can you find one animal that can be clearly distinguished from the others? What characteristics make it unique? Observe the way they group up. How close do they typically space themselves? Does this change? Let's look for interactions between individuals. Does the behavior of one individual seem to affect others? Let's think about the group as a whole. How might we describe the behavior of the herd or flock?"

7. **If the animal(s) leave, call the group together to record metadata and to complete field notes from memory, adding in details they haven't recorded yet.**

a. "Our observations are not complete without recording the date, time, location, and weather. Add this metadata to your field notes."

b. "In time we will forget the details we do not put down on paper. Think for a moment about details or behaviors you observed that are not recorded in your notes. Let's take five more minutes to fill out the rest of your notes. This could be written descriptions or more details in your drawings."

DISCUSSION

Lead a discussion using the general discussion questions and questions from one of the Crosscutting Concept categories. Intersperse pair talk with group discussion.

General Discussion

a. "Find a partner and discuss the things you saw that were interesting or surprising. Compare your journal entries and approaches to note taking to see what you can learn from each other."

b. "Place your journals on this picnic table [or on the ground], open to your last entry. Circulate around and look at the way your classmates recorded their observations. There are many ways to do this kind of work. Find at least three observations that you missed or did not record in your journal. Also look for creative ways that other people recorded their observations. Can you find some journaling ideas that you could use in future observations?"

c. Ask students to bring up any interesting questions they had, and follow up with discussion of them if students are interested.

Patterns

a. "What patterns did you observe?"

b. "What does that remind you of? Where else have you seen similar patterns? Where would you expect to see similar patterns—for example, in structure, color, or behavior?"

Cause and Effect

a. "How did [organism 1] affect [organism 2]? What is your evidence?"

b. "Did you find any evidence that the [observed animal] may be affected by living or nonliving things in the environment?"

c. "How do you think the [observed animal] might affect the living or nonliving things in the environment?"

d. "How might the interactions we observed be affected by the time of day, year, weather, or location? What kinds of things might cause this animal's behavior to change?"

ENERGY AND MATTER

Note: These questions are appropriate for students fifth grade and above.

a. "Let's construct a partial food chain based on our observations. What did you see eating what? Now expand your food chain to a web based on what you have seen in this area, your prior knowledge, and your best guesses about other relationships between animals."

b. "Now trace the cycling of matter through the parts of the food chain you just described. Use arrows to show which direction matter is moving among the organisms you observed."

Structure and Function

a. "Study your notes and drawings of the [observed animal]. Do any body parts seem specialized to do specific things or functions?"

b. "How did you see the animal moving? How did its specific structures help it move? Connect your explanations to the environment, thinking about how the organism's structures help it survive in this specific context."

FOLLOW-UP ACTIVITY

Using Reference Material

Encourage students to look up more information about the species they observed in the field. Can you find references to behaviors they observed? Can you find details that would be fun to look for the next time the group finds this animal outside? Remind them that all the information in their reference books or online resources originated in the same way: someone making careful observations and recording them in their field notes, just as they had done.

SPECIES ACCOUNT

Students choose one species that they can readily observe, and document as many details as they can about it through direct observation.

Much of what we know about nature started with direct observation and experimentation. There are many species that have been deeply studied, but there is always more to learn and discover. Each observation, if recorded and shared, becomes part of a growing understanding of the world. Species accounts are a common approach to cataloguing organisms and building a database of information. In a species account, the observer attempts to learn as much as they can about the type of organism, using words, pictures, and numbers to record details about structures, behaviors, and location in and interaction with the surrounding environment. Once students learn and develop an approach for doing a focused species study, they can apply the skills anywhere they go.

NATURAL PHENOMENA

Any plant or animal that can be observed for a sustained period can be used for a species account. If you think an animal might scamper away, use the *Animal Encounters* protocol instead. Students don't necessarily need to focus on the same organism, unless you want the whole group to build a base of observations to use to reach specific learning goals. Find an area with enough plants, animals, or fungi that individual students could choose their own subject to observe. Plants are very cooperative and will not walk away. Animals are fun because they exhibit behavior that can also be recorded. Encourage students to choose animals that will not crawl or fly away halfway through the observation period. Catching small insects, macroinvertebrates, or other critters in clear plastic cups is a way to deal with this issue. Captive animals are often easy to observe, but may exhibit behavioral and structural differences compared to wild animals.

PROCEDURE SUMMARY

1. Record as many observations and questions about this species as you can, using words, pictures, and numbers.

2. Include information about how the organism looks, its behaviors and feeding habits, where it was found, and the like.

3. Focus on specific observations, not explanations.

DEMONSTRATION

When the whiteboard icon appears in the procedure description: As you suggest things to include in a species account, create a sample page that reflects those suggestions. Do not worry about making a pretty picture. Your bunny can be a circle with two lines for ears. Demonstrate making more than one sketch, to show different

views; changing scale; using words, pictures, and numbers; and including metadata.

PROCEDURE STEP-BY-STEP

1. **Tell students that they will get the chance to learn as much as they can about a specific species by studying it.**

   a. "We are about to practice our observation skills by using them to learn as much about [ants, this type of tree, these worms, a species of students' choosing, etc.] as possible."

   b. "This is your chance to do an in-depth study and become more familiar with this part of nature. We can learn a lot when we focus in on one species and give it directed attention."

2. **Explain that the goal is to describe the species in as much detail as possible using words, pictures, and numbers; and students can use "I notice, I wonder, It reminds me of" to help them focus on specific observations.**

   a. "You don't need to make a pretty picture of this species, but you do need to record as many observations as you can."

   b. "Use words, pictures, and numbers together to describe what you see, relying more on whichever approach is most comfortable."

   c. "You can use the frame 'I notice, I wonder, It reminds me of' to help guide your observations and what to write down."

3. **Encourage students to be specific with their observations and language.**

   a. "When you make observations, be as specific as possible. Don't just say 'The leaf is green'; rather, say 'It is deep blue-green at the base, shading to yellow-green within two millimeters of the edges.' It's important to come up with as accurate a description of this organism as possible."

4. **Tell students to focus on making observations (e.g., "There are yellow leaves at the ends of the branches"), not assumptions or explanations ("The leaves at the ends of the branches are dead").**

   a. "Sometimes when I record observations, I make assumptions or explanations for what I see. I might say, 'Older leaves on the tree are yellow,'

but I don't really know that those leaves are older. I saw the yellow leaves and without realizing it assumed that they were older. This is an explanation for why the leaves are yellow. My assumption that the yellow leaves are older may be wrong."

   b. "When you record your observations, try to avoid assumptions or explanations. Just describe what you see, not what you think is going on."

   c. "I could say instead, 'Larger leaves that are farther from the branch tips are yellow.'"

5. **Ask students for suggestions of observations they could include that go beyond just describing how the organism looks.**

   a. "We can do more than just record observations to show what the organism looks like. What are some examples of the types of observations we can make?"

   b. If students don't mention any of the following, bring them up: notes about its location, notes about other nearby organisms, behavior if it is an animal, similarities or differences compared to nearby individuals of the same species, evidence of where else the organism has been, feeding behavior, contextual information such as nearby soil, the weather conditions, associated species, or a small map showing the area of study.

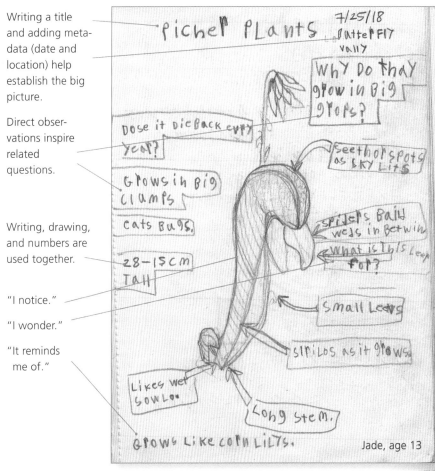

Writing a title and adding metadata (date and location) help establish the big picture.

Direct observations inspire related questions.

Writing, drawing, and numbers are used together.

"I notice."

"I wonder."

"It reminds me of."

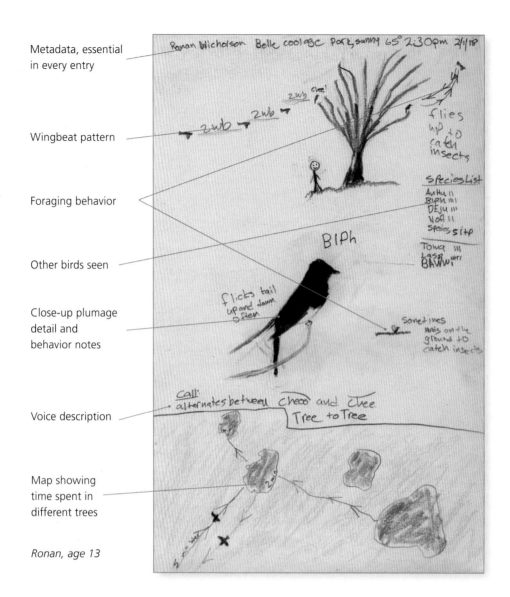

Metadata, essential in every entry

Wingbeat pattern

Foraging behavior

Other birds seen

Close-up plumage detail and behavior notes

Voice description

Map showing time spent in different trees

Ronan, age 13

6. **Encourage students to record questions that occur to them as they work.**

   a. "Put a question mark somewhere on the page, and record any questions that come to you as you work."

7. **Offer specific strategies for drawing the subject students will focus on, as appropriate.**

8. **Send students out to observe.**

   a. "You have nineteen minutes. Are there any questions? Begin."

DISCUSSION

Lead a discussion using the general discussion questions and questions from one of the Crosscutting Concept categories. Intersperse pair talk with group discussion.

General Discussion

   a. "Let's share some of our observations. Look at your notes and find an example where you provided rich and specific details. Let's share some of these with the group."

   b. "How did recording observations in a journal help you learn? Why do you think using journals might be a useful practice in science?"

   c. "What are some of the limitations of gaining knowledge by journaling? Are there other observation tools that would help you learn even more?"

   d. "What were some observations that were more easily shown with a drawing? What observations were easier to record with writing?"

   e. "Were you able to focus on observation and avoid going into explanations in your journal?"

   f. "What could you do next to discover more about this individual organism, or type of organism?"

   g. "Let's go around the circle and each complete this sentence: 'From my direct observation, I now know that…'"

Patterns

   a. "What patterns did you observe? These might be patterns in shapes, colors, growth, behavior, or the location of species."

b. "What does that pattern remind you of? Where else have you seen similar patterns?"

c. "What are some possible explanations for one of the patterns you observed?

d. "Do you expect that the patterns you saw might be different during a different season or in a different location? Why or why not?"

Cause and Effect

a. "Were there any interesting or unique markings you noticed while doing your species account? What are some possible explanations for them?"

b. "When we study an organism, we can notice where it is located, what it is doing, its shape or growth pattern, or any evidence of interactions with other species or the environment. This is all valuable information, and each category can be thought of as a mystery to explain. Look back at your notes and discuss these questions:"

- "Where is the organism located? What are some possible causes of this?"

- "Did you find any evidence that the [observed organism] may be affected by living or nonliving things in the environment?"

- "How do you think the [observed organism] might affect the living or nonliving things in the environment?"

- "How might the interactions we observed be affected by the time of day, year, weather, or location?"

- "Did you see any interactions between the [observed organism] and the environment? What effect might they have had on each other?"

Structure and Function

a. "What were some of the structures you noticed while studying this species? Describe them in detail."

b. "Pick a structure and think about how it might function or work to help this organism survive in this environment. Connect your explanation with a specific description of the structure and how it works in this environment. For example, don't just say 'Its fur helps it hide.' Say, 'The brown spots on the fur look like they might help it blend in to the dead grass or the hillside.'"

Systems and System Models

a. "How do the [observed organism] and the [other thing or organism in area] affect each other?"

b. "What connections between the [observed organism] and other parts of the ecosystem did we observe? How many others can you think of?"

Energy and Matter

Note: These questions are appropriate for students fifth grade and above.

a. "Let's construct a partial food chain based on your observations. What did you see eating what? Now expand your food chain to a web based on what you have seen in this area, your prior knowledge, and your best guesses about other relationships between animals."

b. "Now trace the cycling of matter through the parts of the food chain you just described. Use arrows to show how matter cycles through different parts of this ecosystem."

FOLLOW-UP ACTIVITIES

Conducting Further Research

Encourage students to supplement their personal observations with research. Have other scientists seen the same patterns or behaviors that your students observed? Offer resources such as field guides, research papers, or contact information for local scientists so that students can answer their questions and extend their studies. The chapter Teaching Science and Inquiry: A Deeper Dive, following the activities, includes ideas for how students can engage in future research.

Looking at Others' Field Notes

Grinnell and his colleagues at the Museum of Vertebrate Zoology in Berkeley, California, made detailed species accounts across the western United States to gather distribution and natural history data. The types of information they collected for each species can provide a framework for what students pay attention to whenever they encounter a new species. Students could study these examples of species accounts and use them to guide the observations they make the next time they take notes.

FOREST KARAOKE: TRANSCRIBING BIRDSONG

Students describe bird songs in their journals using writing, drawing, diagramming, and numbers.

Time

Introduction: 10 minutes
Activity: 10–20 minutes
Discussion: 10–15 minutes

Materials

☐ Journals and pencils

optional

☐ Watch with second hand

Teaching Notes

Because you can't control when birds sing, this might be a "start-and-stop" activity that occurs as you and your students explore a natural area. Give the instructions at the beginning of the outing, and when you hear a bird start singing, tell students to start recording right away. If the bird stops singing before you've finished the exercise, pause, have students finish recording what they've noticed so far, then resume if the bird starts singing again.

If you can see the bird as it sings, you may be able to identify it. If it's not visible or identifiable, students can transcribe its song and call it "mystery bird #1" or come up with a descriptive name like "buzzy buzzy bounce." You might find the bird singing from an exposed perch later, and students might be able to pick its song out as well because they've paid attention to it—an exciting discovery!

As students listen to birdsong in accurate detail, they will be able to record sound in their journals. This is an experience relevant beyond listening to birdsong, as it offers an approach to describing any novel auditory phenomenon. Using these multiple modes of recording sound data can change the way students think about and interact with sound. Sound is another viable and valuable way of learning about a place.

NATURAL PHENOMENA

This activity can be conducted anywhere, but is richest in a complex and diverse natural area where students can hear different species of birds, as well as other natural sounds, such as creeks and rustling leaves. Mornings in early spring often are the best times to listen for birdsong. If possible, find a bird singing the same song again and again. Some species, such as mockingbirds, do not repeat a stereotyped song but sing a continuous babble. Avoid these songs when you are introducing this technique and instead focus on species that sing a simple song, pause, and repeat the same song. This activity could also take place in more controlled conditions with recorded bird songs (available on many websites), but it is more fun and relevant in the field.

PROCEDURE SUMMARY

1. Draw birdsong on paper, using longer lines to show longer notes and shorter lines to show shorter notes.

2. Use rising and falling lines to show changes in pitch, and heavier lines to show louder sounds.

3. Describe the song using words (*buzzy, harsh, bouncy,* etc.).

4. Time how long it takes for the bird to sing and the intervals between songs.

DEMONSTRATION

When the whiteboard icon appears in the procedure description: Show the steps of transcribing birdsong as you describe them to students. Draw the "shape" of the birdsong, using lines to show pitch and length of notes. Write the words to the birdsong. Use descriptive words to describe the quality of the song. Time the length of the song and the interval between songs. Describe the context of the song (where, when).

PROCEDURE STEP-BY-STEP

1. **When you hear a bird singing nearby, help students listen to the song by instructing them to "conduct" the song in the air with their hands.**

   a. "Listen—a bird is singing! Close your eyes and focus on the sound of the bird."

   b. "Gently lift your hand in front of you and move your arm and fingers as if you were conducting or controlling the bird's voice."

   c. "Raise your hand higher when the bird sings high notes and lower when the bird sings low notes. Wiggle your fingers when it buzzes or warbles quickly."

2. **After a minute or so, tell students to imitate the song using words and nonsense babble, and then to compare their imitation with a friend.**

   a. "Now quietly imitate that song to yourself. Make a set of noises that as accurately as possible mimics the bird's song."

   b. "Compare your imitation with a friend's. Modify your song if you heard someone else give a better impression than what you initially did."

3. **Demonstrate how to transcribe the birdsong into rising and falling lines on paper, using longer lines for longer notes, rising and falling lines to show changes in pitch, scribbles for buzzes, and thick marks for loud sounds.**

   a. "Now let's draw the birdsong on paper! Use longer lines to show longer notes, and shorter lines to show shorter notes."

   b. "To show changes in pitch, use rising and falling lines. Buzzes can be scribble marks. Loud notes can be heavier lines."

4. **Give students a few minutes to transcribe the birdsong using lines and to add words to describe the imitation they came up with.**

   a. "Give it a try!! Record the birdsong in your journal."

   b. "Above (or below) the lines showing the song, use words to add the imitation you came up with."

5. **Tell the group to use descriptive words to transcribe the "quality" of the song, offering a list of words that describe birdsong (e.g., *quiet, loud, musical, mechanical, buzzy*).**

   a. "Let's use a few other descriptive words to describe this birdsong. Is it loud or soft? Is it the same volume all the way through the song, or does it get louder or softer at the end?"

   b. "Here's a list of terms that describe some bird songs. Write down any that match the song: melodic, mechanical, bouncy, harsh, clear, nasal, sweet, buzzy, slow and relaxed, fast and energetic, piercing."

   c. "Describe the speed. Is it a slow series or a fast trill?"

6. **Tell students to time and record the length of the song and the interval between songs, and to add this information to their journals.**

   a. "Now let's count the seconds that it takes for the bird to sing one song. A second lasts about the time it takes to say 'one hippopotamus.' Record the time in your notes."

   b. "Count how many seconds there are between songs. We will measure the time between the next four songs to see whether they are all the same."

7. **Tell the group to describe the context (location, time of day and month, and any other important details about the surrounding environment) of the birdsong.**

   a. "To make these notes more accurate and more relevant to science, let's record some of our observations about the surroundings and the scene."

   b. "In short sentences or in list format, describe where the bird is singing. Does it sing at the top of a tree or in the middle of a bush?"

   c. "Now describe the surrounding habitat." (Examples might include: meadow, field, dense forest, schoolyard, park.)

How many ways can students describe the sounds they hear? Each approach helps them hear in a different way.

This is a sketch of a squirrel peeking over a log.

Samuel, age 8

These are observations of a California ground squirrel making alarm calls from behind a log. Note the different modes used to record the noise, including music notes and rests, written words, and a graph (pitch over time). The sound was first assumed to be a bird call, but was then discovered to be a squirrel. Note the correction at the top of the page—it's OK to change your mind in the face of evidence!

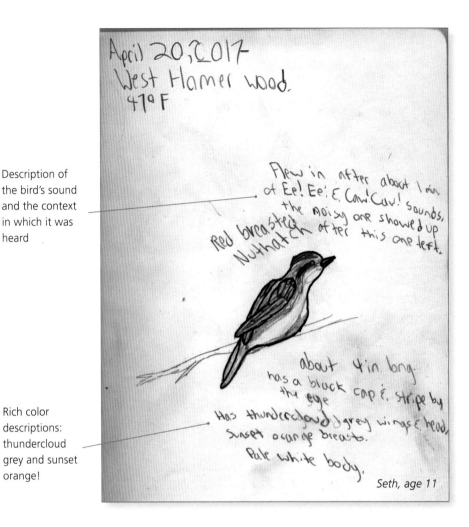

April 20, 2017
West Hamer Wood.
47° F

Flew in after about 1 min
of Ee! Ee! & Caw Caw! sounds,
the noisy one showed up
after this one left.

Red breasted
Nuthatch

about 4 in. long.
has a black cap & stripe by
the eye
Has thundercloud grey wings & head,
sunset orange breast.
Pale white body.

Seth, age 11

Description of the bird's sound and the context in which it was heard

Rich color descriptions: thundercloud grey and sunset orange!

---

d. (If you know the name of the park or area you're in, prompt students to record that, too.) "Let's add the city [or town or wild area] and state names."

e. "Now let's add some other metadata. What time is it? What is the date? Weather?"

f. "Is there any other important big-picture information that we should have in our field notes? This might be anything that could help us think more about the bird and its song in the future."

DISCUSSION

Lead a discussion using the general discussion questions and questions from one of the Crosscutting Concept categories. Intersperse pair talk with group discussion.

General Discussion

a. "We used words, diagrams, and timing to describe the bird's song. What kinds of information can be communicated in each of these note-taking approaches?"

b. "How might you include these note-taking approaches in future sound studies, or as we record information about other subjects?"

c. "Most people do not have a good vocabulary for describing sound. What are other things we can sense, but that are difficult to describe?"

d. "While studying birdsong, did anyone find any questions, mysteries, or things that seemed strange?"

Patterns

a. "Each species of bird makes a different song. Each time we hear this kind of bird today, let's notice where it is singing—such as the type of foliage, the height, or the distance from water—and see whether we can determine a pattern to where we find them singing and where we do not."

b. "We recorded three different bird songs today. Are there any patterns you can notice among their songs, such as the length of notes or the presence of trills?"

c. "What are some possible explanations for similarities or differences in bird songs?"

d. (If you heard many birds throughout the day) "How would you group these bird songs based on the commonalities or the differences between them?"

Cause and Effect

a. "What might be some of the reasons birds sing or call? How might it help them survive?"

b. "Were we able to observe any clues about why birds sing or what affects when and how they sing? This might include behaviors we saw them do while singing, or when they were singing more or less."

c. "Are there any ways that singing might impact birds in a negative way?"

d. "What are some possible explanations for the different types of bird songs that we heard?"

FOLLOW-UP ACTIVITIES

Exploring Pitch

Adults easily grasp the idea of high pitches being "high" and graphing them higher on a page than "low" pitches. However, there is nothing objectively low or high about pitch. They are different frequencies of sound waves. We just think of them as low and high because we have always heard them described that way and have seen music scores that show higher-pitch notes displayed above lower-pitch notes. If students do not have musical training, this is not always intuitive.

To introduce the idea of conceptualizing pitch to students, play music with clear pitch changes and let students "conduct" the music with their hands. Teach them to bring their hands up higher for high-pitch notes and down lower for low-pitch notes.

It is also fun to watch classical conductors and study the way that their hand movements reflect the music. Conductors use the movement of their hands to convey rhythm, and don't control the pitch through the height of their hand. They do often give visuals about dynamics (volume) and intensity to an orchestra. Observing this can help students be more refined in how they "conduct" the music themselves, and prepares them to convert hand movements to lines on paper.

Then play music clips or sounds and have students transcribe the sounds to a graph. Make a simple graph with time on the horizontal (x) axis and pitch on the vertical (y) axis. Search online for audio spectrograms and bird sonograms, similar graphs generated by a computer.

Twinkle, twinkle, little star

Studying Birdsong

Use this activity as a catalyst to inspire research on birdsong. Why do birds sing? How does singing help a bird survive and reproduce? Do all birds sing? What are the dangers or costs of singing? What is counter-singing? An interesting book to extend the study of birdsong is *What the Robin Knows*, by Jon Young.

SOUNDSCAPE MAPS

Students listen to the soundscape around them, then diagram and map the soundscape using symbols, different colors, and other ways to graphically represent sound.

Time

Introduction: 5 minutes
Activity: 10–20 minutes
Discussion: 10–15 minutes

Materials

☐ Journals and pencils

☐ Colored pencils

Teaching Notes

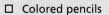

Practice transcribing birdsong with the *Forest Karaoke* activity before asking students to make soundscape maps. Once students have experience listening to and describing a single bird, they are ready to explore the soundscape and listen to the forest as a whole.

Soundscape maps can be constructed in physical space (an overhead map of an area with points for sounds) or in acoustic space (a graph that shows the relative pitches and loudness of different sounds). They will prompt different kinds of observations and will stimulate different kinds of questions. Students could make either or both types of maps.

We seldom pay deep attention to what we hear. When students slow down to listen, they are often surprised by how many sounds surround them! This activity enhances students' perception of sound and brings them into contact with place. There is significant ecological information contained in the sounds of a place. Just as radio frequencies must be calibrated so as not to interfere with each other, so too with natural sound. If two birds sing at the same frequency, their voices will interfere with each other. By using different frequencies and timing, birds can share the same acoustic environment without interfering with each other. Sound frequency also has implications. Both very high and very low sounds are difficult to locate. Low-frequency sound is not as easily absorbed and can be heard farther away. Making a soundscape map will tune students in to another layer of information in nature, offering more to explain and wonder about. It also challenges students to describe novel phenomena in creative ways.

NATURAL PHENOMENA

This activity can be conducted anywhere, but is richest in a complex and diverse natural area where students can hear different species of birds and other natural sounds, such as creeks and rustling leaves. Mornings in early spring often are the best times to listen for birdsong. You may also map the sounds of an urban setting or an urban–wild interface. Compare natural and disturbed systems by repeating the activity in different environments.

PROCEDURE SUMMARY

For a sound map:

1. Listen to sounds and map their location.
2. Put yourself in the center of the map, then start with the most distant sounds, putting them at the edge of your paper, then working in.
3. Find creative ways to show sounds, using symbols and diagramming along with words and sketches.

For an acoustic space map:

1. Make a diagram using lines to show sounds in the environment.
2. Use different colors to show the biophony (sounds from living things), the geophony (sounds from natural, nonliving things), and the anthrophony (sounds from humans and things humans have made).

DEMONSTRATION

When the whiteboard icon appears in the procedure description: Draw yourself in the center of the page and add representations of sounds that you hear, starting with the most distant noises and working your way in.

PROCEDURE STEP-BY-STEP

1. **Begin the activity by asking students to listen deeply, drawing their attention to some of the more subtle noises around them.**

   a. "Sit comfortably so that you do not need to move or rustle the leaves or branches around you. Close your eyes and breathe deeply and slowly. Listen in silence to the sounds around you."

   b. "What draws your attention first? What else do you hear? Listen beyond that to other, more subtle sounds around you. Hold up your hands and lift one finger to count each of the different kinds of sounds around you."

   c. "Let's listen for one minute and see how many we can hear. [Let students count.] When you're ready, you can open your eyes. What did you hear?" (Collect student responses.)

Sound Map

2. **Tell students that they will make a map of the sounds around them, spreading out so everyone can listen well.**

   a. "We are going to make maps of the locations of the sounds around us."

   b. "In a moment, we will spread out so that you will not be distracted by the small sounds made by people sitting too near you. Then you will sit, listen, and begin to draw and describe the locations of the different sounds you hear."

3. **Demonstrate how to begin by adding metadata and adding a North arrow, then placing yourself in the middle of the map.**

   a. "Let's start with the metadata. Write the date and our location on the bottom of the page."

   b. "You will sit facing north (point north). On your map, north will be at the top of your page. You can

make a little North arrow (pointing up) if you wish. Make a little drawing of you sitting and listening in the middle of the page."

4. **Suggest that students start by placing the most distant sounds at the edges of the paper, then working inward.**

   a. "Start with the most distant sounds that you hear and put them around the edges of your paper. Then slowly work your way in."

5. **Tell students that they will need to get creative and come up with symbols to show sounds, then ask them to share some ideas about symbols with the group.**

   a. "You will have to be creative to show some of the sounds you hear. You can use words, pictures, and diagrams."

   b. "How might you show the sound of leaves rustling in the trees all around you? How about the sound of the creek over there? You may invent distinctive symbols for sounds you hear in several locations. Label any symbols you make."

6. **Ask whether there are any questions, set up a signal for students to return, remind them to be quiet, and send them out to make their maps.**

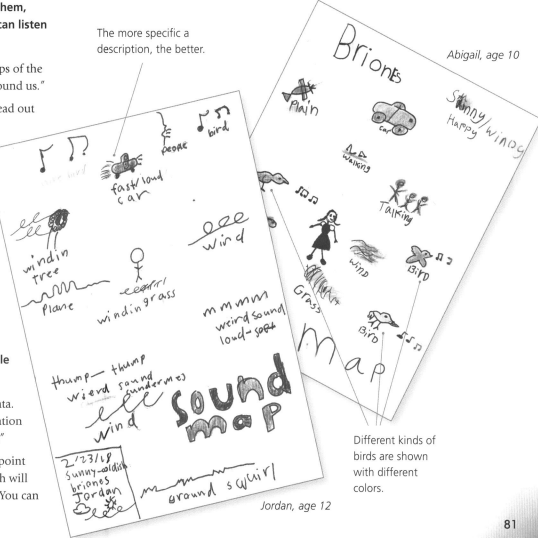

The more specific a description, the better.

Abigail, age 10

Jordan, age 12

Different kinds of birds are shown with different colors.

a. "You will have nine minutes to complete your map. Come back when you hear my whistle."

b. "As you work, stay still and be as quiet as you can. Do not move from your spot during this exercise. Even the sound of your shoes in the leaves can disturb other people."

7. **After the time has elapsed, call the group back together and facilitate discussion about how students chose to record sounds and about interesting or surprising observations.**

a. "Find someone and compare your maps. What kinds of things did you hear? How did you decide to record them in your journal?"

b. "Was there anything surprising that you heard?"

DEMONSTRATION

Acoustic Space Map

When the whiteboard icon appears in the procedure description: Make a graph of the sounds you hear using different colors

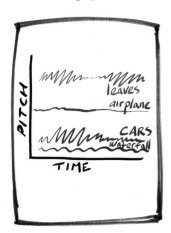

for living and nonliving natural sounds, and sounds from the human world.

1. **Focus students' attention on the soundscape, calling attention to different types of sounds and how they reveal information about the environment.**

a. "A soundscape is like an orchestra. Some sounds are high, some low. Sounds can be constant, rhythmic, or random. Tuning into the sound space will help you notice new things about the environment."

2. **Explain how to classify sounds into the biophony (sounds made by living things), geophony (natural sounds made by nonliving things), and anthrophony (sounds made by humans and machines).**

a. "You can classify the soundscape into three parts. The biophony is all the natural sounds that are made by living things. Can you think of some examples?" (Birdsong, cricket chirps, buzzing bees, etc.)

b. "The geophony is all the natural sounds that are made by nonliving things. What might that include?" (Wind in the trees, a babbling brook, waves on the shore, etc.)

c. "Finally, the anthrophony are all the sounds made by humans and their machines. What examples can you think of?" (Airplanes overhead, people talking, cars driving by, etc.)

3. **Describe how to make an acoustic space diagram by using different lengths, heights, and widths of lines to place sounds on a graph with pitch and time as the two axes.**

a. "You can use writing and expressive lines to describe sound. We are going to make a graph of this sound environment."

b. "Along the vertical axis, we will have pitch (low or deep to high). The horizontal axis will be time."

c. "When we hear a high sound, we'd put it in this part of the graph. A low sound, we will put down here."

d. "We can show volume, or how loud something is, by pressing hard to make a bold, dark mark."

e. "A loud, high train whistle that stays at the same pitch might look like this. Rhythmic waves washing up on the shore might look like this."

4. **Suggest that students begin by recording living and nonliving natural sounds, using a different color for each category.**

a. "Start by recording living and nonliving natural sounds. Use a different color of pencil for each category."

b. "For instance, if I can hear the high-pitched rustle of leaves and the low babble of the creek, I would put them both on my chart with the same color because they are natural, nonliving sounds."

c. "Birds and crickets would be in a second color."

d. "Finally, human-related sounds would be a third color."

5. **Tell students to begin, and to pay particular attention to which sounds stand out from the rest of the noise.**

DISCUSSION

Lead a discussion using the general discussion questions and questions from one of the Crosscutting Concept categories. Intersperse pair talk with group discussion.

General Discussion

a. "Let's look at our maps and diagrams. How are they similar to and different from one another? Can you find the same sound event shown on different graphs in different ways? What are some of the creative ways people showed what they heard?"

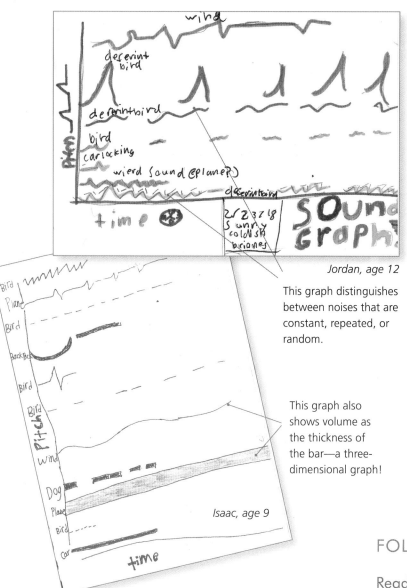

Jordan, age 12

This graph distinguishes between noises that are constant, repeated, or random.

This graph also shows volume as the thickness of the bar—a three-dimensional graph!

Isaac, age 9

b. "Each variable you track on your graph is a dimension. Here we graphed pitch across time. Can you find any graphs that also show volume or the loudness of sounds? These graphs are three-dimensional!"

c. "How many different sounds did everyone hear?"

d. "What sounds stood out from the rest of the noise? What made these sounds stand out?" (Different pitch, rhythm, volume)

e. "What sounds were constant? Which ones were less frequent?"

f. "How does human-generated noise show up in this soundscape?"

g. "What was it like to focus on sound in this way? How did you feel while you made your map?"

Patterns

a. "What are some general statements you can make about the soundscape in this area?"

b. "Did you notice any patterns in this soundscape?"

c. "How might these patterns be different in different places? What kinds of patterns would you expect to see in a desert [arctic, jungle, marine, etc.] soundscape? What do you think would be similar, and what would be different?"

d. "What about animal sounds? What patterns did you notice? You could consider the structure of the sound (pitch, rhythm, and volume), when the animal made the sound, or any synchronization with other sounds."

e. "What kinds of patterns in animal sounds would you expect to see in a different ecosystem?"

Cause and Effect

a. "Did you hear any animal sounds? Were these intentional or accidental on the part of the animal? Describe the animal sounds you heard."

b. "What made the sounds stand out from the background noise? Did the sound expose the location of the animal that made it?"

c. "Why might the animal have made the sound? If the sound made the animal's location apparent, what are the possible benefits and costs of making it? When might the animal make this sound, and when might it not?"

d. "What effects do you think human-made sounds have on the animals here?"

FOLLOW-UP ACTIVITIES

Reading Connection

Read *The Other Way to Listen*, by Byrd Baylor, to set a tone of reverence and reflection.

Repeated Listening

Have students make sound maps at different locations around your campus or other areas (urban, urban–wild interface) to compare the differences among them.

Writing Poems

Instruct students to write poems based on the experience of deep listening.

a. "Think about what you felt like when you were making your sound map. What does it feel like to really listen to a place? Write a poem about what you can hear if you take the time to slow down and listen."

b. "If you don't know what to write or where to start, begin by just saying 'I hear' and listing some things you hear, and then saying 'I feel' and writing about how you feel when you hear these sounds, or as you listen."

INQUIRY, INVESTIGATION, AND SCIENTIFIC THINKING

INQUIRY, INVESTIGATION, AND SCIENTIFIC THINKING

INTO INQUIRY

Intentional curiosity and tools to pursue questions are some of the building blocks of inquiry. Scientific investigation activities build students' skills in asking varied questions, and offer some approaches for dancing with the questions they ask. These practices translate directly to the journal page and will deepen the investigations students make and the understandings they come to.

ACTIVITIES IN THIS CHAPTER

INTENTIONAL CURIOSITY AND INQUIRY

You can train yourself to be more curious and to find mysteries everywhere you look. Using curiosity scaffolds, students will ask richer and more varied questions.

INTENTIONAL CURIOSITY

Children have a natural curiosity and are full of questions. "How come the ocean is blue?" "Why are whales so big?" "Where do clouds go after the rain?" In many instances, adults respond to these questions with impatience, offering quick answers until the child asks a question the adult doesn't know the answer to—which often yields the response: "Stop asking questions!" In school, a focus on standardized tests emphasizes knowing facts over asking questions or sitting with mysteries. Grades and scores evaluate how much students remember, taking the focus off of how they think or how they approach learning. Eventually, our children lose their capacity for wonder. They become the adults who find themselves insecure when they reach the edges of understanding.

But curiosity is objectively important. Curiosity focuses attention. It drives discovery, investigation, and developing intelligence.[22] Curiosity also improves memory and thinking, and promotes flow states in which students perform some of their best work. The pleasant feeling of being curious is a rush of dopamine in the brain.[23] This neurotransmitter improves our ability to remember details, not only of the object of our curiosity but of everything else in the environment.[24]

Our goal as educators is to build a curiosity tool kit—a collection of ways to investigate anything we find interesting. A foundational tool in the kit is the capacity to wonder well and to play with the questions we ask. We want to cultivate a stance of intentional curiosity that supports wonder, exploration, and understanding. The reward of being open to what we don't know is intrigue and mystery around every corner, and the power to learn from evidence.

Mysteries often present themselves as a subtle "That's odd…" feeling that is often ignored. Teach your students to look for that feeling and lean into it when it occurs and to immediately say or write the question before it is lost. We do not need to wait for curiosity to come to us. We can initiate it by deliberately asking questions cued by observation in our environment. Cultivating curiosity takes practice. The more we do it, the better we will get.

Once we slow down to look, mysteries are everywhere. We can attempt to unravel these mysteries, and the process is surprising, interesting, and electrically fun.

THE MYSTERY OF THE FLOATING SAND

Young children aren't the only ones with curious minds. If you can look at the world with the wonder and the intense desire to figure things out of a three-year-old, that openness will make the world an awe-inspiring and mysterious place, full of things to be discovered. Once, with a group of teachers on a beach, we found a nature mystery that kept us occupied for the better part of an hour. We were sitting on the sand next to few large pools of water, well back from the breaking waves.

"Look!" someone said. "Floating sand islands!" We all looked at the pool. Sure enough, on the water's surface, we noticed several small "islands" of sand ranging from around ten grains to upwards of fifty, floating on the surface of the water. "Whoa!" "How are they floating? Sand is rock; it's not supposed to float! How are they doing that?" And so began the mystery of the floating sand.

Immediately, everyone started asking questions, making observations, and trying to figure out the mystery. "How many are there? What can we notice about them? Can we notice any patterns?" "The sand grains don't always appear to be touching." "They're not round spots; they often have a jagged shape. The surface of the water appears distorted next to them."

"Maybe there's something happening on the surface to make the sand grains float?" "Or another explanation is that the grains are really, really light, and just float." We sprinkled some sand grains on the surface and watched as they sank straight through. "Hmm, OK, the evidence doesn't seem to support the idea that the weight is what's causing this. What are some other possible causes we can think of?"

"Maybe it's something to do with the surface tension. What happens if we touch them with a stick?" As soon as we touched the island, the grains separated and sank to the bottom. "Huh, so maybe something with surface tension, but how does it actually work?"

"Do these occur in other pools around here?" The group broke and ran around to the three other pools in the area. There were no floating sand grains. "Is this something to do with density, and the composition of the water? Don't things float better in salty water? Maybe the water in this pool is different."

"Could this be ocean water that got stuck up here after a high tide? Does anyone have any instruments we could use to test the water?" We shook our heads. "We could taste it!" one person said, then bravely stuck a finger in the pool, then licked it. "It's a little salty, I guess, but that could just be my brain thinking it's salty because it would support our idea that there's something about the water composition." We continued playing around for a while, and agreed that we thought the effect had something to do with surface tension, but that we'd need more understanding of this concept to fully understand what was going on.

Although we didn't fully solve the mystery of the floating sand grains, we made careful observations, learned about our surroundings, and had fun working together as a team. This group also demonstrated a high level of scientific rigor—posing multiple plausible explanations, weighing evidence, changing our minds when the evidence contradicted an explanation, and looking out for confirmation bias.

SPINNING THE WHEEL

Beloved professor and marine biologist Ken Norris calls this kind of open inquiry—the shifting back and forth between observation, question, and explanation—"spinning the wheel." Interacting with nature from this place of intentional curiosity is powerful. What does it take to get here?

A key part of instilling a curiosity mindset is giving students permission to be creative, playful, and wrong. We can have rigor and creativity at the same time. The process is less engaging and dynamic if we follow a cookbook set of steps every time, or if the only goal is finding the "right answer." Often, we will never find an answer to our mysteries, but we still learn plenty in the process. When we are focused on exploring and following our curiosity, we are freed up to notice, wonder, propose explanations, make mistakes, and change our mind. The activities and ideas in this chapter will help students build scientific inquiry skills they can apply in their journals and beyond.

> Let me keep my distance, always, from those who think they have the answers.
>
> Let me keep company always with those who say "Look!" and laugh in astonishment, and bow their heads.
>
> —Mary Oliver, from "Mysteries, Yes"

BEST PRACTICES

In their journals, students

- Include metadata (date, time, location, weather, etc.).

- Make observations ("I notice…"), ask questions ("I wonder"), and make connections ("It reminds me of…").

- Intentionally ask varied questions, including questions from observations or from curiosity scaffolds such as "Who, What, Where, When, How, and Why" (what we call the 5W's + H).

- When possible, pursue questions that can be answered through observation in the moment.

- Make explanations, including "Why webs," which present multiple plausible explanations to answer a "Why" question.

- Distinguish between observation and explanation. (This could include icons or symbols for each type of information.)

- Support claims or generalizations with evidence. Students can change their minds if evidence suggests it, and identify changes and shifts in their thinking.

- Cite sources when information comes from a place other than the students' observation or thinking (including people, websites, and books).

- When collecting data, consider bias; describe or diagram methods; and when it's possible to do so, collect a large and random sample rather than small or selected samples.

CURIOSITY SCAFFOLDS OR QUESTION GENERATORS

Children's initial questions about animals and plants often focus on "factoid" or "baseball card statistic" information, such as "How much does it weigh?" "How fast can it go?" "How old is it?" or "How tall is it?" We can train ourselves and our students to ask more and deeper questions that lead us to interact directly with nature mysteries.

Curiosity scaffolds are prompts that help students generate rich and varied questions from their observations. These observation-based questions are also readily investigated through further direct observation. The scaffolds are open ended and encourage creative thinking instead of cookie-cutter questions. Using scaffolds can become part of regular journaling routines by having students write them onto the inside of the journal cover or attach the "cut-and-paste" tool kits (see the appendices) to the back page of students' journals.

You can use specific concepts or questions to focus students' attention while they journal. Such priming can be done by simply asking them to write the word *patterns* or *change* on a page before starting an activity. In this book, we've frequently used the NGSS Crosscutting Concepts to direct discussion, but there are other models for guiding students that you can use. Here we also offer two others we find useful; they may be appropriate for directing inquiry during nature journaling outside of a science context.

- Crosscutting Concepts. These reflect big ideas in science and are widely applicable across different areas of science; they are already adopted in NGSS states.

- Who, What, Where, When, How, and Why (5W's + H). This model is a good baseline for questions for students of all ages; it can be used with either of the other scaffolds.

- International Baccalaureate Key Concepts. These are phrased using intuitive language. There are some overlaps with the NGSS Crosscutting Concepts; they also include thinking about social implications and epistemology; they are cross-disciplinary and could easily be used in history or social science.

THE CROSSCUTTING CONCEPTS

The Crosscutting Concepts are one dimension of the Next Generation Science Standards. They are "big ideas" useful across many disciplines of science, and they reflect how scientists think.

Some of the concepts are easier to understand than others. Introduce one at a time. Begin with Patterns, Cause and Effect, or Structure and Function. These are easier to apply in the context of nature study. Students can typically integrate them into their journaling with minimal explanation even if they haven't encountered the concepts before. The concepts Matter and Energy; Systems and System Models; Scale, Proportion, and Quantity; and Stability and Change are more abstract. Students will be more successful in using these to drive their questioning if they have had previous experience using the concepts in the context of a wider learning activity.

These are some of the kinds of questions you can ask students to think about before an activity to prime them to focus on one concept as they journal. Through the activity *Questioning Questions,* you can also coach and encourage students to ask these questions themselves as they are journaling or exploring nature.

PATTERNS

Ask: What patterns do you notice? How can you describe the pattern? Are there exceptions to the pattern? What might be causing the pattern? What does the pattern remind you of?

CAUSE AND EFFECT

Ask: What happened here? Why is it like this? What might be causing the effect you observe? What are other possible explanations? What might happen if…? Is this causation or correlation? How do you know?

SCALE, PROPORTION, AND QUANTITY

Ask: At what scale have you explored this phenomenon? Would you think about it differently if you zoomed in or out? Can you make a model that helps you understand nature at this scale? How can you measure change at different scales? What can you quantify at this scale, and how can you measure it accurately?

STRUCTURE AND FUNCTION

Ask: What is it like? How might this structure work to help the organism survive in its environment? How does it work? How is this structure like others you have seen? How is it different? How might those differences impact the function of the structure?

SYSTEMS AND SYSTEM MODELS

Ask: What are the boundaries of this system? How does the system work? What are the parts of the system? How do the parts interact? How is this system affected by other things? How does this system affect other things? What would happen if X were removed? What would happen if Y were added? Is there feedback? What was it before? What will it be next? How can you model this system?

ENERGY AND MATTER

Ask: Where does the matter in this system come from? How does it change within the system? Where does it go? How do matter and energy interact in this system? Where does the energy in this system come from? Where does it go? What does the energy do in this system?

STABILITY AND CHANGE

Ask: What causes change in this system? What is stable in this system? What isn't? What changes quickly in this system? What would happen if X were different?

CROSS-DISCIPLINARY CONCEPTS

Big ideas can guide students' learning and thinking in many disciplines. Using Patterns and exploring Cause and Effect look different outside the sciences, but these concepts could be used as anchors for students' observations, and as frameworks for asking questions in any number of disciplines. For example, "What patterns do you notice?" could direct students' observations of art, historical documents, or engineering plans, and teasing apart causes and effects is a key part of learning about history. Systems thinking and considering such questions as "What causes changes in this system?" can apply to social systems in addition to scientific ones.

NEXT GENERATION
SCIENCE
STANDARDS

WHO, WHAT, WHERE, WHEN, HOW, AND WHY

The "5W's" (and one H) used to guide students in journalistic writing can also be a helpful framework for establishing a baseline of observations about natural phenomena. Each question frame will focus students' attention in a different place, tuning them in to spatial and temporal relationships, causation, and identification.

WHO

Identification and classification.

Ask: Who is it? Who was it? Who will it be? Who made these tracks? Who could have made these holes in this leaf? What bird is that?

WHAT

Questions about a phenomenon. Focuses on cause and effect.

Ask: What is happening? What happened? What will happen next? What does it do? What will the caterpillars do when the leaves fall from the trees (prediction)? What causes the darkened sky above the rainbow?

WHERE

Space, location, and biogeography.

Ask: Where is it? Where was it? Where will it be? Where is the bird's nest? Where is the current fastest? Where do frogs go at night?

WHEN

Timing—how long?

Ask: When did it happen? When will it happen? What is the timeline? When did this tree start to grow?

HOW

Mechanism—how does it work? Focuses on form and function.

Ask: How do cracks in mud form? How do beavers survive under a frozen pond? How did that tree grow from such a tiny crack in the rock?

WHY

Meaning, cause and effect, structure and function.

Ask: Why did this happen? Why is it this way? Why do small birds fly in a tight flock and not a V? Why are the leaves darker green on one side and pale on the other? Why is the plant hairy?

INTERNATIONAL BACCALAUREATE KEY CONCEPTS

These concepts enrich questioning in International Baccalaureate schools. They are similar to the Crosscutting Concepts, but offer a slightly different focus.

FORM

The understanding that everything has a form with recognizable features that can be observed, identified, described, and categorized.

Ask: What is it like?

FUNCTION

The understanding that everything has a purpose, a role, or a way of behaving that can be investigated.

Ask: How does it work?

CAUSATION

The understanding that things do not just happen, that there are causal relationships at work, and that actions have consequences.

Ask: Why is it like this?

CHANGE

The understanding that change is the process of movement from one state to another; it is universal and inevitable.

Ask: How is it changing/stable? What was it like before? What will it become? What is the rate of change?

CONNECTION

The understanding that we live in a world of interacting systems in which the actions of any individual element affect others.

Ask: How is it connected to other things? How is this a part of a system or systems?

PERSPECTIVE

The understanding that knowledge is moderated by perspectives; different perspectives lead us to put our attention in different places and make different observations. Our perspectives also affect our interpretations, understandings, and findings; perspectives may be individual, group, cultural, or disciplinary.

Ask: What are the points of view? What is another perspective on this?

REFLECTION

The understanding that there are different ways of knowing and that it is important to reflect on our conclusions, to consider our methods of reasoning and the quality and reliability of the evidence we have considered.

Ask: How do I know? How strong is the evidence? How reliable is the source of my information?

RESPONSIBILITY

The understanding that people make choices based on their understandings, and the actions they take as a result do make a difference.

Ask: What is my responsibility?

Using Form, Causation, Change, and Connection leads students to describe a phenomenon and attempt to explain it by thinking about causes and effects, determinants of change, and systemic interactions. The last three prompts—Perspective, Reflection, and Responsibility—each have distinctly different effects.

The idea of perspective catapults us beyond our own way of making sense. Entertaining multiple perspectives is an essential part of being engaged members of our communities. Getting in the habit of seeking out different perspectives helps us be better communicators and collaborators. This prompt can be a reminder to step back and consider multiple perspectives. In nature journaling, as students attempt to figure out or make meaning of their observations, this idea can remind them to shift their thinking and intentionally build nuanced understandings.

Reflection is an invitation into epistemology. A student or class might make what are seemingly solid explanations of the way a system is changing or why an animal behaved the way it did, only to realize that the information used as evidence was faulty. This prompt keeps us connected to evidence and reminds us to assess the quality of our sources of information.

Thinking about responsibility embeds us within the system or phenomenon we are studying. This concept can be an ideal lens for students to turn outward from a nature journaling entry to consider their place within their community, or their impacts on their surroundings. To begin to think about responsibility in this way, students could ask such questions as "What is my part in this? Am I a part of this system or phenomenon I am studying? If so, how do I or did I impact this situation? How might I make changes to my behavior to lessen my negative impacts? Is it my responsibility to attempt to make changes to my behavior or to listen to those who have had a different experience than I have?" Asking and answering these questions are an essential part of being an engaged and thoughtful community member.

HOW DO YOU ANSWER YOUR QUESTIONS?

We can do a lot more with questions than just look up the answers. Learning to engage with nature mysteries through authentic inquiry can lead to rich, dynamic learning opportunities.

SCIENTIFIC QUESTIONS

When we ask a question, we can begin by figuring out which discipline is most appropriate for its inquiry. Science is a system of creating and evaluating testable explanations of natural phenomena. It is limited to the study of what can be directly observed or inferred from observation. Scientific explanations must be testable, or able to be supported or contradicted by concrete evidence. A scientist should always ask, "How would I know if I were wrong?" If there is no way to test whether an explanation is wrong, it is not open to science. Questions about the soul, ethics, or spirituality are outside the realm of scientific inquiry. This is because they cannot be observed or measured, not because they are bad or unimportant things to wonder about. Exploring questions through philosophy, religion, spirituality, and literature is a valuable part of life, but when we're exploring nature with our "scientist hats" on, they are beyond the scope of our inquiry.

APPROACHING ANSWERS

It's easy to rush to resolve our questions. When we are uncomfortable with ambiguity, we tend to seek immediate answers and accept the first one we come to. Our ability to access information at the click of a button on the internet reinforces this tendency.

The best way to teach a scientific mindset is to model it yourself. Ask questions, find mysteries, look for clues, make exlanations, try to prove yourself wrong, and change your mind in the presence of evidence.

But we can train ourselves to sit with mysteries and use methods to approach deeper understanding. The process takes patience and requires humble, deliberate thought. By these means, we circle closer to the truth and enhance our practical understanding of the world.

INTO INQUIRY: PLAYING WITH MYSTERIES

Here are a few approaches we can keep in our "back pocket" to explore nature mysteries.

Let's Go See

Sometimes we can observe an answer to one of our questions right away. If the answer to the question can be directly observed in the moment (e.g., "Do these same kinds of holes appear on other leaves on this tree?"), we can encourage students to do it. If it's possible to answer the question through observation but it will require additional equipment or time, we "wait and see," gathering the data needed to answer the question later.

Could It Be, Maybe

If we can't directly observe an answer to one of our questions (e.g., "Why are all the leaves on this half of the branch brown?"), we can come up with possible explanations for it. In our journals, we can create "Why webs," proposing several possible explanations to a nature mystery or "Why" question. We can also weigh one explanation against others using evidence to infer which ones are least plausible and, perhaps, which one is more likely to be right.

Look It Up

Books, research papers, and local experts can hold a wealth of knowledge about living things and natural processes. If we can't answer a question through our own observations, we can look up the subject or see whether additional information will provide evidence to deepen our understanding. Because we live in a world where so much can be immediately Googled, we want to encourage students to play with their mysteries first instead of turning straight to an outside resource, and to mindfully consult outside sources after they've thoroughly explored the mystery and have reached the limits of what they can observe and figure out on their own. As students look up answers, they should always ask, "What is the credibility of this source? What is the strength of the evidence for a claim?" Tracking down the original study that a claim is based on is a good way to follow the evidence.

THE SCIENTIFIC "MYTH-OD"

The scientific method is common in science fair projects and classrooms. This step-by-step process is often taught as the single "official" approach to conducting scientific studies, but it is a far cry from the approaches of working scientists. The scientific method originated from a paper written by Oreon Keeslar in 1945. He interviewed scientists, asking them how they approached their work, and the scientists shared many different practices and activities. Keeslar came up with categories for common techniques, then wrote about them in a paper to describe scientific methods that would be appropriate to teach secondary school students. Science textbook companies got ahold of the information and published a "scientific method" with steps in the order that the information was presented in the paper. For decades, students have been taught that this is how science is done.

We can help our students toward a more accurate understanding of scientific inquiry if we teach scientific methods, not one scientific method. Scientists use a variety of activities and practices, and do not do so according to a specific order. Teaching science the way it is represented in the Next Generation Science Standards focuses students on engaging in many practices of science, not one linear approach.

In designing investigations, we can coach students to conduct systematic observations without generating predictions first. From there, they can attempt to explain the results of the systematic observation, and generate predictions they can test in further observational studies. This approach leads to studies less likely to be influenced by confirmation bias and more reflective of how scientists work. It is also more fun, interesting, and creative to take part in than writing a hypothesis at the outset, and it makes space for divergent thinking. As a result, students gain a more accurate understanding of science.

Let It Be

Science is a system of creating testable explanations of natural phenomena. It is limited to the study of what can be directly observed or inferred from observation. If the subject can be neither observed nor inferred, it is outside the realm of science. When we are teaching students to explore nature through a scientific mindset, if they ask questions about the soul, ethics, or spirituality, we can encourage students to "let it be" and sit with the wonder of the mystery, or explore it through another discipline.

A Tool for Each Mystery

In the face of any mystery, we can ask questions, make observations, construct explanations, and look things up, and we don't need to do anything in a specific order. We might ask a question, then come up with a possible explanation, then make an observation that contradicts our explanation, which leads us to another question. We could consult a source to find some key piece of evidence, then go back out to observe more. This whole process can take place in discussion with peers or on the journal page.

The activity *Mysteries and Explanations* is designed to introduce students to finding and exploring nature mysteries. *Phenomenon Model* engages students in proposing visual explanations at a deeper level. The content and activities in this chapter are not meant to be a full course in scientific inquiry methods. Taking the time to go deeper and teach scientific survey methods prepares students to design longer investigations and be critical thinkers and thoughtful consumers of information.

ADDITIONAL RESOURCES

The student activities *Exploratory Investigation* (http://beetles project.org/resources/for-field-instructors/exploratory investigation/) and *NSI: Nature Scene Investigators* (http://beetlesproject.org/resources/for-field-instructors/nsi-nature-scene-investigators-2/) from the BEETLES Project support students in learning how to make explanations and engage in scientific inquiry.

"What a Lousy Bird," a chapter in *Mountain Time,* by Ken Norris, chronicles the story of a group of college students as they begin to "spin the wheel" and explore nature mysteries.

ASK QUESTIONS AND THINK LIKE A SCIENTIST

Asking questions is an important part of science. Recording observations in a journal and using big ideas to guide thinking can help students generate lots of them.

Ruth Heindel

"The drawing helped me focus my observations, while the writing allowed me to express questions that would be difficult to convey with drawings alone. The questions I included on this sketch became the foundation for my dissertation. Some of the questions are ones I'm still thinking about and would like to continue working on."

Thinking Tools: Cause and Effect and Systems of Interacting Parts

Teasing out cause-and-effect relationships is a key part of science. Thinking of an area or phenomenon as a set of parts that influence one another is a way to begin to understand how it works. Using these big ideas as thinking tools helps us ask varied questions about a phenomenon.

Multiple Plausible Explanations

We don't want to get tunnel vision, seeing only the first explanation that comes to mind. Coach students to stay flexible and to entertain many possible explanations. This is a key part of science.

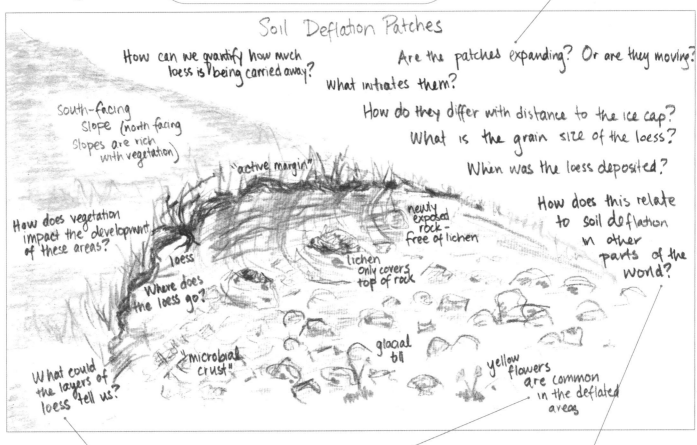

Observations That Lead to Questions

Noticing that there are different layers of soil leads to a question about the information the layers might hold. Using writing and drawing to record observations is a great way to spark questions.

Noticing Patterns

Training students to look for and record patterns in their nature journals can clue them into underlying processes and help them ask questions.

Connections to Other Processes

Thinking about similar processes or using "It reminds me of" can be a way to learn about the phenomenon being studied. It can also lead to more questions.

PRACTICED PRECISION

In a scientific approach to journaling, it's important to provide both the big-picture context and explicit detail. Deliberate word choice, drawings, maps, and diagrams create an unambiguous, vivid picture.

Robert Stebbins

"Although observations may begin without a particular question in mind—soon what is observed will raise questions, and ones that appear to have a chance of being answered can be pursued, scientifically. Children, in particular, are often very insightful in their questioning."

The Grinnell System and Key Details

Joseph Grinnell developed a rigorous system for recording journal entries, species accounts, and collection records that is used by many scientists. This system offers a structure for recording the environmental conditions around observations. Students don't need to follow the strict Grinnell format, but they can and should record the context of their observations.

Map and Cross Section

Stebbins constructs a map and a cross section across the long axis of a shallow pond. He creates a key to show how newts and frogs use different parts of the pond. His written notes compare his observations this year with previous years (this is a long-term project), speculate on the historic height of the pond based on marks on nearby trees, and record the camouflage and behavior of the frogs. A curious mind at work.

Words, Pictures, and Numbers

Words and pictures together describe important details that can be used for classification or other approaches to analyzing observations with a scientific lens. Look for measurements, estimations of distances, temperature, time, depth, and counts of egg masses. He also zooms in to show the structure of a single egg mass and zooms out to a map of the lake that shows the position of each egg mass. His written notes identify the plants in the area and even the light levels.

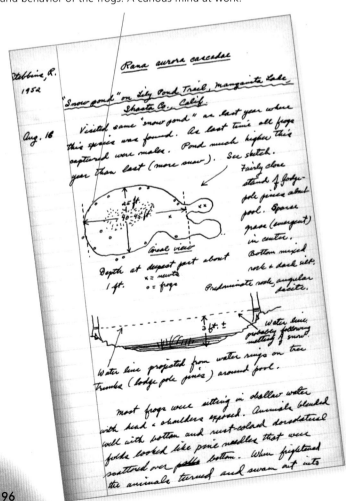

NATURE PUZZLES

Trying to figure out why or how something works is creative and fun. This form of old-school tinkering is an essential part of a scientist's or engineer's tool kit. We can show students that a journal is a safe place to wonder and speculate about what might be going on.

Thinking on the Page

Making visual explanations takes our thinking a level deeper. Observing natural processes and attempting to explain the seen and unseen forces behind them often bring us to the edge of our understanding, offering opportunities for learning.

John Muir Laws

"The pages of my journal focus and curate my discoveries, mysteries, moments, and musings. It is where my thoughts engage and crystallize into meaning. This is my brain on paper. Any questions?"

We Do Not Have to Be Right!

When you and your students find a strange phenomenon, first describe it carefully, then try to figure out what makes it tick. As students create possible explanations, they can label them with the scientific language of uncertainty: "Perhaps," "Could it be," "Maybe," or "I think." This distinguishes their observations from their explanations.

The Undiscovered Country

The most interesting mysteries are the ones we don't understand. Giving students permission to explore and explain mysteries can support their scientific understandings; the more unfamiliar the mystery, the greater the opportunity for learning.

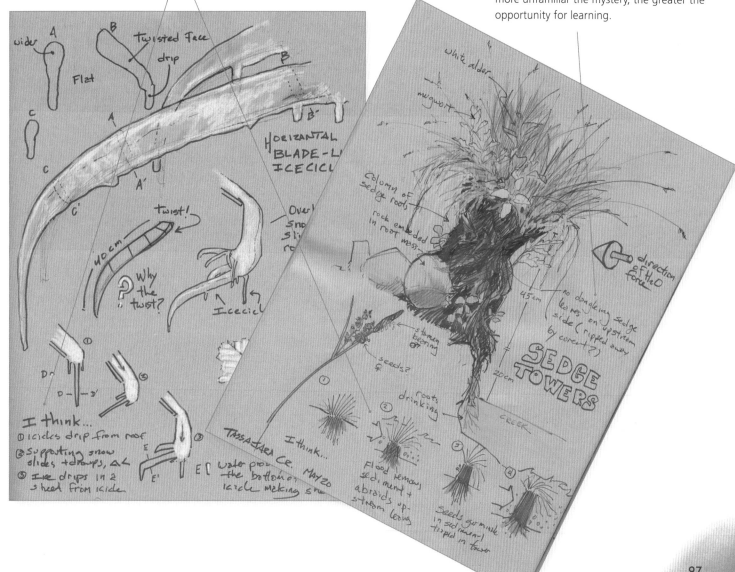

FOLLOW CURIOSITY CHAINS

The most interesting questions are rarely the first ones asked. Being deliberately curious and following a train of thought on the page can lead to learning, wonder, and insight.

Fiona Gillogly

"I am curious by choice. I don't go out in nature and wait. I look for what is out in nature waiting for me. Once I find the mystery, I start asking basic questions that lead to more and more complex ones. It can be difficult to think up an interesting question right at the start, so I use the basic questions to pave the road."

Wonder Wildly

What is the question *behind* the question?

> **Observation-Question-Observation-Question**
>
> Learning to follow the train of thought, wherever it goes, is a way to deepen inquiry. Encourage students to let their wonder loose.

Use Icons

Icons, like exclamation points and questions marks, highlight key discoveries and make the page easier to scan.

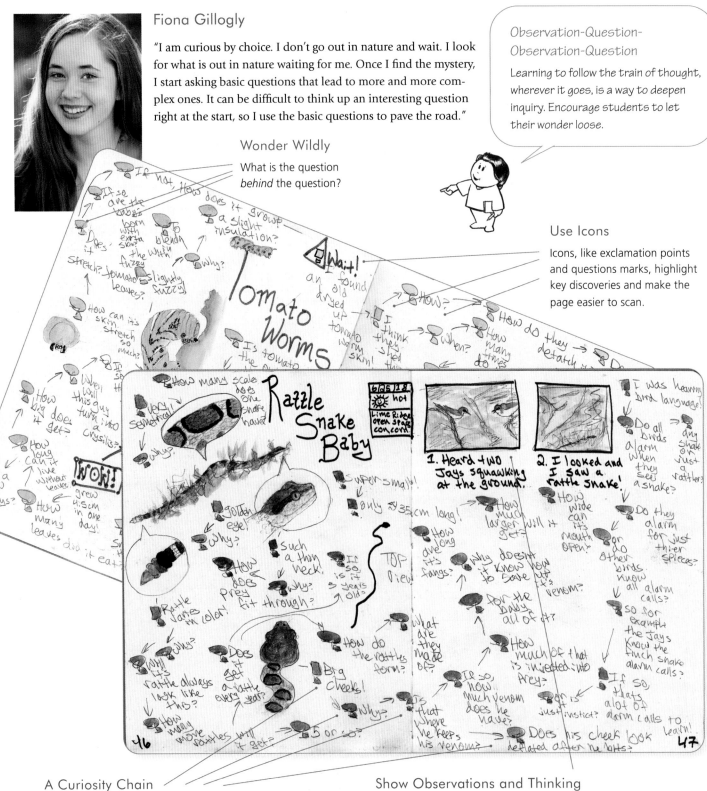

A Curiosity Chain

This set of observations and questions begins with an unexpected observation. This stimulates questions which lead to more observations, which lead to more questions.

Show Observations and Thinking

A set of comic panels describes the interactions between jays and a snake. Inset drawings capture specific details. Words and arrows track the flow of observations and ideas. Including different ways of showing observations and thinking leads to a dynamic entry.

MYSTERIES AND EXPLANATIONS

Students explore nature while digging deep into curiosity. They propose and refine questions and pose possible explanations for what they see, without the need to be right.

Asking questions, finding mysteries, and making explanations are fun and dynamic ways to be creative in nature. Once we start to think of nature as a world of endless small mysteries, we can become fascinated wherever we go. What's causing the pattern of flowers on the hillside? Why are the clouds moving so quickly today? Who is making holes in the stump at the edge of the schoolyard? This activity sets students up to start making explanations about their questions and observations, and gives them a mindset and skills for interacting with nature mysteries wherever they go. Teaching students to be curious, ask questions, and make explanations gives them basic scientific inquiry skills. Wonder also adds to the delight of being alive.

NATURAL PHENOMENA

For the first part of the activity, you can focus students' attention on a small part of nature they can hold in their hand, such as a leaf, a pinecone, an acorn, or a seed pod. Or, if there is an intriguing nature mystery the whole group can interact with—for example, a series of holes in the ground or an animal carcass—focus the group's attention on this. When students get time to practice using the prompts on their own, any natural area that is safe for them to explore will work.

PROCEDURE SUMMARY

1. Make observations.

2. Ask questions and look for mysteries.

3. Make multiple possible explanations about your observations or mysteries; support explanations with evidence.

Note: No demonstration is required because this is a discussion-based activity.

PROCEDURE STEP-BY-STEP

After students have had experience with the activity *I Notice, I Wonder, It Reminds Me Of*, offer these steps:

1. **Gather students, then explain that they'll start by focusing on asking questions.**

   a. "We've had some practice using observation skills to learn, ask questions, and make discoveries."

   b. "Now, let's focus on asking questions."

2. **Explain that students will make observations of a leaf (or other small part of nature you've chosen, or the phenomenon the whole group is focused on), and then will ask questions based on their observations.**

   a. "Start by taking a couple moments with a partner to make observations of a leaf, or statements about what it reminds you of."

   b. "Then ask questions based on your observations or 'It reminds me ofs.'"

   c. "For example, you might notice holes in a leaf, and wonder how they got there. Or you might see a bird's beak that reminds you of a tool, and wonder whether it works in a similar way."

3. **Explain that asking questions based on what we observe is a way of finding nature's mysteries.**

   a. "There are mysteries everywhere in nature. Why is one half of a pond covered in ice in the morning, but not the other? What animal is burrowing and leaving these tunnels? What is causing the patterns and colors on a decomposing leaf? Why are there so many ladybugs on one side of the plant, but not the other?"

   b. "Making observations and asking questions is a way to discover mysteries other people might never notice."

   c. "How many questions can you ask and mysteries can you find?"

4. **Give students a few more minutes to work with a partner finding mysteries on their leaf (or other small part of nature you've chosen, or the phenomenon the whole group is focused on).**

5. **Ask a few pairs to share interesting questions or mysteries they came up with.**

6. **Explain how students can try to answer questions by making more observations.**

   a. "What do we do with the questions and mysteries we find?"

   b. "We might be able to answer some of our questions right away by making more observations, like 'I wonder if other leaves on this tree have this same spotted pattern.'"

   c. "If you ask questions that you can answer through making more observations, go for it."

7. **Tell students that it can be fun and interesting to come up with explanations for the observations, questions, and mysteries they find in nature.**

   a. "There are going to be lots of mysteries you find in nature that you likely won't be able to answer right away."

   b. "There's a chance you will find mysteries that no one may have thought about before."

   c. "It can be fun and interesting to make explanations and try to figure out nature mysteries."

8. **Offer some examples of explanations based on evidence.**

   a. "For example, if we noticed that two halves of a puddle had very different freezing patterns, we could come up with some explanations, like 'Maybe it's shallower in the frozen part, so there's less water and it froze differently there.' Or 'Maybe that part with the lines all in the center was disturbed in the freezing process, so it looks different than the other part."

   b. "Or if you found scratch marks on a tree, you might try to figure out where they came from, saying, 'Maybe it was from a deer's antlers! Or maybe it was a squirrel scratching up the bark to use in a nest!'"

   c. "You could then try to make more observations and see whether there is evidence that supports or contradicts either of your explanations."

9. **Tell students to base their explanations on evidence.**

   a. "We can't just make up explanations. We need to base our explanations on evidence."

   b. "So we wouldn't just say, 'I think these holes were made a long time ago' and leave it at that."

   c. "We would need to share the evidence and reasoning that make us think that, such as, 'I think the holes were made a long time ago, and my evidence is that the dirt is falling down in them and there are plants grown over them.'"

10. **Tell students to use the language of uncertainty and to come up with many possible explanations.**

    a. "When we make explanations, we're coming up with some possible ideas about what *could* be happening and making more than one explanation."

    b. "We don't know for sure what's going on, so we should come up with lots of different possible explanations for our mysteries."

    c. "We also want to remind ourselves to stay open minded to different possible explanations by using what scientists call 'the language of uncertainty.'"

    d. "This means using words like 'Maybe' and 'Possibly' as we say our explanation."

11. **Pick a mystery that a student came up with earlier in the session, then give students about 2 minutes in pairs to make two or more explanations about it.**

    a. "Take a moment and make explanations about the leaf [or the phenomenon the group is focused on]."

    b. "Make sure you use the language of uncertainty, and base your explanations on evidence."

12. **Ask students to share a few of their explanations with the group. Listen to what they say, and coach students to use the language of uncertainty and base their explanations on evidence.**

13. **Tell students that they'll have a few minutes to find nature mysteries and try to explain what they see.**

    a. "In a moment, you'll get to spread out, find nature mysteries, and try to explain them."

    b. "You will get to go wherever you want to within these boundaries [state the boundaries], trying to find cool mysteries and figure them out."

    c. "You can go back and forth between making observations, asking questions, and coming up with explanations. If you're able to answer a question through observation right away, go for it! Then ask another question."

d. "You don't need to be rigid about it, or always go in that same order. You might ask a bunch of questions, then come up with many different explanations. Your explanations might lead you to make more observations, then come up with another question."

e. "You do not need to be right, but do think carefully and be sure to support your explanations with evidence. Be creative and have fun with it!"

14. **As students work, take time to circulate, support, and troubleshoot, and coach students to use the language of uncertainty and make more than one explanation for a phenomenon.**

15. **If a group of students finds an especially interesting or cool nature mystery, consider bringing the class together and facilitating a whole-group discussion focused on trying to explain the mystery.**

16. **Offer time for students to share interesting or exciting explanations they came up with, then discuss in pairs which explanation seems the most likely based on the evidence.**

    a. "Scientists try to come up with the best possible explanation based on all the available evidence."

    b. "Pick a question or mystery you explored and that you came up with multiple explanations for. Which explanation has the most evidence that you can see to support it?"

    c. "Share your reasoning and thinking with a partner."

17. **Explain that scientists go through this process on a much longer time scale.**

    a. "Scientists come up with questions, explore mysteries, and make explanations, too. It's part of how they learn about the world."

    b. "Science is about trying to find the best explanation based on all the available evidence, so scientists spend a lot longer, usually months or years, doing research and testing and considering different explanations."

    c. "When scientists come upon evidence that contradicts their explanation, they stay open minded and are willing to change their thinking and listen to ideas."

DISCUSSION

Lead a discussion with a few of the general discussion questions. Intersperse pair talk with group discussion.

General Discussion

    a. "What was it like to observe, explore nature mysteries, and come up with explanations?"

    b. "What helped you learn during this activity? What skills do you feel like you got better at?"

c. "You can find nature mysteries and come up with explanations wherever you go. It is a fun way of learning about the world around you. What are some other places you'd want to explore using this mindset?"

d. "What are some reasons it might be important for scientists to use language of uncertainty, such as saying 'Maybe' or 'Perhaps,' when they come up with explanations, and to come up with multiple possible explanations?"

e. "A good scientist should always be ready to change their mind if there is strong evidence against an old idea and in support of a new one. We accept an idea in proportion to the strength of evidence that supports it. How do you think we can get better at changing our minds in the face of evidence?"

FOLLOW-UP ACTIVITIES

Making Explanations, Testing Predictions

Use the information in the chapter Teaching Science and Inquiry: A Deeper Dive to guide students into scientific inquiry, through which they come up with alternate explanations, predict what they'd expect to see if the explanation were true, and test those predictions against what they observe. This more rigorous approach to asking questions and making explanations will strengthen students' understanding of scientific processes and methods.

Learning to Evaluate Evidence

To strengthen students' skills in using evidence to support explanations, do the BEETLES Project classroom activities *Evaluating Sources* and *Evaluating Evidence* (http://beetlesproject.org/resources/for-classroom-teachers-2/).

Making a "Why Web"

Some answers cannot be directly observed but may be inferred from evidence. "Why" questions are a good example of questions that can only be inferred. Inference starts by thinking of possible alternate explanations for a phenomenon. A "Why web" is a useful way to visualize this. Choose a question for which you can think of more than one explanation (ideally a question from one of your students) and write it in the middle of a whiteboard. Then build on students' skills in making explanations by encouraging them to make a "Why web" in their journal.

    a. "'Why' questions are challenging and fun to explore. The trick is not to get stuck with the first possible explanation that comes to you. What are some possible answers to this question?"

    b. "Take a few minutes to write down possible explanations in your journal. You do not have to be right. Be creative." (Students generate answers.)

    c. Listen to students' explanations.

d. "All of these are possible. There may be other possibilities that we have not thought of, so I am going to add "something else" to remind us of that possibility. From here we would look for ways to test explanations for evidence against them so that we could cross ones that do not work off the list."

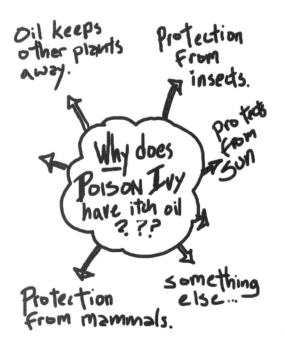

Oil keeps other plants away.

Protection from insects.

Why does POISON IVY have itch oil ? ? ?

protects from sun

Protection from mammals.

something else...

QUESTIONING QUESTIONS

Students observe a natural phenomenon and explore a process for asking varied and interesting questions.

When students are told to ask questions in nature, they often resort to fact-based inquiries, such as "How long does it live?" "How much does it weigh?" or "How fast can it go?" With practice and support, students can develop the tools for more intentional inquiry, asking deeper questions that they can then explore through observation or critical thinking. This activity takes a group of students through a "curiosity crash course," beginning with asking questions from observations, then using a question generator to ask more varied questions. Students put their skills into practice by asking questions about any part of nature that is interesting to them. Students can apply the frameworks they learn in future journal entries, outdoor explorations, science experiences, and learning in other disciplines.

NATURAL PHENOMENA

Students can practice questioning with any part of nature. To find a subject for this activity, try to find a phenomenon that you are curious about or that your students are curious about, or something that makes you think, "Huh, that's weird." Your authentic curiosity will rub off on your students. Make sure you are in a rich area with different natural features, so that in the second part of the activity, students have the opportunity to practice their questioning skills while looking at something interesting to them. If you have specific learning goals or want to use this activity to design an investigation around a specific phenomenon, then use one that is appropriate.

PROCEDURE SUMMARY

1. Use words, pictures, and numbers to record observations, leaving space around the page.

2. Use the question generator "Who, What, When, Where, How, and Why" to ask as many questions as possible, and do not worry about answering them.

3. Start by asking questions about the phenomenon you sketched. Then, if you get curious about something else, you can shift your attention there.

DEMONSTRATION

When the whiteboard icon appears in the procedure description: Make a few quick drawings in the center of the whiteboard and sets of parallel lines to represent blocks of written notes. After the first observation period, write, "Who What When Where How Why" (or whatever scaffold you are using) down the side of the page, then make question mark icons with straight lines next to them to indicate questions. Where appropriate, draw lines to the part of an illustration or note that relates to the question.

Time

Introduction: 10 minutes
Activity: 30–50 minutes
Discussion: 10–15 minutes

Materials

☐ Journals and pencils

optional

☐ Cut-and-Paste Journal Strategies (p. 273)

Teaching Notes

The procedure introduces the curiosity scaffold "Who, What, Where, When, How, and Why" as a tool for generating questions. The follow-up section explains how you could substitute one or more of the Crosscutting Concepts or question scaffolds from the International Baccalaureate Key Concepts p. 92 into the activity. It's important to get to the questioning part of the activity before students "run out of steam," so if your students are only able to focus on journaling for a short amount of time, make their initial observational sketching time shorter.

Consider printing out copies of the Cut-and-Paste Nature Journal Essentials on p. 273 and having students paste thm to the back page of their journals. This will help them remember the list of "Who, What, When, Where, Why, and How" questions.

PROCEDURE STEP-BY-STEP

1. Focus the group's attention on your chosen phenomenon and ask them to share some observations out loud.

a. "Look at this. What do you notice?"

b. "What else do you see? Can anyone add to these observations?"

c. "What does it remind you of? What have you seen before that is similar to this?"

2. Tell students to record observations of the phenomenon using words, pictures, and numbers in the center of their journal.

a. "We're going to observe this in more detail."

b. "First, observe this and sketch it in the center of a new page. Use numbers and words to show your observations, too."

c. "Leave a border of blank space around the outside of your notes."

d. "You will have seven minutes to record your observations."

3. After 7 minutes have passed, call for the group's attention, then explain how to generate questions from observations.

a. "Being curious and asking questions are important parts of what scientists do and how they think. We can use strategies to help us think of questions."

b. "We can always ask questions based on our observations. For example, if you observed a hole in a leaf, you could wonder how it got there, ask a question about its shape, or ask a question about whether other nearby leaves also have holes."

4. Explain that "Who, What, Where, When, How, and Why" is a question generator that can help us come up with lots of questions.

a. "You can also use thinking tools to help you think of questions, and you will learn about one right now."

b. "First, write 'Who What Where When How Why' down the side of your page."

c. "You can use these words to help you ask more varied and interesting questions."

5. Explain that "Who" and "What" can focus us on identifying things and describing a process, then ask students to brainstorm some example questions.

a. "'Who' and 'What' can remind us to ask questions to determine what kind of species or process we see, like 'Who left these tracks?' or to think generally, 'What happened here?' using questions to try to get at a process that is happening."

b. "Take thirty seconds to come up with some 'Who' and 'What' questions out loud with a partner about the phenomenon we're looking at."

6. Explain that "When" and "Where" get us thinking about space, location, and timing, then ask students to brainstorm some example questions.

a. "'When' and 'Where' can remind you to wonder about time, space, and location: when things have happened or will happen, where things have happened or will happen, or the significance of where things are in relation to one another."

b. "Share some 'When' and 'Where' questions out loud with a partner about the phenomenon we're looking at." (If necessary, add examples such as "When did this sapling sprout?" or "Where did this seed come from?")

7. Explain that "How" and "Why" questions focus on figuring out how things work or looking at cause and effect, then ask students to brainstorm some example questions.

Start with observations. These are the backbone of nature study.

When students are new to journaling, it is easier to separate observation and questioning into discrete steps.

As students become more advanced, questions will occur to them as they observe. They can integrate observations and questions as they work.

The level of detail of observation depends on how close and cooperative the subject is. A resting bird gets a close-up. A bird on the move gets a silhouette.

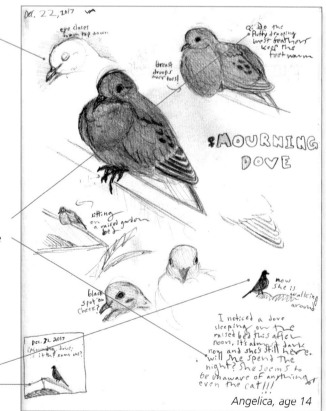

Angelica, age 14

104

a. "'How' and 'Why' can help us ask questions about how things work, or why they might be happening."

b. "Share some 'How' and 'Why' questions out loud with a partner about the phenomenon we're looking at." (If necessary, add examples, such as "Why are there holes on only one side of the leaf?" or "How do the spikes on the leaf help it survive?")

8. **Tell students that they will practice using the question generator to come up with as many questions as possible.**

a. "In a moment, your job will be to ask as many questions as possible. Use your observations and the thinking tools we just talked about to come up with questions."

9. **Tell students that they should start by asking questions about the subject they sketched earlier, but can shift their attention if their interest changes.**

a. "Start by focusing on the sketch you did earlier. Write your questions around your observations, near a part of the drawing that relates to the question. You may also draw a line from the question to the part of the drawing that you are wondering about."

b. "If you start wondering about other things in nature that aren't what you drew originally, you can write down those questions, too. You could shift your attention entirely and focus on a different subject if you would like."

c. "If you are stuck, and feel as though you can't come up with any more questions, you can just say 'I wonder…' to yourself, and see what pops out."

10. **Tell students that they don't need to be rigid with the words that start their questions; the goal is just to get the questions out the door.**

a. "You don't need to be rigid with these words. Sometimes a question that starts with 'What' can accomplish the same goal as a 'Who' question, like 'What animal made this mark?'"

b. "Don't worry about the first word of the question, or 'doing it right.' The goal is just to get the questions out there."

11. **Tell students that they will not need to worry about answering the questions at the moment—their goal is to be curious and to get the questions out there.**

a. "You will not need to answer these questions. Right now, it is just about putting them out there. See if you can catch yourself getting curious about what you see."

12. **As students work, take time to circulate and support those who might be struggling to use the scaffold to ask questions.**

DISCUSSION

Lead a discussion with the general discussion questions, and the topic from the Science and Engineering Practices category. Intersperse pair talk with group discussion.

General Discussion

a. "Let's hear a few of the interesting questions you came up with."

b. As students share, ask follow-up questions, such as "What did you find interesting about that question?" or point out patterns in the group's questioning, such as "Isn't that interesting—many of our questions started with the words 'How' and 'Why.'"

c. After a little while, shift the discussion to reflecting on the question prompts, using questions such as "How did having the 'Who What When Where How Why' reminders impact your questioning?" or "What helped you ask questions?

Asking Questions and Defining Problems

a. "You just practiced an approach for asking varied and interesting questions. You can use these strategies to help ask questions anywhere and to help you become more curious in other settings. Talk with another person about how you might be able to use these question prompts in another setting, like in a different subject in school or in another situation."

EXTENSION

Sorting Questions

1. **Lead students through the process of briefly categorizing questions based on how they could be answered or investigated.**

a. "Let's look back at our questions now."

b. "Draw bold question marks next to your most interesting questions."

c. "Does anyone have a question that we could probably answer with more observation and research, with the tools and time frame we have right now? How would we approach answering the question?"

d. "Does anyone have a question we might be able to answer over a longer period of time or with different tools than we have available right now?"

e. "Does anyone have a question that you don't think can be answered with observation and research? Why not?" *Note:* Some questions might not be investigable because there are not tools available to make the types of observations that would be necessary. Other questions that do not relate to the observable, natural world are outside the limits of what we can study as scientists. Point this out to students, depending on the types of questions they ask.

ALTERNATE QUESTION SCAFFOLDS

Other question scaffolds, such as the Crosscutting Concepts (p. 90) or the International Baccalaureate Key Concepts (p. 92), could either take the place of the 5W's + H steps or be introduced afterward in a follow-up session.

To introduce a Crosscutting Concept as a framework for questioning, give a brief summary of the concept (you can use the information on p. 90 to do this). Then offer some examples of questions that the concept might lead to, and give students time to practice using it to generate questions. The following are some examples.

Patterns

a. "There are patterns everywhere—where leaves come out of the stem on a plant, the shapes of clouds in the sky, the relative location of holes in the ground. These patterns can be clues to underlying processes at work, and we can begin to understand them if we intentionally ask questions to define and describe the pattern."

b. "Ask questions like 'Are there any patterns here?' 'How can I describe the pattern?' 'What does it look like?' 'Are there any exceptions to the pattern?' or any other questions that come to mind when you are using the idea of patterns to guide your thinking."

Cause and Effect

a. "All around us in nature, we see the effects of things that have happened in the past. The things we see happening right now will have impacts that we could observe in the future."

b. "When we use this Crosscutting Concept, we can think of what we can observe as 'effects' caused by events that happened in the past, and we can wonder about this—for example, 'What caused these holes to be here?' or 'Why are there more yellow leaves toward the bottom of the tree?' We can also wonder about what will happen—for example, 'If the slug eats all the leaves on the plant, how will it affect the plant as a whole?' or any other questions that come to mind when you are using the idea of cause and effect to guide your thinking."

Structure and Function

a. "How something looks, its shape, texture, and color, is connected to how it works. The shape of a hawk's wings allows it to soar in the air without flapping. A hummingbird's wings must flap constantly to keep it airborne, but their shape and size enable the hummingbird to make very precise movements."

b. "In nature, we can observe in detail the body part or structure of an animal or plant, then ask questions about how it might function or work in the context of the environment. For example, you might ask, 'How can I describe the structure of this leaf?' 'Why might this leaf by shaped this way?' 'How might the hairs on the leaf help it to survive here?' or any other questions that come to mind when you use the idea of structure and function to guide your thinking."

The other Crosscutting Concepts and the International Baccalaureate Key Concepts could be introduced in a similar way.

MAPPING

Students study patterns in the landscape and create maps of vegetation patterns, wildlife evidence, landscape features, and other characteristics.

Plants, geological features, and animal evidence are not randomly distributed. They occur in patterns on landscapes in response to environmental conditions and such factors as light exposure, soil type, geological forces, or disturbances. Because the focus of the map is showing distribution and not drawing exact representations of a part of nature, creating maps helps students quickly identify patterns in distribution and sets students up to think about possible causes for the distribution. It also sets them up with a new skill and frame to include in future journaling experiences; once students know how to make a map, they can record a small one for context in journal entries focused on other subjects.

NATURAL PHENOMENA

Plan ahead of time what students will map. You don't find ideal subjects for mapping everywhere. Look for phenomena with strong patterns. Any interesting part of nature that has a spatial component (e.g., otter trails at the edge of a marsh, vegetation zones around a small pond, spider webs in a bush, erosion and deposition patterns where a stream feeds into a lake, fresh and old gopher mounds on a lawn, areas of high- and low-intensity burn after a forest fire, a portion of a creek showing alternating zones of ripples and pools with an overlay of animal evidence, or ant trails in a kitchen).

To determine a focus for mapping, you could use your learning goals to decide what phenomenon you would like students to map, then scout good nearby locations for it. Or you could go explore with mapping in the back of your mind and see what features have strong patterns. Identify the boundaries of the mapping location and major landmarks within it as you scout, so that you can point them out to students as you give the group instructions. After students have some practice with mapping, they could select the features for mapping themselves.

PROCEDURE SUMMARY

1. Make a map of the subject in this area within the boundaries described.

2. Use symbols to show where things occur, and make a key to what each symbol means.

3. Include a North arrow and a scale.

4. Start by recording a few landmarks.

DEMONSTRATION

When the whiteboard icon appears in the procedure description: Draw a simple map. Model using landmarks to guide the placement of major features on

Materials

☐ Journals and pencils

optional

☐ Compass

☐ Measuring tape

☐ Examples of maps of your area (vegetation, geology, roads, etc.)

Teaching Notes

Mapping is a complex skill. Give students practice using a journal outdoors before you do this activity, then start with mapping smaller areas (a few square meters of ground) in an area with clear landmarks. Students can make accurate maps without a scale, but you can use a scale if you like.

Ahead of time, figure out whether it will make most sense to orient the map toward a North arrow or to a prominent feature in the landscape.

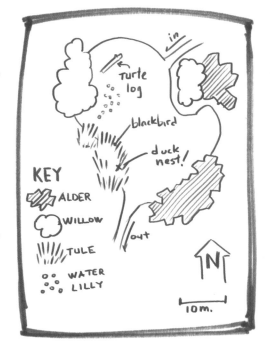

KEY

ALDER
WILLOW
TULE
WATER LILLY

turtle log
blackbird
duck nest!
in
out
10 m.
N

it. Then create a key and add the elements of the key to the map. Add written notes to call out a few important features (e.g., duck nest). Include a North arrow and, if students are making a scale, a scale bar.

PROCEDURE STEP-BY-STEP

1. **Show students an example of a map of a local area and ask them to observe it, discussing the kind of information they notice on the map.**

   a. "What do you notice about this map? What kinds of information do you see included?"

   b. "What kinds of information are not included on this map? For example, you might see a line representing a bridge, but the actual shape of the bridge is not included."

2. **Explain that the purpose of a map is mostly to show the location of things in an area, not to show accurate pictures of each object in that area.**

   a. "Maps don't show everything in the landscape."

   b. "They show the location of certain features (like roads or rivers), but not the details of every single thing in the area."

3. **Direct students' attention to the map legend, pointing out how it identifies the symbols used on the map.**

   a. "This map key, or legend, shows every symbol that appears on the map."

   b. "When you are reading a map, if you see a symbol you don't recognize, you can look for it in the key."

4. **Tell students what the subject of their map will be, and explain that they will focus on showing the location of that subject and anything relevant to it.**

   a. "We are going to be making a map of gopher holes [pinecones beneath a tree, a small stream, rocks, etc.]."

   b. "So we'll record the location of gopher holes [the stream, rocks in the field, etc.]."

   c. "Is there anything else that might be helpful to include?" (If students don't bring it up, suggest details relevant to the subject of your map, such as vegetation, water, or rocks.)

5. **Help students determine the map orientation by directing them to north or a prominent landscape feature they can mark on their map.**

   a. "Our map area will be from [name easily visible landmarks to give boundaries for the map area]."

   b. (If you are orienting the map to north) "This direction is north; please put an arrow at the top of your page with an 'N' under it to show the orientation."

   c. (If you are orienting the map to a prominent landscape feature) "We're going to orient our maps to the flagpole. Put the flagpole in the top right-hand corner of the map, and orient your map around that landmark."

Abigail, age 10

A simple key explains the symbols on the map. This saves time, is easy to read, and takes pressure off drawing.

6. (Optional) Describe how to designate scale, using 1 inch or two squares on graph paper on the map to represent 10 paces. If students already know their pace, you may skip this step and instruct them to correlate 1 inch or two squares on graph paper to a given number of paces on the map.

a. "Now, we will determine and designate the scale. Use your paces to show how far things on your map are from each other; for example, one inch on the map might equal ten paces. Although this should be as accurate as possible, it does not need to be exact." (You can have a compass and a measuring tape on hand if students want to add greater precision to their maps.)

7. Describe how to use symbols and letters to show where the subject and other important features occur on the map, and how to make a key, emphasizing that students should not make detailed drawings of everything they see.

a. "Add organisms or other things you find to the map. When you want to add an organism to your map (like that tree, for example), first make and label a key symbol."

b. "Key symbols can be letters or simplified shapes. They should not be detailed drawings, because the goal is not to show the details of every single thing; it is to show their location."

c. "The symbols should be easy to draw and distinct from one another. They could also be letter codes (such as GH for gopher hole). You can add additional key elements as you go."

d. "You don't need to know the names of species of plants or other features to add them to your map. You only need to be able to tell them apart from each other."

8. Suggest that students use hatching lines or a grid to show features (such as grass) that cover a large area on the map.

a. "If there is a plant, such as a grass, that covers a very large area, you can use hatching lines or a grid to show this."

9. Help students pick out a couple of landmarks to put on their map before they begin.

a. "Before you begin, pick out a few important landmarks to add to your map. Pick things as reference points that could help you place other features on the map."

b. (Or, for younger students) "Let's pick out two landmarks to add to our maps together. We'll pick these as reference points that will help us place other features on our map. Let's add that tree over there, and the bench at the end of the schoolyard. Please pace out to show how far they are from one another." (*Note:* You can also use a pile of backpacks or other equipment and locate it on student maps to get them started, essentially a "You Are Here" sign.)

10. Tell students that they can add questions to their map as they work; make sure they are ready to begin, and send them off.

a. "If any questions occur to you as you make your map, add them in as you go."

11. As students work, take time to circulate, offer reminders, or support students who are struggling.

a. (After half the time has elapsed) "We're about halfway through our time for mapping. If you've only focused on one area, it's time to move on so that you can be sure to complete your map in time."

DISCUSSION

Lead a discussion using the general discussion questions and questions from one of the Crosscutting Concept categories. Intersperse pair talk with group discussion.

General Discussion

a. "What did you learn about the distribution of [your chosen map subject] through making your maps? Where were they, and where weren't they? Were there any patterns you noticed in terms of what your subject was next to?"

b. "What might have caused some of the patterns of distribution you observed? Why do you think that the things you saw occurred in some places and not others?"

c. "What was it like to make a map? What was fun or challenging about it? If you were to do it again in a different area, is there anything you would do differently?"

d. "Was there anything you learned through making your map that was unexpected, anything that surprised you?"

Patterns

a. "When you made your maps, you were noting patterns. Where was there a lot of [subject of map]? Where were they? Were they in groups or spread evenly? Where were they missing?

b. "Are there other patterns you noticed while making your map? What could be some possible explanations for what you observed?"

Cause and Effect

a. "What are some possible explanations for the patterns we observed, of where certain [map features] were and where others weren't?"

b. "What seen and unseen processes or forces could impact where different things occur or don't, or grow or don't grow?"

c. "What kinds of forces could cause a change in the distribution we observed while making our maps?"

Stability and Change

a. "Do you see any evidence of changes or events that happened a long time ago?"

b. "What causes change in the distribution of the phenomenon we mapped, and what is your evidence for this?"

c. "What do you think this area was like _____ ago? What will it be like _____ from now?"

d. "At what kind of time scale are these changes happening? Quickly? Slowly?" "We'll return to this area in [some amount of time] to see how it has changed."

FOLLOW-UP ACTIVITIES

Making a Cross Section

Draw a landscape cross section (see next activity) between two points on the map that would reveal vertical zones of plants or animals such as you might find along the seashore or at the edge of a pond.

Mapping over Time

Return to the same area weeks, months, or years later and make another map, then talk about what changed.

LANDSCAPE CROSS SECTION

Students map shifts in plants or animals across an area with a gradient of species and environmental conditions, such as an intertidal zone, the slope of a hill, or a transition from shore to pond.

A straight line that cuts across an object or area is called a transect. A cross-section diagram is a side view showing the species or features that occur along a transect, as well as other details and observations. (A cross section could also show a phenomenon that has vertical structure, such as a species of moss or lichen on a tree.) Making cross sections of a transect is a powerful tool for noticing patterns of organisms or other features within an ecosystem. The approach prepares students to think about how patterns in organism distribution are related to environmental conditions, and helps them make detailed observations of landscape features.

NATURAL PHENOMENA

Look for areas that transition between environmental conditions (wet to dry, sun to shade, direct to indirect sunlight, one soil type to another, elevations, etc.) with associated changes in the distribution of plants or animals. Or look for an area with variation in patterns of distribution of plant or animal species (even if you're not sure about what types of changing conditions might have caused them).

This might include intertidal zones (on vertical rock faces or wharf pilings, or as you walk from shore to sea among exposed tide pools); a recently fallen tree, along which there are changes in lichen and moss species; a stream profile showing changes in vegetation on either side (being sure to make your area wide enough to show the transition to riparian vegetation along the stream bank); a line across a boulder; the edge of a marsh showing a transition from wet to dry along the bank and emergent or floating vegetation zones in the water; a transect of a valley or a hill on a north–south transect line (north-facing slopes in the northern hemisphere are more lush and damp; they get less direct sunlight than south-facing slopes); vegetation zones up a tall mountain showing plant communities at different elevations (best done on a road trip where you can drive between zones and stop periodically to journal).

PROCEDURE SUMMARY

1. Make a cross-section diagram to show the distribution of plants and other living things in the transect.

2. Begin by drawing a profile of the ground and making rough outlines of any zones or patterns of species you see.

3. Use symbols to show where plants and other organisms occur, making a key to show what the symbols mean.

DEMONSTRATION

When the whiteboard icon appears in the procedure description: Draw a line representing the contour of

Materials

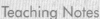

- ☐ Journals and pencils
- ☐ Examples of cross sections pulled from the internet or other resources

optional
- ☐ Compasses
- ☐ Measuring tapes
- ☐ Rulers

Teaching Notes

Students can make a cross-section diagram to a part of their map as a follow-up to the activity *Mapping*. Once students know this technique and the type of insight it provides, they can include a small cross section in other journal entries. Use the discussion questions to help students see the technique as a transferable investigation tool.

Depending on the size of your group and the available area for study, decide whether students should all walk the exact same transect (spread out slightly to avoid being in one another's way) or each walk their own transect.

111

the ground. Then add simple symbols for different kinds of trees and bushes. You can label the plant symbols on the page or create a key to your symbols on the bottom or side of the page. Add a cardinal direction arrow and a stick figure of a person for scale. Represent written notes and questions with sets of lines.

PROCEDURE STEP-BY-STEP

1. **Tell students that they'll be exploring the patterns of distribution of organisms, then tell them to walk along the transect and observe patterns in small groups.**

   a. "We're going to be making a side view or cross-section diagram to show the patterns of plants and other living things along a straight line between _____ and _____."

   b. "Before we use our journals, let's take five minutes to check this out in teams of three or four. The transect begins [name landmark] and ends [name other landmark]. Walk in between those two points and look for patterns of where things occur."

   c. (After students return) "Please find someone in a different group from you and discuss: What patterns did you notice? What else did you observe?"

2. **Explain how cross sections are used to show zonation and patterns and give students time to observe examples of cross-section diagrams in small groups.**

   a. "Scientists and engineers sometimes use drawings called cross-section diagrams to explore and explain vertical zones and patterns. Let's look at a few examples." (Show students printed examples from the internet or other sources.)

   b. "As you look at these examples, talk with those around you about how information is shown in this diagram. How did the author record their ideas and observations?"

   c. (After students have had some time to look in small groups) "What kinds of approaches to recording observations did you see?" (Students might say: symbols are used to show distribution of species instead of detailed drawings; the contour of the ground is included; the cross section includes a key for symbols; written notes of observations are included; a simplified map shows where the cross section occurs in the landscape.)

3. **Demonstrate how to begin making a cross-section diagram by putting in a profile of the ground surface, then blocking out the zones of different organisms.**

   a. "Here's a useful way to approach your cross section. Begin by drawing an approximate profile of the ground (and water) surface, and a stick figure of yourself (noting your height) to show scale."

   b. "Then, using light lines, block in the boundaries of the zones or patterns of organisms you observed."

4. **Describe how to show distribution of specific organisms using symbols, and how to create a key for the symbols.**

   a. "Add organisms or other things you observed to the zones. When you want to add an organism to your cross section (that tree, for example), first make and label a key symbol."

   b. "Key symbols can be letters or simplified shapes. "

5. **Emphasize that students don't need to make detailed drawings of every species and should use symbols above and below the line of ground to show where organisms occur.**

   a. "Do not make detailed drawings, because the goal of a cross section is not to show the details of how a species looks; it is to show where they are distributed relative to each other and the environment."

   b. "Use the key symbol to show where a species occurs in the cross section. Then add the next species."

   c. "If there is not enough space above the line of the ground on your cross section to show all the organisms, you can label them below the ground on your diagram."

6. **Ask students whether they have any questions about the procedure, then spread them out to begin their cross section, offering reminders on timing, and, throughout the process, supporting students who are struggling.**

   a. (About halfway through) "We're about halfway through our time. If you've only just begun filling in symbols on your diagram, be sure to focus on that part of the activity now."

   b. If a student finishes early, approach them and ask, "Are there any more details you can add about the environmental factors that could influence the distribution of these species?" (or) "Is there one species you could map in more detail or record other observations about?"

7. **Reconvene the class in a location suitable for a discussion. Have sutdents make a scale, add cardinal directions and metadata, and make a small inset map.**

   a. "Include a scale bar to show distance, and a small inset map to show how the transect fits into the larger landscape."

   b. "Be sure to add in the date, location, and weather conditions."

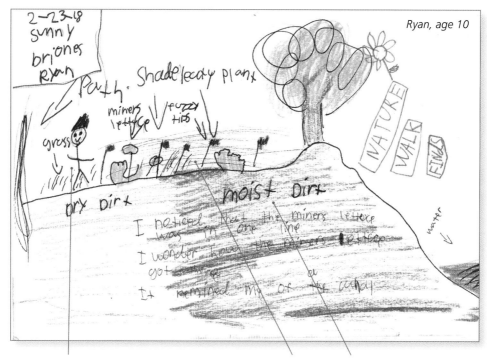

Ryan has drawn himself to show scale.

This labeled drawing holds a lot of information: sun vs. shade, wet vs. dry soil, and plant distribution.

DISCUSSION

Lead a discussion using the general discussion questions and questions from one of the Crosscutting Concept categories. Intersperse pair talk with group discussion.

General Discussion

a. "Find a partner and talk about what you noticed through making your cross section. What patterns did you notice?"

b. "What types of things were higher up or lower down? Were certain organisms or things in some places and not others?"

c. "What are some of the natural forces that influence or affect the organisms you recorded on your cross section?"

d. "Add these factors to your diagram using labels, arrows, drawings, and icons."

e. "How might these natural forces have influenced the patterns of distribution you recorded in your cross section?"

f. "When might a scientist choose to use a cross section in their work? What are cross sections for?"

g. "How did making a cross section help you learn about this landscape and these species?"

h. "What are some other situations where you might want to make a cross section or include one in a journal entry?"

Patterns

a. "What were some of the patterns of distribution in the species you recorded in your cross section?"

b. "Pick one species and discuss its distribution in detail. Where was it? Where wasn't it? What did it grow next to? What kinds of patterns of growth did it show—for example, in clumps? all next to each other? only one in a given area?"

c. "What are some possible explanations for these patterns? This might include living things or nonliving environmental factors or forces."

Cause and Effect

a. "Look at the patterns of species distribution you observed. What are some possible explanations for why certain things grow or live in certain areas?"

b. "What seen and unseen forces may be behind patterns we observed?"

c. "How might the distribution of species along the cross section change depending on the season, weather, or other factors?"

Stability and Change

a. "What factors do you think influence the distribution of species and other natural factors along this cross section?"

b. "If one of these factors changed—for example, if there were more sunlight or more water—how might that affect the distribution of these species?"

c. "Do you see any evidence that the distribution of species or the environmental factors might be changing?"

d. "What might have the patterns of species looked like in this area _____ years ago? What might they look like in the future?"

e. "Where do you see different ages or sizes of plants? How do you think the distribution of the plants along the transect might change in the future?"

FOLLOW-UP ACTIVITIES

Combining Map and Cross-Section Views

Do *Mapping* before this activity, or afterward. On the map, show where the cross section cuts across the landscape.

Looking Back

Look through old journal entries to identify where cross sections would have been helpful in previous investigations.

Any gradient—such as forest to meadow, dry to wet, shade to sunlight—or changes in soil type or agricultural management (plowed or watered) create zones that lend themselves perfectly to investigations with a cross section.

114

PHENOMENON MODEL

Students observe a phenomenon and create a diagram to describe it. Then they think about and label seen and unseen forces that may affect the phenomenon, and use this model to make possible explanations for how the phenomenon works.

The world is filled with phenomena and mechanisms that we take for granted and often do not understand. What are the mechanisms behind the patterns of ripples on the surface of a pond during the rain? Why is a heated pot of water loudest before it is at a full boil? How does a refrigerator work? We can come up with some ideas about how these things happen, but making a model is a deeper approach, one that externalizes our thinking and shows us what we do not yet understand. Modeling a complex phenomenon calls on our background knowledge of science concepts from, for example, physics, ecology, or biology, and sends us to additional sources of information to deepen our understanding. Modeling can change the way we think and how we approach problems, but it is not always an intuitive skill. This activity offers an approach to modeling that students can apply in future journal entries and in other disciplines. Students will also observe a phenomenon in detail and begin to understand it, setting them up for future learning experiences and deepening their knowledge of key science concepts.

Modeling like this makes our ideas visible. It doesn't matter whether our explanation is right or wrong. We model to clarify our thinking. Once the ideas are down on paper, we can more easily see connections, gaps in our thinking, and strengths and weaknesses of our ideas.

NATURAL PHENOMENA

This activity can be done with any interesting phenomenon that can be observed by the whole class. Look for things that make you wonder "How does that work?" or "What causes that?" The more authentically mysterious a phenomenon, the more that students will be drawn to explore it. Look around you for the phenomena that are occurring constantly outside the field of your attention. You may also be able to find aspects of a familiar phenomenon that are new to you, such as the yellowing of leaves in the fall, or steam rising from a wooden fence on a cold day. In this description, we use a mackerel sky (clouds formed in narrow parallel rows) as a phenomenon. These instructions would change slightly for a different phenomenon. Note that the demonstration is about a double rainbow, not the clouds the students will be observing. This is intentional, because it allows you to discuss recording observations without telling the students what they should notice about the particular phenomenon.

If students come to you with a question about something they have found and you can bring the whole class out to see it, take their suggestion and run with it. You can also choose a phenomenon based on the science subject areas that you and your class are working with.

PROCEDURE SUMMARY

1. Observe the phenomenon, writing observations and creating a diagram to show what you can see.
2. (Later) Make possible explanations for how the phenomenon might work, using words, pictures, and numbers to show your thinking.

Time

Introduction: 10 minutes
Activity: 35–60 minutes
Discussion: 10 minutes

Materials

☐ Journals and pencils

☐ Colored pencils (at least two per student)

Teaching Notes

This activity is meant for fifth grade and up.

Models are important tools in scientific inquiry. When we think of a model, we often envision gluing together a plastic plane or making a cell out of marshmallows and macaroni in school. Scientific models (such as those referred to in the NGSS) do more than represent an enlarged or miniaturized object; they are tools to explain the behavior of a phenomenon. By this definition, a labeled drawing on its own is a description, not a model. By adding the interaction of seen elements and unseen forces and making possible explanations for how they produce the observed phenomenon, students move from diagramming to modeling.

This activity can be used as an introduction to modeling. Students do not need to solve or figure out the mysterious phenomenon. Be clear about this at the beginning, telling students to instead focus on generating many possible explanations and opening up ideas.

3. Include seen and unseen forces, using labeled arrows to show how you think they might affect the phenomenon.

4. Revise your model after you share ideas with your classmates.

DEMONSTRATION 1

When the whiteboard icon appears in the procedure description: Draw a line one-third from the bottom of the page. Above this line, demonstrate how a student might take notes about a double rainbow. Do not erase this drawing once it is done; you will add to it later.

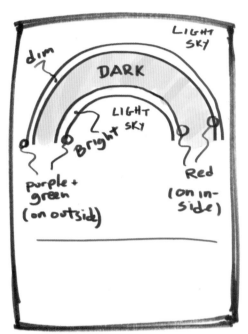

PROCEDURE STEP-BY-STEP

1. **Describe how students might record observations of a phenomenon, and demonstrate on a whiteboard how the notes might look.**

   a. "Have you ever seen a double rainbow? Let's say we wanted to make detailed observation of this phenomenon and record it in our journals. We might observe that the colors are in the opposite order in the rainbows. The lower rainbow has red at the top, and the upper one has red at the bottom. The lower rainbow is brighter than the upper."

   b. "We might also notice that the sky between the two rainbows is darker than the sky outside the rainbows. When you record observations about a phenomenon, your goal is to record every detail you notice. The best way to do this is with words, pictures, and numbers."

2. **Tell students to record observations of your chosen phenomenon using words and pictures.**

   a. "Those are amazing cloud formations [stream currents, ice patterns, etc.]. Let's take seven minutes to record as much information as we can about them."

b. "Start by drawing a light line across your paper, one-third of the way from the bottom.

c. "Use words and pictures or diagrams to describe everything you can notice about the clouds. Use the top two-thirds of the page. Do not worry about spelling or making pretty pictures; just try to make your notes clear and easy to understand."

3. **After students have had some time to record observations in words and pictures, encourage them to try to incorporate numbers as observations. For example:**

   a. "If you can, use numbers to deepen your observation. Is there is a repeating shape or formation? If so, how many do you see? Are they the same size or different sizes? Is the distance they are away from each other consistent? What are some estimations you can come up with of distances between parts of the interesting formations?"

   b. (If you are far away from the phenomenon) "We can't get real measurements because the clouds are so far away, but can you think of a way to more accurately estimate their size or spacing from down here? For instance, how does cloud width compare to the spaces between them?"

4. **After 10 minutes have passed, call for the group's attention. Encourage students to ask questions about the phenomenon and why it appears as it does.**

   a. "What questions do these observations stimulate?"

   b. "Can you ask questions about how or why the clouds [currents, shapes, etc.] appear as they do?" (Examples: "What makes the clouds form rows?" "Why are some of the rows broken into puffy balls while some are continuous?" "Why are the rows smaller in that part of the sky?")

5. **Call for the group's attention, ask them to share their observations, and generalize these to some concise statements about the phenomenon. Keep the conversation focused on observations, not explanations.**

   a. "Let's see if we can turn this large set of observations into a concise set of statements describing this phenomenon."

   b. Elicit observations from the group, encouraging students to add on to or modify what has been described. (For example: "Who can get us started?...Great, can anyone add to or modify that?...What else can we say about what is happening here? What else did we notice?")

6. **Give students a moment to write a summary of what the group has observed in their journals.**

   a. "In your journal, write your own summary of what is happening here."

   b. "Be careful to only write observations. This is what we can directly see, not our explanations of what we can see."

7. **Give students a moment to ask questions and record questions about the phenomenon with a partner, then share them with the group.**

   a. "What were some of your most interesting questions you came up with earlier?"

   b. "Take a moment to share your questions with a partner, then see whether you can come up with any more questions and write them in your journal."

   c. "Let's hear some questions. What are we wondering about?"

8. **Model how a student might create a visual explanation for a phenomenon they observed by first proposing possible explanations.**

   a. "Let's go back to that demonstration of observing a rainbow. We could develop several questions, but let's say we are most interested in what causes the dark band between the rainbows. Here is where creative thinking comes in. What are some possible explanations for this band? Think creatively; you do not have to be right." (Students respond.)

   b. "At this point we are not worried about whether an explanation is right or wrong. Instead, we generate possible ideas."

DEMONSTRATION 2

Direct students to use a different color of colored pencil and the lower third of the page to create a model of a possible explanation for the phenomenon. They can use arrows to show forces at work or stages of a process. They should consider how to show both seen and unseen forces. The explanation does not have to be right.

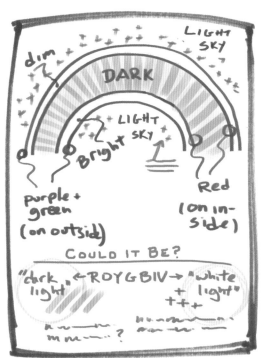

9. **Demonstrate how a student might create a visual explanation for a phenomenon they observed by picking one possible explanation and making a model diagram of it.**

   a. "You can then choose an explanation that is interesting and make a model of the idea to help you think about it. The explanation does not have to be right. In the case of the double rainbow, let's say we chose the explanation that there is scattered 'white light' beyond the violet end of the rainbow and some kind of scattered darkness or 'dark light' beyond the red end of the rainbow." (Study demonstration 2 for an idea of how to draw this.)

   b. "Getting the idea on paper helps you think about it more clearly. After you lay out an idea like this, you can more easily see holes in your thinking or places you could modify the idea."

   c. "In this case, we might see that the idea of 'dark light' was problematic and discard this explanation in favor of another idea. If the model helps you think about an idea, it is a success whether the idea is right or wrong."

10. **Divide students into groups and set them up to generate several plausible explanations for the mechanism behind the phenomenon they observed (e.g., cloud formations), focusing on sharing ideas without worrying about being wrong.**

    a. "Organize yourselves in groups of four and choose a record keeper for your group."

    b. "With your group, take five minutes to come up with ideas about what could cause the clouds to make these shapes."

    c. "Come up with several plausible explanations. Your explanation does not have to be right; we are just playing around."

    d. "Be creative, and make sure you get everyone's ideas. The record keeper will write the explanations down using language of uncertainty, such as "Perhaps…,' 'Could it be…,' 'I wonder whether…,' and 'Maybe….'"

11. **After a few minutes, share and discuss possible explanations with the entire group, emphasizing that the explanations do not need to be right, just plausible.**

    a. "Time is up. As a group, choose the two explanations you find the most interesting. The record keeper will report these out in a moment. We are not interested in right or wrong, just in listening to the possible explanations."

12. **Facilitate the process of sharing explanations. After all groups share, discuss and explore the thinking behind some of the explanations without evaluating whether they are right or wrong, using follow-up questions such as "Say more about that" or "What makes you think that?"**

How might these rock features have formed? Could you draw a series of diagrams to lay out one explanation? What features would your explanation account for? You do not have to be right. Be wrong boldly. Interesting is better than safe.

13. Explain how to make a model of the phenomenon in a journal: Make a simple diagram of what can be seen, then show an explanation for the causes of the observed features, labeling seen and unseen forces, using arrows to show interactions, and using text to explain what is happening.

a. "Choose one of these explanations and create a model diagram that explains how it works. Don't worry about being right; just focus on explaining your thinking clearly."

b. "Use your notes to help you draw a simple, clear, labelled diagram that explains the phemonenon."

c. "Think about the unseen forces that interact with what you can see to create the patterns. Make the diagram of your explanation with a different color of pencil. Add arrows to show forces or movement, lines to show connections, and text to explain what is happening. You can draw steps in a process or a single diagram. You may use the empty space at the bottom of the page."

14. Tell students that they can have fun with the process, coming up with interesting ways to represent their thinking on the page.

a. "This can be a fun and creative process. See whether you can find interesting ways of representing your thinking, explaining the process, and showing interactions between seen and unseen forces."

b. "Are there any questions? You have ten minutes."

15. After students have had time to model their explanations, regather the group and let them to share any new insights, questions, or explanations they came up with, or instances where making the model brought up things they realized they didn't understand.

Note: If students have new ideas or want to add to explanations the group discussed earlier, this process could go on for a little while. Stay with it as long as students are interested. This is a huge opportunity for students to struggle with ideas, construct new understandings, and learn.

a. "When we move from making a verbal explanation to making a model, it often brings up more questions or reveals to us places where our explanation might not be as good as we thought."

b. "What did we learn through making our models? What questions do we have? What do we think we understand, and what don't we understand?"

c. "What do you want to know more about in order to better understand this phenomenon? What questions do you have?"

16. Review and share student work by placing journals on tables and giving students time to circulate and see how their classmates showed their thinking and explained the phenomenon.

a. "Circulate around the room and take a look at the ways other people described this phenomenon. Look for creative strategies and solutions that you could use if you were to do a similar project again."

b. "Also look for places someone else might have been more accurate or might have better explained what is happening here."

17. Give students a little more time to modify their diagram based on feedback or ideas they saw in other students' work.

a. "Take four minutes to add any other details or elements to your diagram that would make it more clear or that would better show your understanding of the phenomenon. It's OK to use ideas you saw in someone else's journal."

18. (Optional) (Upon return from the field) Direct students to sources of information that could offer more evidence to explain the phenomenon.

Caleb, age 6

"So it started as a baby." The diagram depicts a sapling being struck by wind and bending away from the force.

This is an illustration of the same tree, now older, permanently bent by the wind. Curled lines indicate the strongest wind.

Circles on the leeward side of the tree show clusters of *Trentepohlia* algae "popping" up on protected surfaces.

DISCUSSION

Lead a discussion using the general discussion questions and questions from one of the Crosscutting Concept categories. Intersperse pair talk with group discussion.

General Discussion

a. "Today you created a model to explain cloud formations [stream currents, ice patterns, etc.]. What was the process you used to create your model?"

b. "How did developing and using a model help you learn?"

c. "What are other situations or phenomena that modeling could help you learn about?"

d. "What were some approaches your classmates used for showing their thinking in their model on the page?"

e. "We used different colors of pencil for our observations and our explanations. Why do you think we make such a clear distinction between them?"

f. "What is the value of making a model if you aren't sure your explanation is correct?"

Patterns

a. "Describe the patterns you observed."

b. "There is often a mechanism or a force that creates patterns we see. What are some possible explanations of how these patterns were formed?"

c. "What does this pattern remind you of? Can you think of other things or phenomena that create a pattern that is similar to what we observed? Consider things on a larger or smaller scale."

d. "What forces might be behind these similar patterns?"

Scale, Proportion, and Quantity

a. (Follow-up to Patterns question 4) "Forces act differently on large and small objects. How might scale and size change the way forces interact and create patterns in phenomena?"

Systems and System Models

a. "You can think of the clouds [stream currents, ice patterns, etc.] and the forces that combine to make them as a system of interconnected parts that produce behavior, in this case the patterns we observed. What are some of the parts of this system?"

b. "What system behavior did we observe?"

c. "How do the parts interact? What are the inputs to the system and the outputs of the system?"

d. "Where would you draw the boundaries of this system?"

Energy and Matter

a. "How might energy flow into this system?"

b. "How might sunlight energy affect this system?"

c. "Can you think of ways that these patterns might be related to cycles in nature? How?"

d. "How does matter change form in this system?"

Stability and Change

a. "How long did these patterns last?"

b. "When the pattern changed, what came next?"

c. "What forces might make these patterns more stable? What would disrupt them?"

FOLLOW-UP ACTIVITIES

Conducting Additional Research

When students finish this activity, they may have more questions than answers. Many of the phenomena you will encounter will be outside the realm of what students can understand or figure out on their own. Brainstorm resources beyond Google with students that might help them begin to better understand the phenomenon. This might include books, internet resources, or local scientists in the discipline of study of the effect you were observing. The results of the research may or may not support the students' explanations. Either is interesting and useful.

Help students process their current questions about the phenomenon and brainstorm next steps about how to continue to build understanding. Offer instruction and new information that they can integrate into their models, then offer time for modification of explanations. This can be a process that drives an entire unit of study. The chapter Teaching Science and Inquiry: A Deeper Dive offers some approaches for extended investigations.

Studying Models

Models are powerful tools that scientists use to explore, test, and explain the world. They range in complexity and are used in all scientific fields. Search the internet for examples of scientific models. Ask students: "What is the purpose of this model? How is it intended to be used? Who is intended to use it? What are its limitations? Are there any ideas that you could use to help you model or think about other systems?"

TEAM OBSERVATION

Students observe the same subject and work together to discover as much as possible. As they do, they see the variety and depth of the observations that can be made.

Science is a team sport. Rarely does a lone scientist, working in isolation, contribute much to our understanding of the world. Scientists work together to solve complex problems. They share ideas, read one another's work, replicate one another's studies, and build on what is already understood. Everyone has different strengths, and having more eyes on any problem means there will be more ways of thinking about potential solutions. We can see examples with NASA (it takes a village to get to the moon—and back) or any group engineering effort that has produced many of the modern products we take for granted. Any lab at a university comprises a cohort of graduate students at various levels, and just about any scientific paper was written by a team of authors. Your students will go further in their understanding of science if it is a collaborative effort, not a competition to see who gets the best grade. Use this activity to set the tone of teamwork and enable your students to see how much they can learn when they work together.

NATURAL PHENOMENA

Find a natural object, organism, or species that all students can simultaneously observe. The possibilities are limitless. Students could focus on an abundant species of flower in a field; a colony of ants; a tide pool; a large, spreading tree; a small pond; or a stream. A tall, narrow tree (such as a pine) is more challenging for a large group. Even though everyone would be able to see the tree, it might be difficult for everyone to get up close and examine the bark or lower branches. Choose the phenomenon, or set it up so that the group of students can decide what they want to explore. If you want to use this activity to help students build understanding of specific science concepts, then choose the phenomenon and suggest a couple of categories for observations that meet your learning goals.

Time

Introduction: 5 minutes
Activity: 30–40 minutes
Discussion: 10–15 minutes

Materials

☐ Journals and pencils

optional

☐ Loupes, hand lenses, or magnifying glasses

Teaching Notes

Some students go into class wanting to "win" the observation competition. Students will cooperate more with each other if it is clear that there is room for more than one person (and, in fact, everyone) to be successful. Weave this activity in with community building or social emotional learning in your class or group, and reflect afterward on what they could do next time to be a more effective team.

Many sets of eyes observing will lead to a rich experience. Give plenty of time for groups to share what they learned. This information your students worked hard to collect can be a rich source of data, but only if you use it beyond this activity to further the whole group's understanding of the phenomenon. Use subsequent experiences to deepen students' learning.

PROCEDURE SUMMARY

1. Collect interesting observations for your category with your assigned team.

2. Make your own journal entry of observations, but talk to your teammates and work together as coexplorers.

3. Before you begin, discuss with the team what should be observed, and make a plan for how the work will be shared.

Note: There is no demonstration for this activity because different groups of students will focus on collecting data on different parts of this phenomenon.

PROCEDURE STEP-BY-STEP

Note: In this example, we use a tree as the focal object of the study. The procedure would be modified slightly for other subjects. Here we imagine a group of twenty-four students who will be working in groups of four (or six groups). Adjust accordingly for your needs.

1. **Explain that science is a team effort, and scientists think through problems and ideas together, helping one another solve problems and try to come up with the best possible explanations based on the available evidence.**

   a. "In movies and TV, scientists are often shown as loners who work in isolation. In reality, most of science is a team effort."

   b. "Groups of collaborators share their observations and talk about different possible ways to approach a problem or answer a question."

   c. "When they have trouble understanding something, they ask someone who knows more than they do, or they talk with peers about things they have read."

   d. "They work together to come up with the best possible explanation based on all the available evidence. In working together, they come to a deeper understanding than any of them could have done on their own."

2. **Tell students that they will work together like a team of scientists to study a nearby tree [or ecosystem, stream, etc.], emphasizing that the goal is collaboration, not competition, and that each person has a role to play.**

   a. "We are going to explore how working in teams helps us learn as we try to discover as much about this tree as we can, collaborating like a team of scientists would."

   b. "Our goal is to collaborate, not to compete. When we all observe and work together, we learn more than any of us could by ourselves."

   c. "There can be a role for everyone. Some people might be more drawn to counting and working with numbers; others of us might be thinking big picture and connecting what we see to science ideas. Another person might be great at keeping track of the needs of the group, making sure we all have what we need to stay focused. All of these are important roles that will help us learn."

3. **Explain the procedure: The group will come up with categories for the types of observations they think are important to make, then divide into teams of study.**

   a. "First, we need to come up with categories of the types of observations we think would be important to make about this tree."

   b. "One category could be studying the tree's leaves."

   c. "What other interesting structures or categories can you think of?"

4. **If students have trouble starting a list, prompt them toward other categories.**

   a. "Try looking at the shape and color of the trunk, branches, and leaves or looking for evidence of animals or plants living on the tree; another group could study the soil around the tree, or anything else you think might be important."

5. **Divide the group into teams (counting off is a useful strategy for this) and assign categories.**

Note: Older students or those with more experience in the field could be set loose to create their own study groups and categories for observation. Offering this autonomy will increase student buy-in.

6. **Tell students that each team of four will be responsible for collecting interesting observations about their category, recording what they find in words, pictures, and numbers.**

   a. "Your research team is tasked with collecting the most interesting and relevant observations about your category. You will have seven minutes to collect information relevant to your category. Although you will work as a team, you will each record notes separately in your journals, using words, pictures, and numbers."

7. **Suggest that teams take a minute to make a plan for what will be important to observe and focus on, and encourage teams to continue to talk to each other about observations and questions as they work.**

   a. "Before you begin, take a minute to plan what will be important to record."

   b. "As you all work, talk to the members of your team, sharing questions or exciting observations. Are there any questions? Go!"

8. **As students work, take time to circulate, talk to teams about what they are observing, and support any students who are struggling.**

9. **When the time for observation is up, call the group back together and tell teams to meet and share observations, questions, and ideas.**

   a. "Take a couple of minutes to meet with your team and share observations. Talk about the things that you learned, and any questions you came up with."

10. **Give each team the opportunity to share some of the questions and observations they made.**

   a. "Let's hear a bit from each team about the observations you made. What did you find out? What did you notice? What was interesting, surprising, or unexpected?"

   b. "Which observations are important to share as a group, so that we can better understand this tree as a whole?"

   c. "What questions did you come up with?"

11. **(Optional) Lead a group discussion about an interesting question a team came up with, asking for differing ideas, possible explanations, and perspectives from members of different teams.**

12. **Send students back out to individually collect one specific and special observation, a detail they think no one else has noticed.**

   a. "We learned a lot as a group, but it is also fun to explore on your own. You will have five more minutes to explore, with this challenge: Make one observation that is so specific, particular, and detailed that you are sure no one else in the group has the same information."

   b. (After exploration time is up) "Does anyone want to share their discovery?"

13. **Give students a moment to record metadata for their journal entry, including the names of the members of their team.**

DISCUSSION

Lead a discussion using the general discussion questions and questions from one of the Crosscutting Concept categories. Intersperse pair talk with group discussion.

General Discussion

Use the general discussion questions to prompt students to discuss and reflect on their process of working together as a team.

   a. "Why do you think scientists work together in teams?"

   b. "What are some advantages of working in teams?"

   c. "What are skills that scientists need to have to make them effective team members?"

   d. "What are other fields where you see teams or people organizing to solve problems together?"

   e. "How did your small group do at working as a team? What about our whole class? What could we do to work better together next time?"

Patterns

   a. "What patterns did you find? These might be patterns in structures or growth formations, in the location of different parts of the organism [or phenomenon], in features like holes or scars, or in the general location of the organism [or phenomenon]."

   b. "What are some possible explanations for those patterns?"

Cause and Effect

   a. "What were some of the features or patterns you observed? What are some possible explanations for them?"

   b. "Did you find any evidence that the [observed organism or phenomenon] may be affected by living or nonliving things in the environment?"

   c. "How do you think the [observed organism or phenomenon] might affect the living or nonliving things in the environment?"

   d. "How might the interactions you observed be affected by the time of day, year, weather, or location?"

   e. "Did you see any interactions between the organism and the environment? What effect might they have had on each other?"

Energy and Matter

Note: These questions are appropriate for students fifth grade and above.

   a. "Let's construct a partial food chain based on your observations. What did you see eating what? Now expand your food chain to a web based on what you have seen in this area, your prior knowledge, and your best guesses about other relationships between animals."

   b. "Now trace the cycling of matter through the parts of the food chain you just described."

Systems and System Models

   a. "You can think of this tree as an ecosystem in itself. Within these branches are many species. How many different kinds of organisms or their evidence did you find?"

   b. "Let's create a diagram to show some of the relationships between these species. Start with two that you think may affect each other—for example, predator and prey. Draw and label a small box for each species."

   c. "Draw a line between them and write the relationship between them on or around the line. Now add another organism to the chart and connect it too. If you suspect a strong relationship, draw a heaver line."

d. "What connections between the [observed organism] and other parts of the ecosystem did we observe? How many others can you think of?"

In the case of a landscape feature or physical phenomenon:

e. "What are the parts of this landscape feature? How do the different parts affect and interact with each other?"

f. "How can we explain what is happening here?"

Structure and Function

a. "Look back at the structures you observed. Pick one to describe to a partner in detail."

b. "Let's think about how this structure helps the organism survive [or, Let's think about what we know about how this phenomenon works]. What are some possible explanations for how the structure works? How might the specific shape or texture of the structure help this organism survive?"

c. "Were there any structures you were confused by, or were there any whose function you are not yet sure about? Let's discuss this and see what we can figure out together."

FOLLOW-UP ACTIVITIES

Working toward Effective Collaboration

Look up resources on effective teamwork and work together with your students to improve group cooperation and management. Or, read accounts of how scientists have worked together in teams, and use this information to lead a discussion about how students can improve their group cooperation.

Engaging in Further Research

Make a class list of questions about the organism or phenomenon, then work in the same teams (or new ones) to look at other sources of information to deepen understanding.

WORDS

ARTICULATED THOUGHT
AND STORYTELLING

WORDS: ARTICULATED THOUGHT AND STORYTELLING

WRITING TO OBSERVE, WRITING TO THINK, WRITING TO REMEMBER

Writing is a powerful way to capture our observations. It also pushes our thinking and helps us articulate our ideas. Activities in this chapter offer support for how students' writing can appear on the page, what to write, and how to integrate writing with pictures and other forms of communication.

Writing is also a way to reflect, connect, and find the stories that emerge from our lives. This chapter also includes activities that focus on storytelling—guiding students to build narratives from nature observations, call on their life experiences, and invite themselves onto the page.

ACTIVITIES IN THIS CHAPTER

WRITING TO THINK AND OBSERVE

Well-chosen words can be worth a thousand pictures. The inspiration of writing in nature can unlock the storytelling within as children describe the intricacy, complexity, and wonder before them.

What comes to mind when you think of writing? Crafting complete sentences and cohesive paragraphs? We often focus on the finished product of writing, but few of us think in essay format. Irene Salter of Chrysalis Charter School (known for its focus on science and nature) distinguishes between writing to think and writing to communicate. When students are writing to think, the goal is for them to put their ideas on paper, externalizing their thinking to take it a level deeper. This type of writing might not occur in complete sentences or hold together in a perfect paragraph. It might not be fit for an audience, but that's OK. That is not the goal yet. This is the stage where students are entertaining different ideas and observations as they process their thoughts and make meaning.

Writing to communicate, by contrast, requires planning and careful thought about the purpose of the writing and whom it is for. It occurs when students have enough mastery of a subject to communicate about it, or when they are sharing about their own experiences and feelings with others. Learning how to write to communicate is important, but it is often overemphasized in learning situations to the point where it's all we associate with the term *writing*.

When we frame writing as a tool for thinking, we free students to reap its benefits. Those who have been intimidated by writing because they have always been expected to produce error-free paragraphs will have a very different experience if they see writing as a way to map their thoughts or investigate ideas. Writing is a way for students to articulate their thoughts with specificity, and it enables them to express ideas, questions, or information that would be difficult to capture in a drawing.

Nature journaling typically occurs in the "writing to observe" or "writing to think" phase of learning. The goal of the writing is to gather information, record thoughts and observations, spark questions, or track data. The focus is on thinking and learning, and on capturing information with enough clarity to come back to do more thinking later. This mirrors the work of scientists, whose field journal entries contain detailed writings meant not for publication but as records of key ideas and experiences they will build on later. It is also similar to the brainstorming stage of other types of communicative writing, using writing freely to ask questions and note interesting ideas and connections without worrying about creating a perfect draft.

On a journal page, writing can include descriptions of events, species, structures, or phenomena; the context for drawings; labels to elaborate on or clarify drawings; lists of observations or questions; accounts of methods of study; stream-of-thought narration about thinking or reflection; or written analysis of observations, data, or ideas.

SPELLING AND GRAMMAR

We don't require perfect spelling or grammar in students' nature journals. Doing so can get in the way of students' ability to express their thoughts and ideas. In particular, asking students who are dyslexic or language learners for perfect grammar and spelling on every single assignment puts the focus on skills they are developing rather on than their thinking and ideas. A student who never has had an educator see past a misspelling may never receive recognition for their intelligence and contributions. There is a time to work on fundamental language skills, but journaling in the field is not it. As long as you and the student can understand what they wrote in their entry, it is good enough for this phase of the process. Students will not develop bad habits or begin to think they never have to use proper grammar. They will learn that writing is a tool for thinking, and they will feel supported because you are doing your best to clearly see them and their thinking. Spelling and grammar become important at the stage of publication or communication. If students go back into the classroom to elaborate on what they have done in the field in a follow-up activity where they will do more formal writing, then they can focus on mechanics.

WRITING FORMATS

The way we think while writing a full paragraph is different than the way we think when we make a list or add a label to a drawing. It is not that one way of writing is better or worse. Different ways of using words have different purposes. Encouraging students to include different writing formats in their journals offers them flexibility in their thinking and in their approaches to recording information.

Labels

Labels clarify drawings by identifying and describing key features. Labels can simply name a feature in a drawing (e.g., "antennae" on a bug); offer context for a drawing (e.g., labeling a leaf drawing "underside" or "cross section"); or describe a structure or feature (e.g., "dense fuzzy hairs" referring to a plant stem). Labels can be used to add information and context that would be difficult to capture in a drawing by itself. Adding labels also helps build vocabulary because students can see which elements they have terms for and which they do not. We can use lines and arrows to connect labels to our drawings.

Bulleted Lists

Bulleted lists give just enough structure to help ideas flow onto the paper. After a student writes their first item, they add another bullet below it. Another idea will probably come. If students use open circles for their bullets, they can come back later and fill in the circles next to their most compelling ideas.

Fragments and Complete Sentences

Short sentences and fragments can capture an idea or information similar to what we include on labels, such as adding brief observations about the subject of the journal entry (e.g., "dove is sitting on garden fence"), or quick questions (e.g., "Why the tiny hairs?") We can also cluster short sentences and fragments around one subject: "Fox is gray on top and white on underside. Took four running steps, then stopped. Lifts tail while running. Why?"

Writing takes time, and we can use short sentences or fragments intentionally to quickly capture observations or ideas. This could include lists of questions, observations of a phenomenon, highlights of the day, or signs of the season.

Longer complete sentences are ideal for developing questions and ideas and for describing events in more detail, including more additional information than just a discrete observation. "It's been a week since I've visited this pond, and the oaks have leafed out!" "When the hawk dove past the tree, I heard loud, sharp bird calls coming from the branches." These kinds of sentences can be used to elaborate what is shown in a drawing or to describe events that might be harder to draw efficiently.

Full Paragraphs

A full paragraph calls for deeper, more thorough thinking than a single sentence. "The hawk dove off the telephone pole" is very different than "We watched the hawk sit on the telephone pole for 5 minutes. It turned its head back and forth, and we wondered what it was looking at. Then, it dove! We saw it staring at one place on the ground before it dove, its wings tight against its body, off the telephone pole and out of sight down the hill." Writing full paragraphs lends itself to describing observations of events, narrating thinking processes, and exploring ideas in a more intentional way. It is also useful for describing experimental procedures and methods of data collection. Students can summarize observations, discoveries, insights, experiences, and questions in full paragraphs.

BEST PRACTICES

In their journals, students

- Use labels, lists, fragments, sentences, and paragraphs to describe observations, thoughts, and events.

- Use words to show thinking and explore ideas.

- Use specific and precise language (e.g., *bright light lime-green* instead of *green*).

- Use the language of uncertainty (e.g., "Perhaps…," "Could it be…," "The evidence seems to show…") to indicate explanations or tentative ideas.

- Use writing intentionally (using words to describe observations that would be tedious to draw, recounting an event in complete sentences to capture detail).

- When appropriate, use words to describe methods of data collection.

- Use grade-appropriate scientific vocabulary.

- Include personal thoughts and reflections.

- Integrate words with pictures and numbers.

CREATIVE WRITING AND STORYTELLING

Writing in our journals doesn't need to be purely scientific or observational. Allowing space for students to include personal notes, reflections, and creative writing in their journals can strengthen their connection to the practice and leaves them with deeper memories of their explorations.

Writing is also an opportunity to craft narratives from our observations and invite stories onto the page. Stories are everywhere, if we know how to look. The snake that slithered across the trail is a story. The ants on the schoolyard, where they go and how they interact, a story. How you felt being outside with friends, a story. Telling the stories of our experiences and feelings in nature nurture our connection to ourselves and to the outdoors, and is a valid practice in journaling. Even a fragment of a poem is worth the time it takes to write down the lines. It is a way of finding joy and meaning in the small moments of each day.

The activities in this chapter will attune students to the small stories that surround them in each moment. The activity *Poetry of Place and Moment* uses *I Notice, I Wonder, It Reminds Me Of* as a scaffold for writing poems. Students start by noticing and wondering about their surroundings, then look inward, recording what they notice about themselves or memories that surface. *Event Map* and *Event Comic* focus students on recording external events and landscapes with pictures and words, then crafting written narratives about their journeys and the events they witnessed.

The writing that results from these activities speaks to students' internal and external experiences, and will reflect different ideas and identities. When students put themselves onto the page, they build on their personal narratives, nurturing a sense of self that is connected to nature, as well as a vision of the outdoors that includes them. Sharing these stories and poems as a group can build community, understanding, and respect for different perspectives.

Writing down emotions and reflections alongside scientific observations is also an opportunity for students to practice distinguishing between them. Scientists are supposed to be objective observers, but no human is capable of true objectivity. Instead, we encourage students (and scientists) to include creative ideas and personal perspectives in their journals, as they cultivate the ability to differentiate observations and emotions.

BUILDING WRITING SKILLS

All the journaling activities in this chapter include some scaffolding and suggestions for what students could capture in their journals (e.g., "Use labels to describe things that would be hard to draw"), and they offer some specific support in writing. To further support students in improving the writing in their journals, we can

- Remind students that they can always fall back on the activity *I Notice, I Wonder, It Reminds Me Of*, making a list of observations, questions, and connections as a framework for writing in their journal.

- Show students journal entries of field scientists to see the range of ways that writing can be used to record information. (The example pages from experienced journalers in this book are a good place to start!)

- Have students read grade-appropriate science texts and writing in other technical subjects.

- Point out opportunities for students to integrate writing with other elements on the page, such as by adding text to elaborate on a drawing, or providing context for numbers.

- Honor students' work by reflecting back to them what we learned by examining their entry. This shows they have been effective in recording their thinking.

- Carry a small voice recorder for students who are more comfortable expressing thoughts verbally than in writing. The student can describe their observations and ideas into the recorder, then use the recording to transcribe written notes into their journal. This process gives students more confidence and helps them develop the capacity to go from thinking and talking to writing.

SUPPORTING LANGUAGE LEARNERS

Learning and communicating in a new language are challenging in and of themselves. Developing proficiency with academic language is even harder, and on average, it takes 5–10 years to

do so.[25] Journaling can provide a language experience for all students whereby they are able to use drawings and different strategies to represent their thinking. This is an excellent opportunity for language learners to develop their language proficiency and build visual literacy skills. Students practice key language skills as they study nature, learning new words and integrating text with drawing. Here are some practices to support language learners in engaging in journaling activities:

- Strike a balance between encouraging students to speak and write in their native language and challenging them to develop proficiency with the language of instruction.

- Make sure that language learners know they can participate in journaling and discussion at their current level of proficiency in English (or whatever the language of instruction is). Students should focus on their ideas rather than spelling and grammar.

- Offer processing time after an activity, during which students can describe their journal page in their native language.

- Pair a proficient speaker with a language learner and ask them to compare journal entries, focusing on how they wrote descriptions or made diagrams and on teaching words to each other. Both students benefit from this experience.

- Model engaging with the activity by "narrating" your thinking out loud, showing examples of how a student might explore a question or follow instructions.

- Have students create a vocabulary list or word wall of terms specific to journaling and nature exploration, to which they add throughout their journaling experiences.

- Immediately define "process words" (e.g., *journal*, *pencil*, and *draw*) that students will need to know in order to follow directions; reference a physical example whenever possible.

- Hold up or point to natural objects or other parts of nature as you reference them.

- Offer an example of a finished journal page and reference it as you give instructions.

Language learners are an asset to your learning environment. Cultural and linguistic diversity is a powerful resource if you include it thoughtfully, and it can strengthen a group's learning.

ADDITIONAL RESOURCE

The Writing Strategies Guide, by Jennifer Serravallo

FIND POEMS WITHIN AND AROUND YOU

Direct observations of places and parts of nature can inspire poems. Insight and reflection can emerge from stillness in the outdoors. Sharing this process in a group and listening to others' perspectives can build resilient, responsive communities.

Terry Tempest Williams

"To be whole. To be complete. Wilderness reminds us what it means to be human, what we are connected to rather than what we are separate from."[26]

In this journal entry, Terry Tempest Williams weaves inspiration and reflection from her observations of nature.

José González

"Nature is one of the best literary and creative teachers. It provides abundant opportunities, invitations, and challenges for creative writing with its models, metaphors, and reactions. My creative writing is an expression and reflection of that process: nature as muse, introspection, and educator."

In this poem, José González focuses on processes of transformation in nature. The meditation expands to the process of change, and nudges readers to see similarities between their own transformations and natural processes. Students, too, can expand on their observations and write metaphor poems.

> NEW YEAR'S EVE: The Rim
>
> At the foot of the Cottonwood
> offering my prayers for all
> the gifts — droughts: flood
> highs — and lows — shadow
> and light of 2017 .
>
> So many blessings
> So many heartache
> So much growth
>
> This wise Cottonwood reminds
> me growth is all
>
> transmissions
> transgressions   Truths.
> transformations
>
> Roots . dig down deep

> At the moment the seed became
> a flower
> The caterpillar became a butterfly
> And even when the cloud became
> rain
> There was no going back, stopping,
> or wondering why
> There was simply the moment
> of being
> Of blooming without being vain
> Flying free without attachment
> to past pain
> And letting go, letting go
> without refrain

Build Environmental and Emotional Literacy

When students look at what is in front of them and notice their reaction, they cultivate a relationship with nature and find richness, resonance, and insight. This kind of attention is essential to students' environmental and emotional literacy. Giving students permission to write poems allows this literacy to grow.

Make It Personal

Journals are a place where students can reflect on personal experiences and relationships with nature and the outdoors. This reflection nurtures self-awareness and a connection with nature.

133

REFLECTION, CONNECTION, AND JOY

When we let ourselves onto the page, we can start to see ourselves as a part of nature, not separate from it. When we support students in expressing themselves, we deepen their relationships with their surroundings and themselves.

Sarah Rabkin

"With the variety of line weights and hues available from [my] writing implements, I can record my observations in lettering of various sizes and shapes, which makes the information easier…to view and absorb when I return to the page later on. I can also make visual distinctions among the various threads of thought and emotion that land on my journal pages—in this case, the nutmeg study alongside…some interior musings."

Shifting Points of View

This set of drawings uses four points of view to fully depict the nutmeg. This shifting of perspectives is mirrored in the writing, which examines an experience from different vantage points. Together, they form an exquisite record of a moment in time.

An Invocation

Dropping a quote or idea in the journal before starting an entry can frame the experience.

"A New Favorite Entity"—Letting In Joy

We can let ourselves love what we look at. Journaling is a way to let joy in, and we can welcome students' expressions of enthusiasm on the page.

Making Memories

Insights, conversations, and memories of meals are recorded alongside observations. These are the moments that make up our lives, yet they are easily forgotten over time. Students can write down these details and remember moments from the experience for years to come.

Shifts in Thinking

A narrative description of the causes of shifts in thinking leads to reflection and memory, and a highly personal narrative.

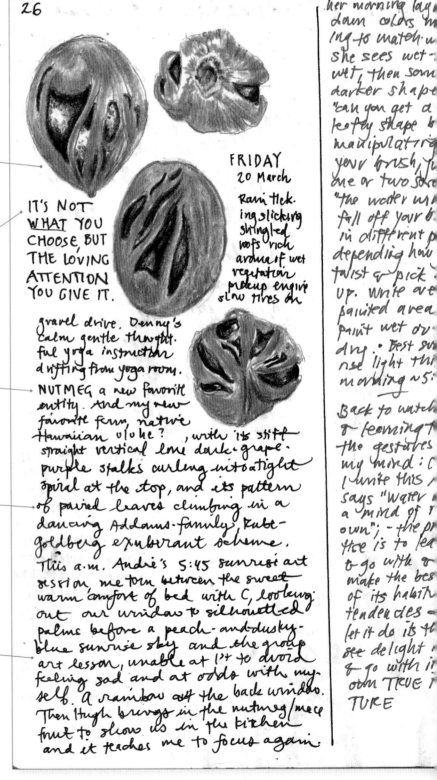

"Sketching and writing about an entity I'm interested in helps me get to know it. I ended up feeling more intimately acquainted with and affectionate toward this nutmeg fruit, and able to remember it more acutely, than if I had simply observed it without making visual and verbal notes. And it's not just my thinking that was affected, but my whole being. As often happens, I came away from this session of alert, relaxed, engaged attentiveness feeling more whole, happy, vibrant, alive."

Awakening Awareness

To be awake to ourselves and in the world is a valid goal of journaling. This entry models the process of shifting attention outward and inward, honestly recording observations and internal responses to them. We can give our students permission to be human, affected by their experiences and reflective on the page. This offers awareness and insight that deepen the experience of being alive.

Lines to Structure Page Layout

A few simple lines on the page can create a structure and flow for the ideas, and give a distinct feel and look to the entry.

Citing Sources

We learn from the people around us. Citing the source of what we record in our journals helps us maintain a clear connection between an idea and where it came from. It is also a way of acknowledging others and how their ideas influence us.

Varied Line Weight

The larger, bolder writing stands out in this entry. Bringing along even one or two larger felt-tipped markers enables you to make subtitles, titles, and headings pop.

(handwritten journal notes, center:)

27

andie on "Effort-less Beau-ty"

painting with twigs as antidote to the painstaking preci-sion of illustration — allowing a wilder more responsive ges-ture. Connect with the idea of watery move-ment — use your whole body to make your mark. Andie uses black sumi ink.

PRACTICE SCIENTIFIC STORYTELLING

Every encounter with nature is a story. Paying attention and recording these encounters lead to learning something real, directly from the source. Crafting nature narratives builds students' storytelling skills.

William D. Berry

"Well, I took up drawing when I was about three. From then on, I spent most of my time either chasing bugs and lizards or drawing pictures."

Comparison

Comparing two objects or organisms side-by-side makes the unique characteristics of each easier to see.

Life, Memory, and Reference

Sketching from reference material and memory can be a helpful supplement to live sketches done in the field. Berry distinguishes these by labeling sketches "life," "memory," and "ref." for "from reference materials." Some are a combination "life and memory." This practice gives context to the observations and helps the reader gauge their reliability.

Writing and Drawing

Which are better, written notes or sketches? Trick question. They are both important, just different. Some observations are easier to show with sketches, others with words. Each mode also leads to different observations and thinking, and they are complementary parts of storytelling and communication.

Event Comic

This numbered sequence of observations describes the strategy of a fox hunting in snow. Approaches like this give students tools to craft narratives about their journaling experiences and notice the small stories that surround them.

Including Humor

Step 6 includes scientific observation and a sense of humor. Journal entries can welcome play and humor in addition to scientific thinking.

Writing the Narrative

Writing out a detailed description of an event in nature can lead students to document it more thoroughly. Significant details that might go unrecorded in a sentence fragment are noticed and preserved in a longer account. Encouraging students to take the time to write paragraphs and explain their thinking will leave them with a richer record of their experiences.

WHAT PICTURES CANNOT SHOW

Words are a powerful tool in journaling. Students can use writing to record what is difficult or impossible to draw, such as animal movements or personal reflections.

Liz Cunningham

"With these drawings I asked: How can I portray the unique life force of these fish? How does that life force express itself in the fluid medium of water? An overarching question was how to translate that 'everything is alive' feeling experienced underwater—the profusion of life—into drawings and words."

Use Analogies and "It Reminds Me Ofs"

Using analogies is a fun way to deepen memory and observation. It can also reveal significant patterns. If the tail structure looks like a fan, could they function similarly?

Describe Feelings

A sketch cannot convey the depth of feeling that students experience in nature. Encourage them to let their thoughts flow alongside their scientific notes.

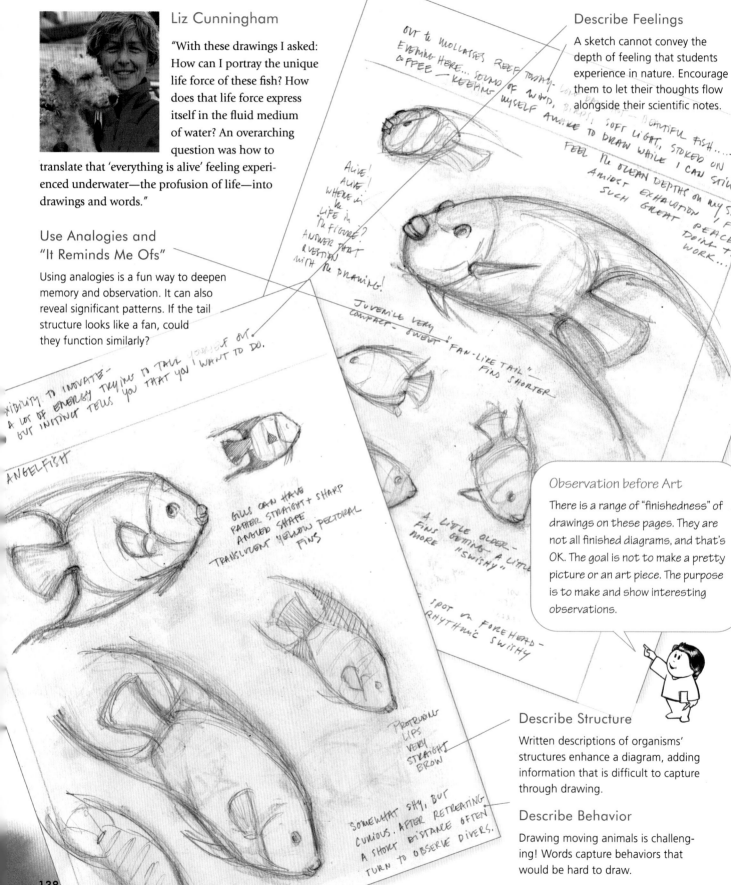

> ### Observation before Art
>
> There is a range of "finishedness" of drawings on these pages. They are not all finished diagrams, and that's OK. The goal is not to make a pretty picture or an art piece. The purpose is to make and show interesting observations.

Describe Structure

Written descriptions of organisms' structures enhance a diagram, adding information that is difficult to capture through drawing.

Describe Behavior

Drawing moving animals is challenging! Words capture behaviors that would be hard to draw.

WRITING TO OBSERVE, WRITING TO THINK

Students focus on a subject in nature and practice using different writing approaches to capture their observations and thinking.

Writing isn't just about the final product. Writing is also a tool for observation and thinking. When students see writing as more than just producing five-paragraph essays, they can start to use it as a tool and a strategy. Students can use many different approaches to writing in their journals. This activity will engage them in thinking about how to best use different kinds of writing, such as labels, shorter sentences, and longer paragraphs to capture information.

NATURAL PHENOMENA

Students can focus on any part of nature during this exercise. Leaves, trees, slow-moving animals, or landscape features will all work. Animals moving quickly through the landscape (e.g., a hawk flying by) aren't ideal, as students will need to focus on the subject for several minutes.

PROCEDURE SUMMARY

1. Make a simple drawing of your leaf (or other part of nature).

2. Add in observations by writing labels, bullet-point phrases, sentence fragments, short sentences, and paragraphs.

3. Add in writing to show your thinking, including connections to what you already know ("It reminds me of/makes me think of…"), questions ("I wonder…"), explanations of observations ("Could it be…," "Maybe…"), and reflections or reactions ("I am surprised by…").

4. Connect your written thinking, observational writing, and drawings together with lines and arrows.

DEMONSTRATION

When the whiteboard icon appears in the procedure description: Draw a simple outline of a leaf (or other part of nature), then model "writing to observe" by adding labels, horizontal lines to represent bullet points, and clusters of lines to represent paragraphs. Leave some space around your drawing and notes. When you introduce "writing to think," add in lines with question marks next to them, sentences with starters like "Could it be" and "Maybe," longer paragraphs, and lines to connect drawings and ideas together.

Time

Introduction: 10 minutes
Activity: 40 minutes
Discussion: 10–15 minutes

Materials

☐ Journals and pencils

☐ Poster board or whiteboard with sentence starters written on it:

Writing to observe: I notice… We saw… I heard… First, then…

Writing to think: I wonder… Maybe… Could it be… It reminds me of… I feel… It surprised me when…

Teaching Notes

Writing mechanics (grammar, syntax, spelling, etc.) are important, but don't emphasize them in this activity. The goal of this activity is for students to practice using writing as a tool for observation and thinking.

This shouldn't be the first journaling activity you do with students. Prior experience making observations, sketching, and thinking on paper will prepare students to engage with this activity as a next step in their growth as journalers.

This activity can be done all at one time or spread out into several lessons.

PROCEDURE STEP-BY-STEP

1. **Ask students to talk with a partner about what they think of when they hear the word "writing," then ask the group to share a few comments.**

2. **Point out that writing can be a tool for learning in addition to a form of communication.**

   a. "We often write to communicate our ideas to someone else, but writing can also be a tool to help us learn."

3. **Explain that writing can be a tool to expand our own observations and thinking.**

   a. "Writing down our observations and thoughts helps us notice more, clarify our thinking, and remember what we see."

   b. "We're going to practice using our journals in this way."

4. **Explain how writing can be a tool for making observations.**

   a. "One kind of writing we can do in our journals is writing to observe."

   b. "When we're writing to observe, we're trying to notice and record: What is here? What is happening? How does it look? What is it doing? What can I notice with my senses?"

5. **Explain that different kinds of writing capture different kinds of information, and students can use labels, bullet-point phrases, sentence fragments, short sentences, and paragraphs to record their observations.**

   a. "We can record our observations with labels, bullet-point phrases, sentence fragments, short sentences, and paragraphs."

   b. "Each kind of writing leads us to record different kinds of observations."

6. **Ask students for examples of observations they could record in labels, bullet-point phrases, sentence fragments, short sentences, and paragraphs.**

   a. "Labels are single words or a few words connected directly to a drawing. What kinds of observations could we record with labels?"

   b. Listen to students' responses. If they don't bring up any of the following ideas, mention them: labels of colors or features that can't be shown in a drawing (e.g., "light lime green" or "fuzzy hairs everywhere"), labels clarifying what a feature is (e.g., "hole"), and labels for parts of an organism (e.g., "feet").

   c. "What kinds of observations might we record in bullet-point phrases, sentence fragments, or short sentences?" Listen to students' responses. If they don't bring up either of the following ideas, mention them: quick notes about what is happening in the moment ("Hawk flew by!"), notes about location ("The frog was next to the pond").

   d. "What kinds of observations might we record in full paragraphs?"

   e. Listen to students' responses. If they don't bring up the following ideas, mention them: longer descriptions of structures or features, narratives of events, and longer accounts of their thinking and ideas.

7. **Tell students they will have around 10 minutes to make a sketch of a natural object, then to document as many observations as possible in writing; do a demonstration on a whiteboard as you introduce the procedure.**

   a. "Start by sketching a leaf, rock, bird, or other part of nature that's interesting to you."

   b. "Just put some basic lines down and don't get too focused on your drawing."

   c. "Then start writing. Write your observations to build on what you show in your sketch. Use labels, bullet-point phrases, sentence fragments, short sentences, and complete paragraphs."

   d. "Your goal is to capture as much information as possible in writing. If you start to get stuck in your writing, look back at your drawing and see whether there are more words you can add to build on what is in your sketch."

   e. "How much can you learn in the next ten minutes?"

8. **Offer scaffolding and sentence starters to guide students' observations.**

   a. "I'll share some guides you can refer to as you're writing down your observations."

   b. "Here are some sentence starters or ideas if you get stuck." (Hold up poster board that says, "I notice…," "We saw…," "I heard…," and "First…then….")

9. **Encourage students to see this as an opportunity to develop their writing skills, and clarify that their writing won't be evaluated.**

   a. "When we're writing to observe, we're writing for ourselves. The goal is to build our own learning, not to communicate with anyone else."

   b. "This is an opportunity to develop your ability to use writing as a tool to notice and think."

   c. "Don't worry about spelling if there are words you're not sure about, and don't get stuck trying to write with perfect grammar. Right now, the goal is to make observations and show them in words. As long as you're doing that, you're meeting expectations."

10. **After 10 minutes have passed, call for students' attention and tell them to share with a partner some observations they discovered.**

11. **Refocus the group, and explain how writing can be a tool for thinking.**

   a. "We can also use writing as a tool to think and get our ideas out on paper."

   b. "When we ask questions about our observations, make connections between ideas, make explanations for what we see, or write about how we are feeling, we're thinking on paper."

12. **Ask students to brainstorm different kinds of thinking they could include on their journal page.**

   a. "What kinds of thinking or ideas might we describe in our journals?"

   b. Listen to students' responses, and bring up what they don't mention: questions ("I wonder…") or connections ("It reminds me of…") related to the observations they have made; possible explanations for what they have observed; longer-paragraph descriptions of the procedure they used to make observations; poems, personal notes, or reflections; notes about how they feel in the moment or in response to the environment; questions about ideas they are not sure about or things they want to look up later.

13. **Offer scaffolding and sentence starters to guide students' thinking.**

   a. "I'll share some guides you can refer to as you're writing down your observations. "

   b. "Here are some sentence starters or ideas if you get stuck." (Hold up a poster board that says, "I wonder…," "Maybe…," "Could it be that…," "It reminds me of…," "I feel…," "It surprised me when…', etc.)

   c. "You can record your thinking in sentences or in more extended paragraphs."

   d. "See if you can build on the observations in your journal, asking questions or trying to figure out what is happening; or write about your response to being in this place. How many ideas can you get on the page?"

14. **Encourage students to follow their interest.**

   a. "Allow yourself to follow your interest and ideas. It's OK to write out your thinking, then shift back to drawing or writing observations."

   b. "Focus on writing your thoughts about what is interesting to you right now. Whether that is making an explanation, or thinking about how the object connects to your life, or playing with questions, that's OK."

15. **Give students 10 minutes to add to their journal entries.**

16. **As students work, take time to circulate, troubleshoot, and ask questions about students' observations and ideas.**

   a. "When we're working in our journals, our goal is our own learning. We don't need to communicate about our ideas yet."

   b. "This means that we're not trying to communicate with anyone in this moment, and we don't need to focus as much on spelling or grammar as we would in an essay, because the goal is learning, not communicating."

17. **Call students back together, and challenge them to write out an extended paragraph describing an explanation, a part of their journaling process, or how they are feeling.**

   a. "Writing full paragraphs can be useful. As we write out our thoughts, we clarify them."

   b. "Getting something down on paper can often make us realize that we don't understand something as well as we thought we did, or can lead us to unexpected insights."

   c. "You'll have five minutes to write out a full paragraph."

   d. "You could write out an explanation for something you saw, using your words to try to figure out why or how something is happening."

   e. "Or you could describe a part of your journaling process, such as by writing about the process you used to make a part of your journal page, explaining the choices you made about how to structure the page layout, or what you noticed first, then second, and so on."

   f. "Or you could write a paragraph narrating your inner thoughts, writing out how you felt throughout the learning experience, or explaining how you feel in this place."

18. **Tell students to take 5 minutes to write a paragraph. Let students know they can use *I Notice, I Wonder, It Reminds Me Of* if they get stuck.**

   a. "Please take five minutes to write out a paragraph."

   b. "If you are stuck, you can try turning the prompts 'I notice,' 'I wonder,' and 'It reminds me of' inward. What do you notice about how you feel right now? What does this experience remind you of?"

19. **Point out how scientists use their journals to write down their observations and thoughts before they communicate about what they've learned.**

   a. "When we write down our observations, we're forced to be more specific than if we just think of observations in our heads."

   b. "Putting thinking on paper helps clarify thoughts, come up with more ideas, and deepen your memory."

   c. "Scientists don't just sit down right away and write a paper on their research or give a talk on what they know."

   d. "They use journals because journaling is a tool for learning, and writing in their journals helps them make more observations and deepen their thinking."

   e. "Scientists spend months using writing, drawing, and other methods of data collection to make observations and to think."

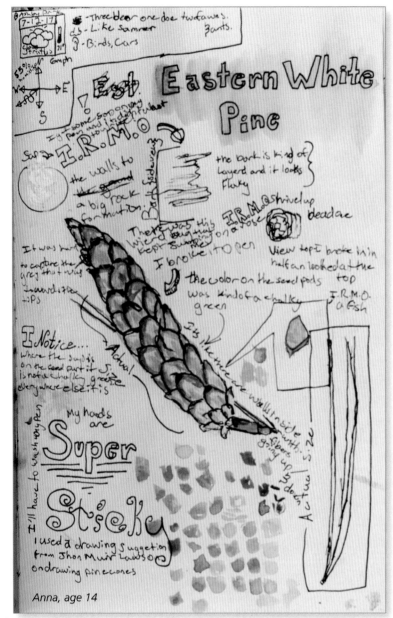

Anna, age 14

You can write in so many ways. The variety of fonts, angles, and modes (labels, sentences, bullet points, titles, etc.) make the page easier to scan and full of information.

> **f.** "Remember to use writing as a tool for observing and thinking, not just as a way of communicating."

20. Explain how students can also use writing to reflect on their learning process or to record their reactions and feelings.

> **a.** "Another way we can use writing is to reflect on our learning or to record our reactions and feelings about our experiences."
>
> **b.** "This is a way of getting better at learning and journaling."

21. Give students a few minutes to write about the following questions in their journals.

> **a.** "What was it like to use writing to capture your observations and thoughts? What felt successful, and what felt challenging? What might you do differently next time?"
>
> **b.** "How did it feel to use your journal to learn as a scientist might?"

DISCUSSION

Lead a discussion with the general discussion questions. Intersperse pair talk with group discussion.

General Discussion

> **a.** Ask students to discuss and reflect: How did writing help you to observe or to clarify your thinking?
>
> **b.** How might you use writing as a tool to help you observe, think, and remember what you journal about in the future?
>
> **c.** Writing is a useful tool for helping us think about anything, not just science or nature. If you could use your journal as a tool to write and learn about any topic or idea of your choice, what would you choose?

FOLLOW-UP ACTIVITY

Aiming for Specificity

Challenge students to look through their writing and add more specific language to their observations. For example, instead of saying "green," they might say "intense yellow-green"; instead of "orange," they might say "the color orange of the mac-and-cheese box." Instead of "There are dots on the leaves," they might say, "There are tiny black dots the size of the dot of an 'i,' and there isn't really any pattern I can notice of where they are." Give students a couple of minutes to work with their writing, then ask if anyone came up with any interesting substitutions.

EVENT COMIC

Students create a series of simple diagrams, much like a comic book or a movie storyboard, to record an event they have witnessed.

Making a comic or storyboard is a fun way to document observations and develop visual communication skills. Laying out panels and showing the flow of action using sound effects, arrows, and other touches give students creative freedom while keeping the focus on accurately documenting what they saw. Then students can deepen their memory and strengthen their storytelling skills by writing a narrative of what they saw. The approach of going from a visual representation or storyboard to writing a story scaffolds the process of writing a narrative.

NATURAL PHENOMENA

If you spend enough time outside, you are bound to witness exciting nature dramas, but you can't schedule them. This activity is best done spontaneously, in response to a cool "nature moment." For example, you might get to see a falcon dive into a flock of shorebirds, blackbirds mob a hawk, a duck put on its courtship display, or a lizard catch prey. These condensed, action-packed events are ideal for making event comics. Also keep your eye out for more subtle nature dramas, such as a snail eating a leaf, or ants overcoming an obstacle in their path; although these events might seem less exciting initially, they are worthy of study and can make fun subjects for comics.

PROCEDURE SUMMARY

1. Describe the action of an event immediately after seeing it.

2. Use a series of panels and visual elements such as arrows, interesting perspectives, and action words to tell the story.

3. Have fun and be creative, but record observations accurately, staying true to the event that happened.

4. After finishing the comic, write a narrative describing the event, including as much detail as possible.

DEMONSTRATION

When the whiteboard icon appears in the procedure description: As students describe strategies for laying out the comic, create a simple replica of it on a whiteboard. Add elements and details that the students suggest, such as sound effects and action arrows. Consider long vertical or horizontal panels or elements that break out of the frame (an exciting comic book effect).

Students don't need to draw detailed portraits of the animals in the comic. The goal here is to tell the story of how the animals interact.

Time

Introduction: 10 minutes
Activity: 30–50 minutes
Discussion: 10–15 minutes
Extension: 15–20 minutes

Materials

☐ Journals and pencils

optional

☐ Examples of graphic novels

☐ Rulers

Teaching Notes

Memories fade quickly. Have students verbally review the event as soon as possible after viewing it to help them remember details long enough to get them on paper.

When we have our "scientist hats" on, we strive to be as accurate as possible in understanding the world, and anthropomorphism (ascribing human motivations and feelings to nonhuman things) sometimes gets in the way of that. It's fun to give a duck a thought bubble, but if we just think about what it would be like for us to be in the duck's position, we miss the opportunity to try to understand what it's like for the duck to be the duck. Students can strive for accuracy and still have fun in this activity by using anthropomorphism consciously, and discerning between their concrete observations (the hawk banked away from the raven) and their interpretations (the hawk was annoyed at the raven).

Any sequence of observations can be made into a comic-style journal entry. Students must break the seamless event into discrete observations that most clearly describe what they saw. In so doing, they learn to parse discrete moments from a continuous event and teach themselves the fundamentals of visual storytelling.

Lydia, age 12

Arielle, age 10

PROCEDURE STEP-BY-STEP

1. **After an exciting "nature moment," prompt students to verbally review the details and order of events they just observed, and ask follow-up questions to keep the discussion going.**

   a. "Wow, we were so lucky to see that hawk catch a snake! Quickly now, before we start to forget, let's review what we saw, what happened in what order, and any details you remember."

   b. "What were other observations we made? Were there any sounds we heard? What did we notice about the way the hawk was moving? about the snake's behavior? about the landscape around it? What is the weather like right now?"

2. **Explain that students will make a series of panels, imitating the format of a graphic novel or comic book, to make a "true-life" nature comic.**

   a. "We are now going to document this event in our journals by making a true-life nature comic."

   b. "In your journal, you will make a series of panels that show the full sequence of events with as many accurate details (not made-up ones) as you can include."

3. **Ask students for ideas on how they could creatively set up their journal page, recording their suggestions on a whiteboard and filling in with your own suggestions.**

   a. "Laying out a story in a comic format can be creative and dynamic. What are some of the elements of comic books you have seen that make them exciting to look at, or effective for telling a story?"

   b. As students generate ideas, lay them out on your whiteboard. Ideas may include interesting points of view (such as the snake's- or hawk's-eye view), long or tall panels, sound effects, close-ups, action symbols, arrows showing movement, and elements that dynamically break out of the frame.

4. **Suggest that students take a moment to write down a plan for the number of panels or "scenes" they want to show, then tell them to begin.**

   a. "You might want to start by writing down a plan for how many panels or 'scenes' you will show in your comic."

   b. "Let's take the next fifteen minutes while we still remember what we saw. When we are done, we can share and compare our work. Are there any questions before we start?"

5. **As students work, take time to circulate, troubleshoot, ask them questions about what details they are choosing to include or leave out, and remind them to include metadata.**

   a. "It's interesting that you're choosing to show the clouds in the sky behind the bird. Why do you think it's important to include that?"

   b. "Remember to include your metadata—the date, location, and time. That provides important context for the story."

6. **Call the group together and have students discuss different ways to tell stories.**

   a. "Making a comic is one way to tell a story, but there are other forms of storytelling. What are some other ways we could tell the story of what happened?" (Students might say making a film, writing a book, etc.)

DISCUSSION

Lead a discussion using the general discussion questions and questions from one of the Crosscutting Concept categories. Intersperse pair talk with group discussion.

General Discussion

When students seem finished, or when it is time to move on, call the group together, tell them to take a moment to add a title to their comic, then to discuss their comic with a partner.

   a. "Add a title to your page that captures the story."

   b. "Find a partner and talk with them about the types of details you both chose to include or leave out. Neither of you is wrong or right; it is just interesting to see how another person told the same story that you did."

   c. "Compare the ways you chose to write your narratives. Are there any approaches someone else used that you could incorporate into other journal entries?"

Patterns

   a. "Have you ever seen animal behavior that was similar to this? What's the closest thing to it that you've observed?"

   b. "What other situations might lead to the behaviors we observed today?"

   c. "Do you think the interaction that we observed between these two types of organisms is common? Why or why not?"

   d. "What are other ways these organisms might interact with other living or nonliving parts of this ecosystem? How might these behaviors be similar or different in other areas or ecosystems?"

   e. "What other organisms do you think might exhibit similar behaviors?"

Cause and Effect

   a. "Why might have the [hawk, bird, snake, etc.] behaved in the way it did?" (Refer to more specific behaviors if possible.)

   b. "What might have happened next after we stopped watching the animal? Why do you think that?"

   c. "Do you think these organisms might have acted differently under different conditions—for example, in another weather pattern, during a different time of year, or in the presence of some other organism? Why do you think that?"

   d. "What are other things you can think of that affect this organism's behaviors?"

Structure and Function

   a. "Describe how the [deer, hawk, snake, etc.] moved. What body parts seemed most involved? Describe the organism's movement in detail."

   b. "When scientists study animals, they often try to think about how their specific structures help the animals survive in their habitat. How might some of the body parts you observed help the organism survive? Be specific in connecting the structure of the body part to its function. For example, don't just say, 'Its claws help it catch things.' What is it about the claws' shape or material that make it ideal for catching things?"

EXTENSION

1. **Ask students to discuss how writing a story describing the event they witnessed is different than making a comic about it.**

   a. "If you were going to write a short story of this event, what kinds of details would you need to include?"

b. "How would it be different to tell the story using only words, without pictures?"

c. "What kinds of visual information did you show in your comic that you'd need to describe in words?"

d. "How could you use the information in the comic to plan or structure your written story?"

2. Discuss students' ideas, and add any of the ideas listed here that they don't bring up:

a. Students would need to describe the setting and the animals in their story, and could look at their comic panels to guide their thinking about how to do this.

b. Students would need to use words instead of arrows or pictures to describe the movement and interactions that took place, and could look at their comic panels to guide their thinking.

c. Students could use their comic as a guide for structuring their story, writing one paragraph to describe each panel and replacing arrows with words like *ran*, *jumped*, *flew*, and so on.

3. Tell students they'll have 15–20 minutes to write out a narrative version of their event comic. Circulate and support students who might be struggling, and ask students about the kinds of details they're recording in writing.

a. If students are struggling with the assignment, ask them to look at the first panel of their comic and describe it to you verbally, including details such as setting, behaviors, movements, and so on; then tell them to write down what they have said to you.

b. You can also offer some sentence starters, such as "First…," "After that…," "Suddenly…," and the like to offer support in writing about an event that unfolds over time.

4. Ask students to share their work with each other and notice differences.

a. "Compare the ways you chose to write your narratives. Are there any approaches someone else used that you could incorporate into other journal entries?"

FOLLOW-UP ACTIVITY

Studying Comics

Let students bring in their favorite comics or graphic novels. Analyze these as a class with an eye toward collecting dynamic effects, layout and design ideas, and elements that move the story along. Give students sticky notes to mark good examples and useful ideas. Then ask them to discuss when they might use these strategies in their journaling.

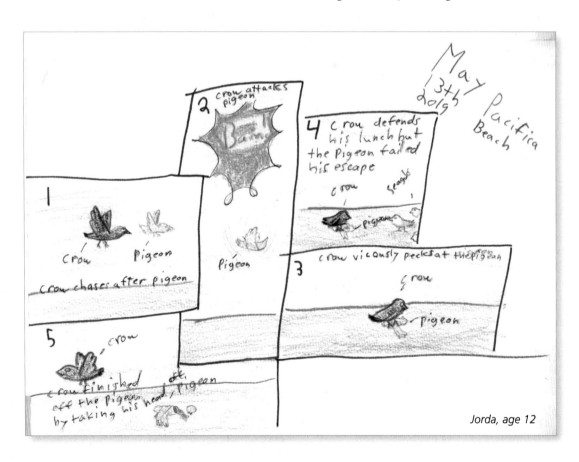

Jorda, age 12

EVENT MAP

Students draw a treasure map as they hike, highlighting discoveries and features along the trail and making quick sketches of plants, animals, or other surprises and writing notes to record memories.

Have your students ever come back from a nature hike and been unable to give a vivid report of their experience? "How was your hike?" "Good." "What did you see?" "Trees and stuff." Externalizing thinking (getting our ideas and observations down on paper) is a good way to enhance attention and memory. Making an event map, or a physical chart of experiences in time and space, is a way of intentionally recording memories. In this activity, students make a map of hidden treasure. The treasure is not buried at the end of the trail; it is all along it. When they return from the hike, students will not only remember the sequence of events along the trail but also have a spatial memory of where they were and the distances between the discoveries. An optional extension guides students to write a narrative of their "treasure hunt." This is valuable practice in storytelling and an opportunity for students to strengthen their writing skills.

NATURAL PHENOMENA

Lead this activity on a level mile- or half-mile-long trail in a natural area. Ideally students should be able to see the whole route from the start and be able to track their progress as they walk. (We are halfway there; we are three-quarters of the way, etc.) The trail should be clearly marked and free from hazards so that students can make the hike at their own pace.

PROCEDURE SUMMARY

1. Make a treasure map of things you find along the trail, using writing and drawing.

2. After each trail treasure, add a dashed line to show the section of trail you traveled.

3. Take your time, and do not disturb others in the group.

4. (In the field, or when students return) Using your map to guide you, write a story of your adventures.

DEMONSTRATION

When the white-board icon appears in the procedure description: Draw a dotted line that represents the shape of the route (loop, winding trail, etc.). Show the locations of a few prominent landmarks

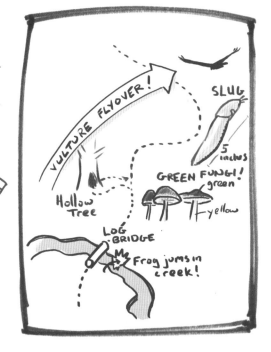

Time

Introduction: 10 minutes
Activity: 40–120 minutes
Discussion: 10 minutes
Writing extension (can occur in the field or when you return): 15 minutes

Materials

☐ Journals and pencils

☐ Two group leaders

Teaching Notes

This activity can stand alone or can "run in the background" as students complete a field day, with the group stopping every now and then to reflect on what they have seen and to add to their treasure maps. Use your knowledge of your exploration site and your students to decide whether to keep the group together or let them explore separately at their own pace. It is ideal for each student to set their own pace, but at times it might be necessary to keep the group together.

If your students will need more structure to be successful with this activity, set it up so that you walk ahead and an assistant or chaperone sends students down the trail after you, separated by about a minute to ensure they have space to wander at their own pace.

In an optional follow-up activity, students look back at their maps and write a story about their adventure or "treasure hunt." This approach is a great way for students to develop written communication and storytelling skills by building on what they've already captured on their maps. Writing a story also deepens their memory of the experience.

to help orient the students both to the landscape and their map. Then demonstrate how you could show events (e.g., a vulture fly-by or frog jump) and discoveries (cool mushrooms or slug). Model using words, pictures, and numbers.

PROCEDURE STEP-BY-STEP

1. **Tell students they will make an event map showing treasures they discover as they walk along a trail.**

   a. "We are going to search for hidden treasure. You will each make your own map. Unlike a pirate map where the treasure is found at the end, this map is of the treasures we discover along the way."

   b. "As we walk, you will add drawings and notes about the treasures you find."

2. **Show students the route, either on a map or pointing it out in the landscape, and demonstrate on a whiteboard how to lightly record the route, making a template that can be added to as they walk.**

   a. "Our route will take us in a loop, through the meadow and over the creek to the pond, then back along the forest trail, again crossing the creek and returning here."

   b. "You will want to make sure you have room for the whole map and what you find by lightly penciling in the creek, trail, and pond like this. These are just guidelines that you will be reinforcing or drawing over later, so make them as light as possible."

   c. "As you go, try to keep a general idea of where you are on the route so that you can add landmarks or discoveries at appropriate distances on the map."

3. **Tell students that they can get creative with their maps by making up place names or making a sidebar of notes about discoveries and a key to where they are found on the map.**

   a. "You can either add the little sketches along the side of the trail or make a sidebar of notes and sketches and use a letter key to show where each discovery can be found on the map." (Add to the whiteboard demonstration.)

   b. "You can also add creative place names for significant landmarks on our path."

4. **Begin by hiking the trail together, and stop early on at an interesting landform or plant to record as an example together. Choose something that everyone in the group can easily see.**

   a. "Here is our first trail treasure. Use words, pictures, and numbers to describe this jewel on your map."

   b. "When you are done, add a dashed line to show the section of trail we have already traveled."

   c. "Keep your eyes open as we walk. Who can find the next treasure?"

5. **If your group is self-directed, continue hiking, stopping as the group wishes in order to record group treasures or allowing students to hike at their own pace and record their own treasures.**

6. **(Alternatively) Set up a solo-hike protocol whereby individual students are sent down the trail at intervals, one leader walking ahead and the other staying behind to send students off and collect stragglers.**

   a. "We each will be finding our own treasure from this point forward. We are now going to continue down the trail separately. I will walk ahead, stopping here and there to add other treasures to my map."

   b. "Every thirty seconds, my assistant will choose the next person to start down the trail to make their map. On the trail, walk attentively and look carefully for treasures. When you find one, stop and add it to your journal. Add at least seven treasures to your map. Keep most stops to five minutes or less. This is a solo walk. You may pass other students who are stopped, but do not disturb them or walk together. When we get to the end of the trail, we will share our discoveries with one another."

   c. Give trail directions and safety reminders as necessary.

A map with inset icons shows you the route and the sequence of discoveries.

Detailed drawings and notes correspond to the labeled icons on the map, enabling the student to capture more information about each find.

7. **(After everyone has completed the hike) Give students a few minutes to add to their maps, adding scale features, embellishments, or place names.**

   a. "Take five minutes to improve your map by adding scientific features such as a scale, direction arrow, or key, or creative pirate-style embellishments or other notes. Be sure to add the metadata of the date and location somewhere on the map. It's OK to use ideas you saw on someone else's map."

8. **Have students lay their maps open on a picnic table or on the ground.**

   a. "Take a moment to look at some of your classmate's maps. What are some features you like?"

   b. "What was the best treasure you found? What surprised you?"

DISCUSSION

Lead a discussion using the general discussion questions. Intersperse pair talk with group discussion.

General Discussion

   a. "What are some feelings you had while walking along and making your map? Add them to your page, starting with the words 'I felt' so that it's clear this is a feeling, not an observation. This will help you remember the experience better later."

   b. "What did you enjoy about making your map? What was challenging about it?"

   c. "How did the process of making the map affect your experience? What was it like to make the map?"

EXTENSION

Writing Narratives

When you have returned from the field trip, either immediately or after some time has passed, gather students with their maps. Tell them that they'll have time to look back over their maps and write a story about their adventures.

1. **Gather students and give them their maps, then tell them to take a few minutes to look back and remember their experiences.**

2. **Give students a minute to think about how they would tell the story of the adventure they had while making their map.**

   a. "Take a moment to think. How would you tell the story of your map-making adventures to someone else?"

   b. "Imagine you were talking to someone who had never been to the place where we made our maps. How would you talk about the setting?"

   c. "What kinds of details would you include? How would you describe the treasures you found?"

   d. "Are there any feelings or responses you had when you found different things on the trail? How would you talk about them in your story?"

3. **Tell students to take 2 minutes to tell a partner the story of their adventure, then switch roles. Encourage students to practice active listening when they're in the listening role.**

4. **Explain that students will now get to write out a story that goes with their map.**

5. **Explain that students should have the goal of telling their story in a way that is meaningful to them, while including details that will help a reader understand their experiences.**

   a. "You're going to write a story capturing your experiences making your map."

   b. "Tell the story in a way that is meaningful to you. For example, you might choose to spend most of the story describing the cool or exciting things you saw. Or you might want to only briefly mention what you found, and spend more time talking about the feelings of excitement, curiosity, or surprise that you experienced along the way."

   c. "Your story will, though, need to make sense to someone who wasn't there on the journey with you, so you will need to include details about the setting and each of your finds."

6. **Offer some scaffolding and ideas about how students could structure their story.**

   a. "If you're not sure where to begin, you could start by writing a short description of the setting, focusing on details like the weather or what the trail looked like."

   b. "Then introduce yourself. You could write in the first person ('First, I saw…') or create an adventure-character version of yourself to write about in the third person ('Aria the Adventurer went off on a journey to…' or 'Welcome to the adventures of Edmund, Curiosity Machine')."

   c. "Share about each treasure. Look at the pictures on your map and use them to write about each find. Explain why you chose to include it on your map and why it was meaningful to you."

7. **Encourage students to have fun, be playful, and get creative with elements in their story.**

   a. "You came up with fun, cool, creative names for many of the events on your map. Use them in your story."

b. "Use fun action words to bring the reader into the story and introduce events."

c. "For example, if a frog jumped into the pond, instead of just saying 'The frog jumped into the pond,' say 'SPLASH! a frog!'"

8. Give students 15–30 minutes to write their adventure stories.

9. As they work, take time to circulate and offer support. If students are struggling to write out narratives, ask them to tell you parts of their story verbally, then coach them to write down what they narrated.

10. After students have finished writing their stories, offer the opportunity for them to share what they've written or to talk about their experiences writing their stories.

FOLLOW-UP ACTIVITY

Map Study

Every map is made for a reason. Some of them help with navigation; others present data. This map was created to record discoveries. Search for examples of maps that are made for different purposes. Ask students, "What was this map made to do, and who is the audience?"

POETRY OF PLACE AND MOMENT

Encouraging students to slow down and notice their response to a place and a moment in time helps them bring their personal perspectives onto the page and create lasting memories.

Turning everyday moments into poems can bring more meaning into our lives. When students include poems and creative writing in their journals, it can be a powerful experience of learning and sharing together. The roots of many poems can begin with the tools fundamental to inquiry—noticing, wondering, and making creative connections. Anything can become the subject of a poem when we observe it, then add memories, questions, or connections to ideas. This scaffold for writing poems gives students an approach they can use again and again to support their creative expression and integrate poetry into their journaling tool kit.

NATURAL PHENOMENA

Any place students can spread out, sit, and write comfortably will suffice. To support students in being able to slow down, reflect, and focus on their own experience, pick a place that is safe and/or familiar to them.

PROCEDURE SUMMARY

1. Sit and look around you, then write a poem inspired by this place and your response to it.
2. Use the sentence starters "I notice," "I wonder," and "It reminds me of" to describe your surroundings. Then, turn the sentence starters "I notice," "I wonder," and "It reminds me of" inward to describe your internal experience, emotions, and thoughts.
3. Alternate back and forth between looking out and describing what you see, then looking inward to write about what you notice about how you feel and what you are reminded of.

EXAMPLE POEMS

Instead of doing a demonstration of a journal page, read these example poems to give students some ideas of what their writing could look like.

I notice roots of a tree growing into the ground
They remind me of elephant legs,
or columns on a building.
I wonder, how many people
have visited this place?
How many birds
have sat on this tree?
I notice
sitting here leaning against the trunk,
with friends around me,
I feel calm,
like I have strong roots, too.

I notice big flowers all around.
They remind me of trumpets,
or the beaks of a hundred birds.
I wonder what it would sound like
if they could sing.
I notice they make me
feel happy.
I wonder, how do their
reflections sing in the water
like a shining star?

Time

Introduction: 10 minutes
Activity: 10–30 minutes
Discussion: 5–15 minutes

Materials

☐ Journals and pencils

Teaching Notes

Just as some students may be nervous to try drawing, others may feel intimidated by writing poems. Scaffolding the process and offering students a place to start give them a way to engage. Ideally, lead this activity after students have done the routine *I Notice, I Wonder, It Reminds Me Of*, so that they are familiar with the prompts they will use.

This is not a time to restrict students' creativity, judge their work, or limit what they write about. Although students are instructed to begin their poems with descriptions of the place around them, wherever students' ideas lead them after this is OK. As students freely notice, wonder, and make connections, their personal experiences and perspectives are validated and invited onto the page. This process leads students to develop a connection to nature and to craft poems that will help them remember that experience for years to come. The shared experience of writing poems in a group can build community and connection, and makes space for students to share their cultural perspectives and stories.

PROCEDURE STEP-BY-STEP

1. **Tell students to take a moment to look at their surroundings.**

2. **Explain that students will write a poem about the place and their experience in it, based on their observations.**

   a. "In a moment, you will write a poem recording your experience in this place."

   b. "Do not be intimidated. Writing a poem, like creating any journal entry, can begin with what we observe around us."

3. **Tell students to use "I notice," "I wonder," and "It reminds me of" to describe their surroundings.**

   a. "To start your poem, all you have to do is write down things you notice about your surroundings—for example, 'I notice ants crawling along the ground.'"

   b. "You can also leave out the 'I notice' part, and say your observation: 'Ants crawl along the ground.'"

   c. "As you observe, add 'I wonders' about what you are noticing, or any other questions that come to mind—for example, 'I notice ants crawling along the ground. I wonder how many times they have walked this path before.'"

   d. "Add in connections, or things you are reminded of. These can be about your own experiences, something that a part of your surroundings physically looks like, or connections you can make between what you see and other parts of the world—for example, 'This reminds me of people walking back and forth, each day treading the same path' or 'Ants, their shiny bodies move like lines of seeds.'"

4. **Tell students to turn "I notice," "I wonder," and "It reminds me of" inward, and write down how they feel and what they think about as they are in this place.**

   a. "As you are describing your surroundings in the poem, turn the prompts 'I notice,' 'I wonder,' and 'It reminds me of' inward to describe your feelings and thoughts."

   b. "What do you notice about what it's like for you to be in this place? What do you wonder about yourself? What does this experience remind you of? When you look at a tree branch, the sky, or another part of your surroundings, what connections can you make?"

   c. "Write that down in the poem."

   d. "After you have gotten some lines down, keep writing. As it feels right to you, shift between what you notice about the world around you, connections you can make to the rest of the world or yourself, and what you notice about your own experience."

> "Poems hide....What we have to do is live in a way that lets us find them."
>
> —Naomi Shihab Nye

5. **Tell students that their poems do not need to rhyme and do not need to follow any exact order, sequence, or topic.**

   a. "Your poem does not need to rhyme or have a specific rhythm, but it can if you would like."

   b. "You also do not need to follow the exact order of 'I notice, I wonder, it reminds me of.'"

   c. "You could have several 'I notices' about your surroundings in a row, then a series of questions. Then you could focus on the connections you feel to your own experiences or memories."

   d. "Anything is fair game. Do not limit yourself. If you are reminded of memories, people, experiences, ideas, or knowledge, that is all welcome. Write down what feels meaningful to you."

   e. "If the poem takes you in a completely different direction, to a totally unrelated topic from where you began, that's OK. Follow it. An exciting part of writing poems is being surprised by where they lead you."

6. **Let students know that if they feel stuck, they can always just go back to "I notice," "I wonder," and "It reminds me of" and write down what they see or sense.**

   a. "If you run out of things to say or feel stuck, don't worry; you can always go back to noticing and describing your surroundings."

7. **(Optional) Encourage students to look back at previous journal entries and incorporate lines or writing that speaks to them.**

   a. "If you want, after you've gotten started, flip back through some of your journal entries and take a look at what you've written."

   b. "If there is a line you wrote down that speaks to you and connects to the poem you're writing, take it and add it in."

8. **Tell students to begin, then give them 10 or 15 minutes to write.**

9. **As students work, follow the instructions of the prompt yourself to model engagement with the activity. Circulate if students need support or are struggling to focus.**

10. **(Optional) If you've done journaling activities such as *Zoom In, Zoom Out* or *Comparison*, make an announcement partway through the activity explaining how students can use these strategies in writing poetry.**

    a. "Last week, we did the activity *Zoom In, Zoom Out*, where you sketched [a tree, a plant, etc.] from close up, at life size, and far away."

b. "Try taking this technique into your poem and shifting perspectives. What do you notice when you look at something up close?"

c. "How does what you notice shift as you pull back and look at it from far away, and look at the context around it?"

d. "If anything interesting comes up, add it to your poem."

11. **When students have had time to write, but before they become disengaged, call the group back together.**

12. **Facilitate poem sharing in a way that works for your group, being clear that students only need to share what they have written if they want to.**

a. Students could share poems in a number of ways. If there is time, individuals could read a whole poem or part of a poem to the group. Students could also pair off and share a poem or part of a poem with a partner.

13. **After each student shares a poem, respond evenly, thanking them for their vulnerability and avoiding judgmental or evaluative statements.**

a. Have an even response after each student shares. Do not, for example, say "OK, thanks" to one student and "Wow, oh how amazing, what a great poem!" to the next.

b. Thank each student who shares for their courage and vulnerability, giving a similar response each time.

c. Avoid evaluative statements. This can shut students down. The goal here is for students to notice and record their experience. As long as they did that, they are successful.

14. **Point out how sharing poems is an opportunity to learn from each other as a community.**

a. "Thank you all for sharing your words with us."

b. "We all have different perspectives, ideas, and experiences. Listening to one another helps us learn from one another, and this strengthens us as a community."

15. **Tell students that they can include short poems in their future journal entries and use this technique in the future, reminding them to distinguish between what they observe and what they feel.**

a. "When you have personal thoughts, ideas, or feelings that arise while you are journaling, you can write them down in your entries along with your observations."

b. "This can help you form deeper memories of your experiences, and can be a way to keep learning about yourself and enrich your experience."

c. "As you do this writing, remember to make note of when you are recording an observation and when you are recording an idea, emotion, or thought."

DISCUSSION

Lead a discussion using the general discussion questions. Intersperse pair talk with group discussion.

General Discussion

a. "What was it like to spend time writing a poem in nature?"

b. "Did anything surprise you as you were writing?"

c. "What were some interesting or unexpected observations or insights you came to?"

d. "How might you include poems in your future journal entries?"

FOLLOW-UP ACTIVITY

Simile and Metaphor Poems

Engage students in writing simile and metaphor poems. Encourage students to look for parts of their surroundings that remind them of themselves, and to write them into a poem. You can say, for example: "Maybe, like the squirrel, you are watching everything closely, storing seeds for a more difficult time. Maybe you feel you are a spider web, because you make connections between many things. Maybe you are like a butterfly, like a cloud, or like a tree. Only you will know! You can include these thoughts in your poem, describing what you see and connecting that to who you know yourself to be."

Sharing Poetry

Look for examples of short, relevant poems to share with students in the field. This may become a part of your regular opening or closing ritual.

SIT SPOT

"Find your sit spot and see what comes." This activity gives students the time and autonomy to connect with nature on their own terms. As simple as it sounds, Sit Spot is one of the most memorable and potentially life-changing journaling experiences.

Time

Introduction: 10 minutes
Activity: 20–50 minutes
Discussion: 10 minutes
With practice, this activity can be expanded to take an entire morning or a day.

Materials

☐ Journals and pencils

☐ Loud whistle

Teaching Notes

Part of building self-motivated learners is giving them the autonomy to direct their own experience. This demonstrates your confidence in them and their ability to work beyond your supervision. The directions for this activity are intentionally open ended, giving students the room for highly personal and creative work.

Your introduction of this routine will be more authentic if you have done it yourself. To inspire students, share stories of your own memories of sitting quietly outside. You do not need to have seen an owl catch a mouse or a coyote walk by. Any memory will do. Model finding wonder in the smallest things, and your students will follow suit.

Read the needs of your group to help you set boundaries and determine a time limit. You might need to shorten the activity for more energetic groups. This is OK. Even 5 minutes of sitting still and quietly in nature can produce profound effects.

The experience of being on your own in nature, with a flexible structure and permission to encounter it on your own terms, is a formula for magic. Some educators fear that if they are not in direct control of a group, students will go off task and will not make productive use of their time. However, trusting students and giving them autonomy open up the possibility for powerful and personal experiences beyond what we can direct. This is an opportunity for students to focus not just on observing their surroundings but also noticing what it is like for them to be there.

In this activity, students find their own place in an outdoor area, then pay attention in whatever way inspires them. They can record their experience in their journal, and share about it afterward. Self-directed experiences like *Sit Spot* can help form a bridge between assigned work and students' own journaling, supporting them in adopting a nature journaling or exploration practice outside of assigned school projects. This is also an opportunity for students to think about their own story and identity, and how to craft a narrative of who they are and what is meaningful to them.

This is not a new activity or practice in nature. Cultures all over the world have and had ways of slowing down to observe their surroundings.

NATURAL PHENOMENA

Initiate this activity in a natural area that is big enough or has enough cover for students to spread out and get a sense of privacy and being alone. Define clear boundaries for the group with general safety considerations in mind (heat, cold, falling hazards, water, snakes, etc.) and an awareness of others who may be in the space (in public areas). Spots that offer concealment and a sense of safety from which students can look out on a broad view are ideal. This activity does not have to be done in a remote wilderness area. Forests, schoolyards, classroom gardens, ancient forests, or waterfronts are all places that can work for sit spots if there is enough space and cover to provide a sense of autonomy.

PROCEDURE SUMMARY

1. Find a special place.
2. Open your senses, sit quietly, and observe using "I notice, I wonder, it reminds me of."
3. Document your experience.

Note: There is no demonstration for this activity because students can construct their journal pages however they like.

PROCEDURE STEP-BY-STEP

1. **Create a sense of anticipation and excitement about the activity by telling students that they will get time to sit by themselves, paying attention to whatever is interesting or important to them.**

a. "We are about to do an amazing activity. It is called *Sit Spot*. This can be a powerful experience where people make amazing discoveries, have close encounters with wildlife, or have insights about their lives. No two people will have the same experience."

b. "In a moment we will break from this area. You will go out, on your own, to find your own sit spot. This is a place that is just for you, where you get to sit quietly and just be."

c. "Once there, you will pay attention to whatever is interesting or important or whatever is going on. That may be something you find, or it may be the feelings and thoughts you have in this place. Then you will record this in your journal in whatever way feels appropriate."

2. If you, the instructor, have had an opportunity to do sit spots before, share some of your experiences or journal entries.

3. Set expectations and state the sit-spot ground rules and boundaries: Find a place that calls to you, sit alone, stay within boundaries, and respect others' solitude by being quiet.

   a. "The first thing you will do is find your sit spot. Go alone and look for a place that interests you or somehow calls you. This may be at the base of a special tree in a wide-open area, or somewhere kind of hidden."

4. State clear boundaries for where students can find their spots, designate any areas that are off-limits, and explain any rules that your group of students might need in order to focus on the experience.

a. "When you get to your spot, the only thing you are required to be is alone and quiet."

b. "Be respectful of others' solitude and quiet. In order to make the most of this experience, you should be alone. You do not want to sit close to other people, as this might distract you or them."

5. Offer guidance (but not requirements) for how students might engage with their surroundings at their sit spot, suggesting that students focus on one sense at a time or on making careful observations.

   a. "Once you have sat down, pay attention. Relax and tune in to the environment around you."

   b. "You could try focusing on one your senses one at a time to see what you notice, or look around at different things next to you."

   c. "There is no wrong way to do this. As long as you are quiet, still, and alone, you can be in whatever way feels right for you."

   d. "If you are drawn to some object or view, experience it as fully as you can, observing, asking questions, and looking for connections. The 'I notice, I wonder, it reminds me of ' observation routine may be useful here."

   e. "Getting up and walking around will distract others, so when you find your spot, stay put. The longer you remain still and quiet, the more the animals might get used to your presence and begin to emerge near you."

   f. "You don't need to work in your journal the whole time you are in your sit spot. You can also just be still, relax, let your mind go quiet, or take in the view."

6. Offer guidance (but not requirements) for how students might engage with their inner world, suggesting that they turn their observation prompts inward to slow down and focus on their state of being.

   a. "You may find that the experience brings up personal thoughts and reflections. If this happens to you, you can focus on and deeply experience these feelings."

   b. "To intentionally turn your attention inward, you can notice and wonder about your own state of being, or think about what this experience reminds you of."

   c. "You may find that partway through the experience you feel done and that there is nothing else to see. This is normal. If this happens to you, just relax and remain where you are. Try focusing on senses you do not

A fall day, looking down a valley toward a distant farmhouse. This is a memory that will be kept forever.

Stefan, age 11

usually use. Notice what you observe after you pass the 'I'm done' boundary."

7. Offer guidance (but not requirements) for how students might record their experience in their journal, suggesting that they write, draw, and record observations or whatever else feels right to them.

  a. "After you have sat for a little while, you may document your experiences and thoughts in your journal in whatever way seems the most appropriate. If you are inspired to draw, draw. If you want to write a poem (it does not need to rhyme), write a poem."

  b. "Whatever approach you use to capture the story of this moment is OK."

  c. "If you are curious about a phenomenon or object you found, you can explore it with writing, drawings, questions, and observations. If you go deep into personal thought, record these ideas and feelings."

  d. "This is your time to be however you want to be, as long as you are not disturbing others. No two people will have the same experience. Let your journal reflect yours."

8. Give reminders about boundaries and expectations, and set a time limit that is appropriate for your group and context.

  a. "We will give this [three, fifteen, twenty] minutes If you have your own watch, you can keep your own time. If you do not have a watch, listen for my whistle. I will blow it two times when it is time to come in. Stay within earshot of the whistle if you do not have a way to keep time."

  b. Share any reminders based on the needs of your group.

9. Let students go, and trust them.

10. Pay attention to the needs of the group as the time elapses. Call the group back early if they are becoming restless, or let them stay longer if they are quietly engaged in the experience.

DISCUSSION

Lead a discussion using the general discussion questions. Intersperse pair talk with group discussion.

General Discussion

  a. "Find a partner and share about your experiences. You may share your journal entries. If you found yourself writing personal things, or do not want someone else to see the entry, you may keep it private."

  b. "Let's come together as a big group. If you feel comfortable doing so, open your journal to the *Sit Spot* page and place them on this picnic table. Take some time to see what other people experienced." *Note:* Sharing journal entries among peers can be a vulnerable act. Only offer this as an option if your students already have experience with protocols for observing others' work.

c. "Let's gather as a group. Think of one word that describes your experience at your sit spot, and share it with the group when you are ready."

d. "What was the sit-spot experience like for you? Does anyone want to share a story of something they experienced or felt at their sit spot? What made this experience special?"

e. "Find a way to craft your sit-spot experience into a story. This could be a story about what you saw or how you felt. If you do not want to share from your own perspective, you could imagine the story of an organism you observed, or the story the land around might tell. Then share your story with a partner or with the group."

FOLLOW-UP ACTIVITIES

Repeated Visits to the Same Spot

Students can repeat this activity throughout the day, week, or year, which will give them insights into their spot at different times. If you have consistent access to the same place for outdoor explorations, you could also institute a routine where students return to their sit spot for 5 minutes at the beginning of each foray outside. This quick "check-in" will be enough to slow them down and notice patterns, and engage in the study of phenology (seasonal changes).

Sit Spots in New Places

Once students understand this routine, you can initiate the activity in different habitats or places you visit just by saying, "Find a sit spot; you have ten minutes." This is a great way of slowing students down in a new place, or offering some time to "reboot" in the midst of a busy or social learning experience.

Longer Sits

Once students are familiar with the activity, they may be ready for longer sit-spot experiences. For longer time periods, students may spread farther apart (for more seclusion and autonomy). They may also want to move around more, as it is hard to sit still for half a day. Think of it as a "sit area" as opposed to a spot. They will still need to stay generally in one place and be careful to respect the privacy of others by staying hidden and not approaching other students.

The Green 15

One middle school teacher assigns students the "Green 15" every week. This is a "homework assignment" that requires students to spend 15 minutes outside at least once during the week. What the students do is up to them. This autonomy and continual practice helps students develop a routine of being outside in a way that feels good for them.

PICTURES

DRAWING AND
VISUAL THINKING

PICTURES: DRAWING AND VISUAL THINKING

DRAWING IS A SKILL

Anyone can learn to draw. Practice and learning illustration techniques help students improve more quickly and help them draw what they see and imagine. This chapter includes drawing exercises, practice skills, and techniques for breaking down and diagramming complex objects.

Combining drawings, diagrams, maps, cross sections, and other visual elements with structured page layouts helps students develop visual communication skills. Activities in this chapter also guide students to think about page layout and structure and how to best show their thinking on the page.

ACTIVITIES IN THIS CHAPTER

OBSERVATIONAL DRAWING

Anyone can learn to draw and to teach drawing (even you!). Visual thinking and communication strategies—such as observational drawing, drawing from memory or imagination, making structural diagrams that emphasize parts or construction of an object, creating mental models, making maps and cross sections, and creating quick sketches—are skills that improve with training and practice.

Drawing improves observational skills. As students draw, they must notice and record specific details, such as the angle of a branch from the trunk, the shape of a flower, or the direction that a heron's legs bend. A drawing also captures details that might be lost if the observer were just using written notes. Any observation not specifically described in writing is lost, but in a sketch, many observations are unintentionally recorded in the course of drawing. While drawing the fore and hind legs of a deer, for example, you also record the distance between them, even if you are not focused on that detail as you sketch.

Although both writing and drawing improve memory,[27] drawing is the more effective memory hook.[28] It does not matter whether the picture looks good or not. The attention required to draw locks a moment into memory. Because writing and drawing engage different processes in the brain,[29] it is reasonable to assume that using writing and drawing together would be even better for memory than drawing alone. In addition, the journal entry serves as a memory anchor, reminding you of key details of an experience through the fog of time.

OBSERVATION BEFORE ART

A nature journal is not an art project, though it could be the initial foundation for one. If creating drawings that are perfect and beautiful becomes the main focus of journaling, it distracts students from observing the natural world. Students' nature journals should be a place where they feel no pressure to make pretty pictures (or spell correctly or use the right grammar). If the pictures end up looking good, that is a bonus, but not the goal.

For students to fully commit to nature journaling—and for journaling to work its observational magic—they must understand that the goal of such drawing is to accurately observe and record data. If the goal becomes simply making a pleasing drawing, the "pressure for pretty" can get in the way of documenting observations.

Students will often hesitate to start a sketch if the words of an inner art critic or memories of disparaging comments from a previous teacher ring in their memory. Children (and many adults) remember the words of a teacher who (even unintentionally) dismissed their drawings with an offhand comment. For people who have been so discouraged, starting to draw again is intimidating.

By contrast, if the goal is to clearly and accurately observe and to record observations without regard to whether it "looks good," the pressure of producing Art is lifted, and students' focus shifts to making and recording observations. Any drawing, however crudely executed, is a success if it enables students to see more clearly or document their observations. When drawing is framed this way, students who do not consider themselves artists are liberated to draw without the pressure to produce a "masterpiece." An interesting side effect is that because students draw frequently in a sustained journaling practice, their work improves with practice.

BEST PRACTICES

In their journals, students

- Strive to show accurate and specific observations through drawings.

CAMERAS AND OTHER MEDIA

Cameras, tablets, and any number of digital field instruments are prevalent in many research labs and classrooms and can be useful tools for you and your students. Photography and apps are tools for gathering information, but they are not a replacement for the unique thinking and learning that happen in a journal. A field scientist may use tools such as a camera or data-recording instruments along with a journal, but the tools are used at specific times for a specific purpose. Student use of technology should also be purposeful. Without structure, a tool, such as a camera, can distract students from making careful observations and actually impair memory.[30]

Be strategic and specific in how you guide students to use cameras or other digital tools along with their journaling. Integrate these tools into the same learning process, either at a different time from journaling or with journaling to record data in a way that a journal cannot. For example, students might take a photograph of a field site they will observe over time, and paste it into their journal to add another layer of information, or use an app to measure sunlight in different parts of a study area. The activity *Photo, Pencil, and Found-Object Collage* gives students the opportunity to use photography as a tool to enhance a journal entry, and to consider the benefits and drawbacks of different media.

- Show multiple views of the same subject (top, side, angled view, etc.).

- Show the subject at different scales (magnified view, life size, zoomed out)

- Indicate scale ("actual size," 4x, 15 mm, scale bar, etc.).

- Use diagrams to emphasize specific details or record information more efficiently.

- Choose a drawing approach that fits the need of the situation—a quick sketch to show the position of an organism relative to a hillside, a detailed illustration to collect structural observations, a simple silhouette to show movement of an organism relative to the landscape.

- Use drawing as a tool for making observations; build on drawings by asking questions; integrate drawings, writing, numbers; and use labels, callout boxes, arrows, titles, and maps.

ANYONE CAN LEARN TO DRAW

Drawing is a skill, not a gift. No one is born knowing how to do it. It is learned, and the more you practice, the better you get. Both adults and children improve rapidly when given the opportunity and motivation to draw regularly. Coaching students to observe more deliberately at the start of a drawing and sharing simple drawing tricks help students improve more quickly.[31] The observational drawing exercises and tricks and tips in the section "Observational Drawing, Step-by-Step" will help students draw quickly and accurately. Offer one or two relevant drawing tricks from time to time before an activity (e.g., offer tips for drawing moving things before an animal observation).

Is There a Right Way to Teach Drawing?

Although there is some common ground among approaches to teaching drawing, there is also a lot of conflicting and contradictory advice. This is one of those areas where intelligent people disagree. There are a lot of successful ways to teach drawing, and very different approaches can work.

Practice and supportive positive feedback seem universally agreed on as important. If your curriculum did nothing more than provide a safe and supportive space where children could regularly practice their drawing, they would improve with no additional guidance. Whatever you do on top of that will probably be successful. Experiment, pay attention, and keep doing what works.

One area of disagreement is whether to teach specific drawing techniques, such as line weight, value, proportion, color theory, or how to draw a bird. Some art instructors feel that teaching children these techniques stifles creativity and gives them the

SHOULD YOU HELP STUDENTS WITH THEIR DRAWING WHILE THEY ARE MAKING IT?

In an art class, helping students with drawing technique is appropriate and expected. Students improve quickly with guided instruction and demonstrations of techniques and drawing principles. This increases their confidence and technical ability. In a nature journaling session, by contrast, do not offer any art critique, even if the student will later use this nature study to create a finished drawing. The emphasis of journaling is on observation, memory, and curiosity.

If you are teaching nature journaling as part of an art class or drawing workshop, students will often ask for or expect help with their drawings. Before offering a suggestion, take time to learn what the student is most satisfied with, where the student is pushing their ability, and what aspect of the drawing does not feel right to the student. Ask, "What parts of this drawing are working for you?" "Why?" "What part of this drawing is giving you trouble?" "Can you be more specific?" Reinforce drawing principles that the student is using that they should use again in other drawings. Try to find a specific suggestion for what the student is struggling with. Don't give feedback on the value range if they are struggling with proportions; it will just feel like one more thing they need to do. Meet them where they are.

Do not draw directly on a student's drawing. This will imply that they have made a mistake that they could not have resolved themselves, and will discourage drawing. One way to help a student with a drawing is to place a piece of tracing paper over their drawing and show how you might change perspective, proportions, or other details. Alternatively, the student can draw on the tracing paper themselves. This is helpful early in the drawing process. Once a student has laid in a lot of detail, it may be difficult to make changes without a lot of erasing.

impression that there is one "right" way to draw. Other art instructors regularly teach drawing principles and methods.

We have found that teaching fundamental techniques of drawing accelerates students' learning. Just as it would be more difficult to learn the piano by banging keys and figuring it all out by yourself, it is hard to learn to draw in a vacuum. This is also true in the world of formal art. The great artists and painters did not figure everything out by themselves. For centuries, artists have innovated and shared drawing techniques to render different subjects, textures, moods, and materials. Techniques and principles that work get passed along. Fundamental drafting methods help artists get what they see down on canvas or paper, quickly and accurately. These methods build confidence and encourage the artist to explore increasingly complex or subtle material.

Students will be most successful with new techniques if drawing fundamentals are taught as tools, not rules. Offer these techniques as a place to start and as an adaptable set of tools to be called on as solutions to any drawing problem. Learning drawing techniques will not make your students mere clone illustrators. Having a toolbox full of drawing and drafting techniques gives students greater agency in making decisions about how to create their drawings.

This chapter outlines observational drawing exercises that can help build students' fundamental drawing skills, then offers a step-by-step approach that can be used as a flexible framework for making drawings in the field. These focused drawing lessons will increase students' confidence and ability to put their observations on the page.

Permission Granted

Is it OK to make abstract or geometrical doodles? Yes.

Is it OK to draw from your imagination? Yes.

Is it OK to use a step-by-step tutorial? Yes.

Is it OK to use photographs as reference? Yes.

Is it OK to copy a single photograph? Yes.

Is it OK to copy someone else's drawing? Yes.

Is it OK to trace a picture? Yes.

Should I be drawing from life? Yes.

Should I be drawing all the time? Yes.

Yes, yes, and yes. It is OK to do whatever works to get the marks down on paper. The more your students draw, in any way, the better they will get.

WHAT IF YOU ARE NOT AN ARTIST?

You can do this. You do not have to be an artist to teach fundamental drawing skills to your students. As you instruct your students and practice yourself, your own drawing ability will

WHAT MOTIVATES STUDENTS TO DRAW

Specific, authentic, and positive feedback. This kind of feedback takes attention and work on the part of the instructor. This attention further validates the work of the student.

Observational drawing. It is more difficult to draw from your imagination than from a real object. Many students have never drawn from life, and think it all needs to come out of their head. They are surprised and delighted when they learn that by closely observing and drawing an object in front of them, they can render it more successfully than they thought they could.

Seeing their own improvement. Looking at old work (at least twenty drawings ago) is evidence of growth and change. Ask students to look for ways that they have refined and improved their drawing skills. When students finish a journal, have them flip back to the beginning and notice ways they have improved.

Having a new tool or technique. A new drawing skill or tool is fun, and students can see their skill set grow.

Combining writing and drawing. Using writing in addition to drawing takes a lot of pressure off the drawing to explain everything, and looks more like note taking than art.

WHAT DISCOURAGES STUDENTS FROM DRAWING

Generalized, insincere, or negative feedback. Students can feel discouraged if they think they drew something wrong. Instead of saying, "You made the head too big," you could ask a question to lead the student to discovery so they can fix it themselves: "How big is the head compared to the body?" or "How many head lengths is the body?" Even positive feedback can hurt if you don't mean it (children usually know when we are lying), as can feedback that is so generic that it's clear that you are not really looking at the drawing ("Oh, that is so pretty").

Being out of practice. To a fifth grader who has not drawn since third grade, starting to sketch is scary. Do not emphasize drawing techniques at the start. Focus on observations and ways to record them. As students begin drawing more, they will naturally become curious about tips and techniques to improve their drawing accuracy and speed.

Having someone else draw on your page. Avoid fixing a drawing for students by drawing directly on their page. This action implies that they cannot do it themselves. Instead, help students observe and thereby make the changes independently. Sketching on another piece of paper or laying a piece of tracing paper over the drawing and letting students draw on that may facilitate the discussion and make it easier for them to see past the lines they've already drawn.

A right way to draw. There is no right way to draw anything. Formulas that lead students step-by-step to the same finished piece are useful only if they are seen as an example of drawing principles (such as how drawing techniques may be integrated or to learn the progression of a drawing from basic shapes to refined edges, to values and details). These demonstrations should not be interpreted as "That is not how you draw a bird; this is how you draw a bird."

blossom. Many adults are afraid to draw, but the more you are able to face this, be brave, and move forward, the more your work will inspire students.

Throughout this book, we emphasize the importance and excitement of teachers' keeping their own field journal, and that applies here too. Sketch while your students do, and they will see sketching and journaling as a valuable practice instead of just an assignment. It's more useful for students to see you struggling to develop new drawing skills than it is for them to see you produce perfect work. If your journal is a record of observations and a space for learning, your work reinforces what you have told the students: that the purpose is not art but observation.

How to Talk to Students about Your Drawings

Be honest with students about your process as you seek to improve your drawing skills. Refrain from belittling yourself with such comments as "My pictures are terrible too," "I can't draw," or "I am not an artist." Instead, reflect a growth mindset in how you discuss your learning process: "Just like you, I'm learning how to do this. Look at how much I have improved in the last [month, year]. Look at what I discovered while making this sketch." Supportive talk shows students that you are also learning and that it is OK to be learning. You are the role model; show them how to be gentle with themselves as they learn a new skill.

As you get better at drawing, students may compare their work with yours. If they have a fixed mindset about drawing—that is, if they assume that they either have the drawing "gift" or don't—they will be intimidated by your ability to draw. To reduce the negative impact of comparison when you show students your work ("Your drawing is so much better than mine, it's pointless to try"), emphasize again that you did not start out knowing how to draw, and that they too are improving through practice. Point out parts of your own drawings that have recently improved, along with areas that are challenging and that you are still working on. Help students play the "long game" by reminding them that they are better at drawing this month than they were last month, and that improvement never stops. If you have been drawing for much longer than your students, help them calculate how many more years of practice you have had. Keep coming back to the point that drawing is not a static trait but a skill that they can improve through deliberate practice.

OBSERVATIONAL DRAWING, STEP-BY-STEP

A flexible step-by-step approach will help students draw confidently and efficiently in the field, and avoid becoming overwhelmed.

Having a system for approaching a drawing helps students jump in fearlessly, and will preempt many struggles they might come up against. This is particularly important in the field, where drawing time is limited. Study and experiment with this workflow, teach it to your students, then help them modify it to fit their individual styles.

1. OBSERVE

The backbone of an accurate drawing is specific and careful observation. Do not take this for granted. It is easy to start drawing too soon and get focused on what you have on the paper or the way you think the subject "should" look, instead of how it really does look. One way around this is to begin by verbalizing your observations before you begin drawing. Don't just look at the subject and silently observe. Say what you see, and be specific. This can be done alone or with a partner. If you make observations about shape, angles, proportions (relative measurements, such as how big the head is relative to the rest of the body), or where structures are attached before you start drawing, it is more likely you will include these elements in your drawing.

2. BLOCK IN THE BASIC SHAPE

With the lightest possible gesture lines, block in the rough shape of the object with a pencil or non-photo blue pencil. This light framework or "ghost drawing" is the scaffolding on which you will hang your details. Start with the central axis or tilt of the object with a fluid stroke. Then add the general proportions (height vs. width, relative sizes of the major parts of the object) with circles and angular strokes. It will help to measure part of

your subject. Use a prominent feature such as head length to make relative measurements (e.g., "This bird's body is 2.5 head lengths"). Use negative shapes (the shapes of space around an object) in and around the object to further block in the basics.

3. CHECK AND MODIFY THE BASIC SHAPE

Now check the basic shape, looking at the proportions, negative shapes, measurements, and angles of your drawing. You can use "through lines" (described later, in the section "Drawing Tricks and Tips: Instruction for Students") to check the alignment of things on opposite sides of the drawing. This step is really important, but is the one that drawers of all levels are most likely to skip. It is so much fun to get in there and draw that people do not stop to check the ghost drawing before they proceed. It is only when the drawing is complete that people back off and discover that they made the head too big or the negative space between the legs too long. These sorts of problems are easily fixed at the phase of a drawing when you have a light framework of pencil lines. You do not even have to erase; just draw a new set of lines over the existing ones. Once you start to draw deliberately, however, you are locked in. Once a hard line is down on the page, you will convince yourself that it is right because you do not want to erase, and the confidence of a deliberate line gives the illusion of accuracy. Remeasure parts of the sketch. Look back at the subject, then at your drawing. Correct any inaccuracies before you move on.

Negative shapes carve the angles between each leaflet.

An oval and a point capture the overall proportions of the leaf.

Radiating lines run along the axis of each leaflet.

Rough ovals and points capture the dimensions of each leaflet.

4. DRAW ON TOP OF THE BASIC SHAPE

If you have been using a non-photo blue pencil, switch to a regular graphite pencil or pen. Draw deliberately on top of the light framework, defining edges, contours, and form. As you draw, let your mind flicker back and forth between seeing the shape of the subject and the shapes of the negative space around and within the subject. As the drawing progresses, use more pressure, adding accents and reinforcing key lines, or parts of foreground objects. Vary the boldness of your lines, avoiding the temptation to make

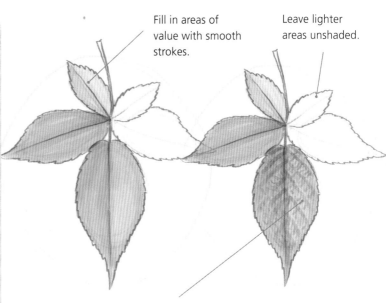

Fill in areas of value with smooth strokes.

Leave lighter areas unshaded.

Create contrasting areas of value to suggest texture and form by erasing highlights and deepening shadows with your pencil.

every line bold. There is no accent if everything is bold. This step is what most people think of as "drawing," but notice how much has come before. It is tempting to start with this step, but just as an essay will be improved by an outline, a sketch is improved by the light preliminary drawing.

7. ADD COLOR

Once your values are defined, you can tint them with color. As you get more experienced, you can start to apply both value and color at the same time.

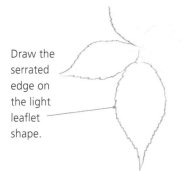

Draw the serrated edge on the light leaflet shape.

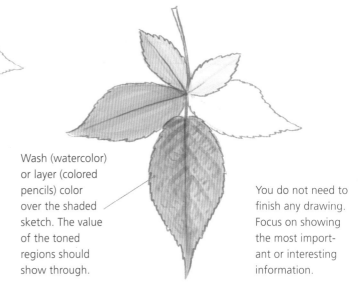

Wash (watercolor) or layer (colored pencils) color over the shaded sketch. The value of the toned regions should show through.

You do not need to finish any drawing. Focus on showing the most important or interesting information.

5. CLEAN UP THE DRAWING

Now pull out the eraser (white vinyl erasers are very good for this) and clean up any extra lines. Note that erasing comes late in the game, very different from the "draw a little, erase a little, draw a little, erase a little" approach. Erasing at the end is about calling attention to the intentional lines you have drawn throughout a slow process. Erasing as you go is not as effective.

6. ADD VALUE

Many artists simply color the drawing from here. This can result in a flat, coloring-book look. Getting the values, the patterns of dark and light, is more important than matching the color. Squint at your subject and your drawing to blur detail and help your brain focus on value. Shade in the darkest area and middle tones.

8. ADD DETAIL

Think of detail as a spice to be added conservatively at the end of a drawing. If you add detail too early in a drawing, it becomes obscured by shading, color, erasing, or the smudging that happens as you draw. Add more detail with a sharp graphite pencil. Put detail in areas of focus and interest, and in the foreground. This will guide the eye of the viewer and give a greater sense of depth. Detail everywhere will flatten your drawing.

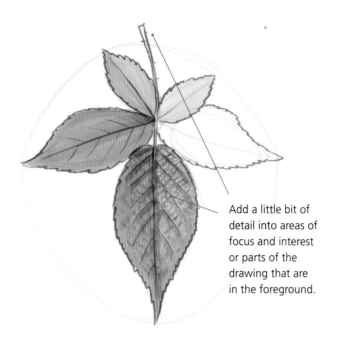

Add a little bit of detail into areas of focus and interest or parts of the drawing that are in the foreground.

9. STOP BEFORE YOU OVERWORK THE DRAWING

It is hard to know when to stop. A drawing that is partially complete is more interesting than one that is overworked. One rule of thumb is to stop drawing before you think you are done.

10. MAKE ANOTHER DRAWING

This is a process, and each drawing is just practice for the next. There is no end, no masterpiece, just open-ended practice and learning.

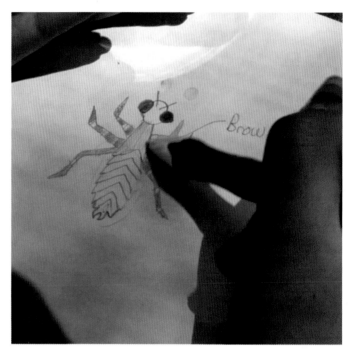

BUILDING NEW SKILLS: BASIC DRAWING EXERCISES

Artists pull from a toolbox of methods that help them transfer what they see onto the page. You can learn and teach these techniques.

These fundamental mechanics of observational drawing are easy to teach and greatly improve students' ability to draw what they see. Even a short time spent on basic drawing skills can help students be able to more accurately represent their observations in their journals and have more fluid visual communication tools.

These art lessons can be given to the entire class or to an individual student who needs support with their drawing and is ready to learn. You could teach all these techniques together in a drawing lesson, or one at a time right before an activity where students will apply the technique. They are presented in a progressive series, but also can be taught on their own to help a student overcome a drawing hurdle. Familiarize yourself with these strategies and practice on your own and with your students. Your challenge is to push students just a little outside their comfort zone—to try something new or take a little risk. Can you spot the next tool that the student needs in their drawing toolbox?

BUILDING EYE-HAND COORDINATION

1. **Explain the connection between drawing and hand-eye coordination.**

   a. "Training your fingers to do what you see in your mind is one foundation of drawing."

   b. "The more pencil miles you put in, the more you train your fingers to respond to what you see."

   c. "We can practice drawing simple shapes (triangles, circles, squares), learning to fill areas with a solid or graded tone, and making light to bold lines."

   d. "These are important parts of every artist's vocabulary."

2. **Instruct students to make straight lines, drawn from the shoulder, and parallel sets of elbow, wrist, and finger arcs.**

   a. "Doodling is great practice. Fill a page with random geometrical patterns."

   b. "Make some lines dark, others light, some thick, some thin. Make variable lines. Make circles, squares, triangles, long straight lines, smooth curves, and elbow and finger arcs."

   c. "Fill in some of the spaces with even, continuous tone, or a smooth graded tone going from rich dark to as light a value as you can make. Fill boxes with hatched lines and cross hatching, or invent textures: a hairy surface, a bumpy surface, a pitted surface, a pebbled surface, dry cracks."

   d. "With this training, you will be able to draw straight lines, smooth curves, and sets of parallel lines, which are great for plant stems."

3. **Encourage students, letting them know that this practice will develop their drawing skills.**

   a. "Think of these exercises as warm-ups and drills. Just as you get better at basketball with practice and repetition of basic skills, so too with drawing."

   b. "The more you practice making marks, the more you will intuitively understand what your hand and pencil can do."

FINGER TRACING AND AIR DRAWING

1. **Coach students to use their fingers to trace an object and practice noticing angles and changes in contour.**

   a. "Before you can draw the object, you must train your eye to see angles and changes of contour. The goal is to draw the shape that is in front of you, not the one in your imagination."

2. **To help students understand this idea, place an object in front of them or have them hold an object in one hand, then close one eye and trace the edges with a finger of the other hand.**

   a. "Describe out loud the new angle each time your finger changes direction. For example, 'Straight down, sharp right turn, short flat edge, starting to curve down, little bump, now straight out at an angle.'"

   b. "Now do this with a pencil, this time tracing the shape in the air with the tip of it. Imagine a line being inscribed on the wall behind the object."

   c. "Each micro change of pencil direction or line-segment length is a decision. You want to notice each decision as you move around the object."

LEARNING TO DRAW LIGHTLY

What happens when you suggest that young students start a drawing lightly? The moment their pencil hits the paper, they begin drawing with full force. Learning to draw lightly is a skill we can help them develop with a little practice. The energy behind this kind of drawing will be used again and again in observational drawing. Use this sequence as a warm-up to drawing, and your students will find it easy and natural.

1. **Encourage students to loosen up and draw with their whole arm.**

   a. "Start by loosening up. Relax your shoulders, arm, and hand, and draw a circle on your paper."

b. "Drop your shoulders, unclench your jaw, loosen up on your pencil grip, and move your pencil grip back a little."

c. "When you draw, move your whole arm, not just the fingers. Draw another circle, and then another."

2. Tell students to fill the page, drawing circles as lightly as possible.

a. "Now lighten up. Draw a circle as lightly as you can. Now see if you can draw one that is even lighter. Fill the whole page with the lightest circles your pencil can make."

b. "Now speed it up. Keep the lines loose and light, but this time the goal is to also make them fast. Make a series of overlapping circles, scribbles, and flowing lines."

c. Make a game out of it. Walk around the room and find examples of really light circles and hold them up as examples. "Do you see this one? No? Exactly! That is because it is so light!"

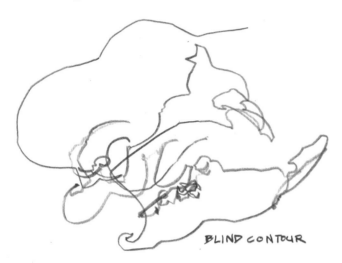

BLIND CONTOUR

CONTOUR DRAWING

Sit the students at tables with interesting objects before them.

1. Explain that contour drawing is a way to train yourself to look at drawing subjects.

a. "The most important part of accurately drawing an object is to look at it carefully."

b. "This seems too obvious to mention, but all too often we rely on our mental image or what we think our drawing subject should look like, instead of observation."

c. "Contour drawing is the most powerful way to train yourself to look at the subject."

2. Tell students to start their blind contour drawings without looking at their paper.

a. "In a blind contour drawing, the point is not to draw but to see. It is a fun exercise that will train the connection between your eye and your pencil."

b. "Stare at the object and slowly begin to draw its shape without looking at the paper. Let your eye crawl slowly along the contour of the object. As you do so, let your pencil creep along your paper moving up or down following the curves and angles that you see."

c. "With every change in angle, let your pencil respond with its own change of direction. Do not lift your pencil or look down to see where you are. Take your time."

d. "When you are done, take a look. The results are comical and fascinating. Look for places where your lines reveal subtle changes or aspects of the real object."

e. "You can keep practicing with other objects. This will train your hands to respond to what your eyes see."

3. Offer instructions for a modified contour drawing.

a. "A modified contour drawing also helps us hone our observations, but results in a drawing that looks much more like the object."

b. "The process is the same, only this time you get to peek. Every now and then, you can glance down at your paper to allow yourself to relate the spacing and size of the lines to each other."

c. "You can also pick up your pencil and move it to another spot. To keep the energy of the contour drawing, keep your eyes on the object as you draw."

MODIFIED CONTOUR

LEARNING TO USE NEGATIVE SHAPES

1. **Define negative shapes and explain how to look for them.**

   a. "Negative shapes are the shapes that occur between the objects we are drawing. In approaching a subject such as a skull, you would probably focus on the shape of the upper and lower jaws."

   b. "The negative shape is the shape of the empty space between the upper and lower jaws. The jaws have height, width, and angles, as does the negative space."

2. **Explain how negative shapes can be a tool for accurate drawing.**

   a. "By drawing the negative space as an actual shape, you may discover that you drew the jaws too close together or too far apart."

   b. "If your negative space does not fit, don't ignore it and move on. You will have found a valuable indication that something is off with your proportions."

   c. "Find out what is wrong and fix these errors before continuing to draw. Using negative space is one of the most powerful but underused tricks in the artist's tool kit."

NEGATIVE SHAPE

GESTURE DRAWING

1. **Tell students to draw a circle.**

   a. "Would you like to draw a perfect circle? Grab a piece of paper and draw a circle with one clean line right now."

   b. "It will probably be lopsided or uneven. Drawing a circle this way is hard. I can't do it."

2. **Offer a technique for drawing a circle and correcting the lines.**

   a. "Let's try an easier way. Lightly and loosely draw a circle. It's OK if it's a little lopsided."

   b. "Now, without erasing, draw over it, correcting some of the imperfections with continued light lines. Overlap five or more circles, slowly correcting the roundness."

   c. "Your brain will gravitate toward the right lines. As it does, you can press a little harder, reinforcing the lines you want to keep. Watch a perfect circle emerge from the page."

3. **Explain the value of using light lines to start a drawing.**

   a. "The key is to begin lightly, make lots of lines, and reinforce those that seem to be most accurate."

   b. "By keeping it light, you let your brain sort between several possibilities as you carve into or add to your original shape."

   c. "If you start with bold, hard lines, you will feel committed to those lines even if they are wrong. Use this approach when starting any subject."

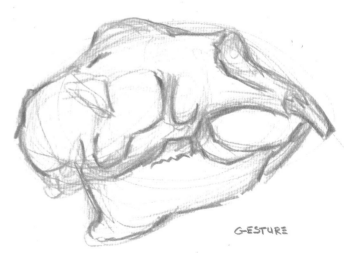

GESTURE

DRAWING TRICKS AND TIPS: INSTRUCTION FOR STUDENTS

You do not need to be an art teacher to teach these practical drawing techniques. Sharing them with your students will help them draw faster and more accurately.

These are useful drawing tricks and "tidbits" that can increase students' proficiency in drawing. Some highlight key tricks or skills, and others focus on components of the overall drawing approach described in the earlier section "Observational Drawing, Step-by-Step." Offer them to students to guide their growth and development, or before an activity where the technique will be especially useful.

BLOCKING IN A DRAWING AT THE START

Whenever you start a drawing, lightly and loosely sketch the basic shape of your subject at the start. Ignore the details and focus on the size and position of the largest masses of the form. Once you have a quick shape down on paper, double-check the proportions (relative sizes of structures) and angles. Refine the basic shape before moving on. It is easy to change a light drawing without detail. It is hard to change your drawing once you add strong lines or detail.

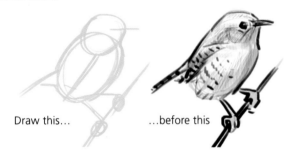

Draw this...   ...before this

USING "THROUGH LINES" TO SEE ACROSS A DRAWING

Knowing where certain features of an animal or plant line up with each other can help us draw more accurately. As you draw, imagine a vertical or horizontal line between key landmarks on your drawing (e.g., the base of the head and the line of the back) to see where it comes out on the other side of the drawing. Try closing one eye and holding your pencil up to the subject to see how things line up. Then do the same on your drawing. If parts do not align, correct the problem before going further.

USING NEGATIVE SHAPES

Negative shapes are the shapes of the air next to your subject. Although you may have an idea of how the leg of a cow looks, you do not have a sense of how the air next to a cow's leg looks.

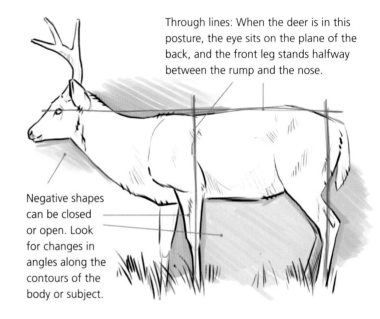

Through lines: When the deer is in this posture, the eye sits on the plane of the back, and the front leg stands halfway between the rump and the nose.

Negative shapes can be closed or open. Look for changes in angles along the contours of the body or subject.

When you look at the air, you are less distracted by how the subject *should* look and can more easily see how it *does* look. When we draw, the angles, size, and position of negative shapes are as important as those of the object itself.

THREE VALUES WITH SHAPES

Value (the range of dark to light) is more important in a successful drawing than color. If the values are correct, the drawing will read well even if the colors are way off. The opposite is not true. It is easier to see the dark-and-light patterns if you squint your eyes. Instead of trying to copy all the shades of gray, simplify the values to three: the darks, the middle tones, and the lights. First identify the darkest shapes and make sure they are appropriately dark on your drawing. Do the same for the shape of the lightest areas. If you capture the shape of the darks and lights, you will automatically shape the middle tones between them. The shapes of the dark and light areas give an object structure and solidity.

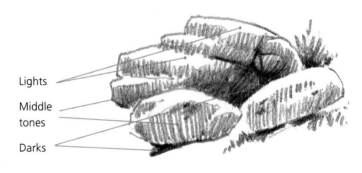

Lights

Middle tones

Darks

DRAW ANGLES, NOT CURVES

If you start drawing a curved line, the hand and the pencil get carried away and overround the shape. Many "curved" objects are actually slightly angled, where the shape is better described with a series of short lines rather than a continuous curve. Look for the slight angles of lines in curves, and emphasize these angles.

HOW TO SHOW DEPTH

Does your drawing look flat? To add depth to a drawing, you can

- Add larger objects in the foreground that overlap smaller objects in the background.
- Add more **detail** in the foreground, less detail in the background.
- Use heavier **lines** in the foreground, lighter lines in the background.
- Use strong, contrasting **values** in the foreground, lighter and less contrasting values in the background.
- Use stronger **colors** in the foreground, and pale, more muted, and blue-tinted colors in the background.

THE BASICS OF ADDING COLOR: MIX WITH CYAN, MAGENTA, AND YELLOW

The primary colors are cyan, magenta, and yellow, not red, yellow, and blue. Magenta and cyan make blue and purple; magenta and yellow make red and orange; and cyan and yellow make green. Notice cyan and magenta as distinct colors, not just hues of red and blue. To make muted colors, add all three primaries together.

Note: Make sure your students have good cyan and magenta colors in their kits. (They probably already have the yellow.) Experiment with color mixing in class before trying it out in the field.

Overlapping layers of color build a variety of hues. Note how the green mixed from blue and yellow is dull, whereas the one mixed from cyan and yellow is bright and clean.

GETTING MORE OUT OF COLORED PENCILS

When you use colored pencils, build up colors with lightly applied layers of different pencils. If you lay down a color too forcefully early on, the paper becomes slick and smooth with wax, and you no longer can add layers of color. By combining light layers, you can make many more colors than you have in your pencil box.

Pass out paint-chip samples to each student and challenge them to match the color and value of the chip by overlapping layers of color.

MAKING DRAWINGS OF MANAGEABLE SIZES

Many people draw very small pictures. It is difficult to see or add detail to these cramped drawings. Instead of making micro pictures, work larger! A good general rule when you draw an object is to make your drawing at least as large as your fist.

At the other extreme, many people make very large drawings. The solution is to draw the subject at a more manageable size. If the subject is large, perhaps a tree or a landscape, try making "thumbnail" drawings, perhaps 2 by 3 inches. These are more manageable, and are fun to make! If there are features that are particularly interesting and do not show at the scale that the drawing has been done, draw an enlargement of the interesting area to magnify it. This allows you to include small features without needing to make the whole drawing too big.

LEARNING TO ERASE LESS

Some students feel pressure to make a perfect drawing. They draw a little, erase a little, draw a little more, then erase that too. Erasers are useful tools, but not in this way. Instead of erasing a bit at a time throughout the drawing, set the eraser aside and just draw over the mistakes. Get several light lines down first, then decide which ones to emphasize. Out of the three or four lines for the curve of the horse's back, your eyes will gravitate toward the best choice. Strengthen this line with a heavier pencil stroke. Once you have chosen the best lines for the whole drawing, then you can pick up the eraser and clean up extraneous marks.

WHAT TO DO IF IT MOVES!

Living animals move, and move again. Prepare your students to work through the challenge of a moving subject with clear strategies:

- When you draw animals, they are very likely to move.
- Work on one drawing until the animal moves to a different position, then start another drawing on the same page. If it moves again, start another drawing. If it returns to a position you have already started to draw, go back to that drawing. The drawing you get the furthest along on will probably capture the animal's most characteristic posture.
- Switch between different animals in a herd or group. If you are drawing a side view of a deer and it turns to look at you, switch your attention to another animal that is in the side view.
- Slow down and focus on smaller details. Any time you feel overwhelmed by too much information, you can dial it back. If comparing birds at a pond is too much, make a study of their beaks or feet.

PAGE STRUCTURE AND LAYOUT

The way you lay out a journal page changes how you think and observe while you work, and affects how you retrieve information later.

A journal entry does not need just to show a drawing in the center of the page with observations written around it. Many phenomena can be represented with more creative approaches to the layout and structure of the journal page. Frameworks—such as panels to show progressions in time, boxed features to distinguish between different species observed, or tables for recording data—help the journaler use and place words, pictures, and numbers on the page. Smaller layout features—arrows, icons, lines, insets, and the like—can integrate these elements and create a precise record of observation and thinking.

TITLES AND SUBTITLES

Page titles and subtitles categorize information on a page, highlight themes, and make the page easier to scan. The process of creating a title also forces the journaler to think about the big picture of what the page is about. Titles may be creative and playful. Let students use colored pencils, block letters, bubble letters, or picture-words (imagine the word *fast* with speed lines behind it) if they wish. Bonus points for students who use bad puns in their titles.

When faced with a blank page, we can be tempted to put a title down just to take up some of the empty space. As students first start journaling, they can title their entry based on the prompt they are given (e.g., "Willow Comparison"). As they grow, coach them to wait until observations and ideas are on the page; at the beginning of a journal entry, you seldom know where the investigation will take you. A theme or title may emerge, one that is playful and speaks to the type of observations you made while creating the entry.

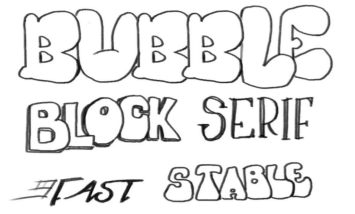

Creative picture-words are also a great way to "prime" students' thinking. Instead of having students add the title at the end of an activity, ask them to spend a few minutes before an activity creating a picture-word for a key concept

(e.g., *system, structure, change,* or *patterns*), which will guide their thinking during the activity.

ARROWS, DIVIDERS, AND BOXES

Once students have their observations on paper, they can use boxes and arrows to connect and relate observations or other components on the page. Dividing lines break a page into different sections. Students could use a colored pencil that matches a prominent color on the page. This extra layer of structure visually "chunks" information on the page and makes it easier to understand.

BEST PRACTICES

In their journals, students

- Integrate words, pictures, and numbers on the page (labels or written notes to elaborate or clarify a drawing, drawings to show what was quantified, all three elements focused on one subject).

- Use format features (icons, boxes, arrows, titles, headings, subheadings) to organize the page and call attention to ideas.

- Include contextual features such as maps or insets.

- Have a cohesive overall structure or layout.

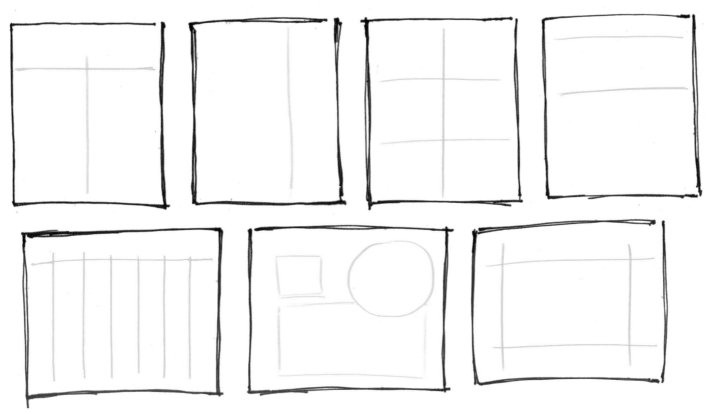

Encourage students to play with different layouts and think ahead of time about page structure.

DEVELOPING STRUCTURE AND LAYOUT SKILLS

Use a whiteboard while introducing activities so as to model roughly how students could choose to structure their journal entry. This support helps beginning journalers envision different kinds of layouts and approaches to page structure.

As students develop, occasionally prescribe a particular page structure for an activity (e.g., "Please structure your journal entry in four panels" or "Show at least three different perspectives in your drawing, and integrate this with text"). Although you should not always direct how students lay out a page, occasional guidance can offer them a new approach or lead them to try something they would not have otherwise. Reflecting afterward on how the approach made them think will help them put it into their visual layout tool kit.

Students can also study the journals of scientific researchers and professional and amateur naturalists, and look for ideas about how to structure a page. Ask students to discuss creative approaches to visual layout and representing information on the page that they want to try themselves. Comic books and graphic novels are another source of information where students can see ways of depicting motion or organizing pages they can adapt to their journaling. In addition, let students see one another's work and look for new approaches to laying out information on the page.

Reflection is a key tool in the learning process. After a journaling experience, give students time to study their page structure. Ask what worked and what didn't, and what they might do differently next time.

Additional Resources

The Sketchnote Handbook, by Mike Rhode

The Naturalist's Field Journal: A Manual of Instruction Based on a System Established by Joseph Grinnell, by Steven G. Herman

The Laws Guide to Nature Drawing and Journaling, by John Muir Laws

Nature Drawing: A Tool for Learning, by Clare Walker Leslie

Keeping a Nature Journal, by Clare Walker Leslie and Charles E. Roth

Artist's Journal Workshop, by Cathy Johnson

Drawing on the Right Side of the Brain: The Definitive, 4th Edition, by Betty Edwards

Use the John Muir Laws blog videos (johnmuirlaws.com/blog) as "draw-along" lectures to build skills in drawing specific subjects.

Collaborate with the school art teacher to teach basic observational drawing skills.

THINK WITH PICTURES

Journaling trains you to think visually. Models and diagrams help students observe, think, and figure out how things work.

Marcelo Jost

"This entry was to document an exquisite flight maneuver I saw a booby doing at a windy beach. The fact that the bird was gliding against the wind was so strange that I was compelled to watch it intently….The maneuver was so weird that it required a top view and a double wave cross section to record it in a way I could understand. Finally I added a very concise explanation that verbally described [my understanding of] the maneuver and wrapped everything together. An awesome lesson in flight!"

Dynamic Diagrams

Diagrams do not need to be a detailed. They are about ideas, not art. Using diagrams to describe processes and movement is a useful strategy for thinking and communication.

Showing Motion: Lines and Arrows

It would be hard to draw the bird at every pose in its flight path. Lines and arrows show direction and motion. This is more clear and faster to draw.

Cross Sections

Using letter pairs (A, A') shows how maps and cross sections relate.

Unseen Forces

In a diagram, arrows can show unseen forces acting on a phenomenon. This arrow shows the direction of the wind relative to the waves and the bird.

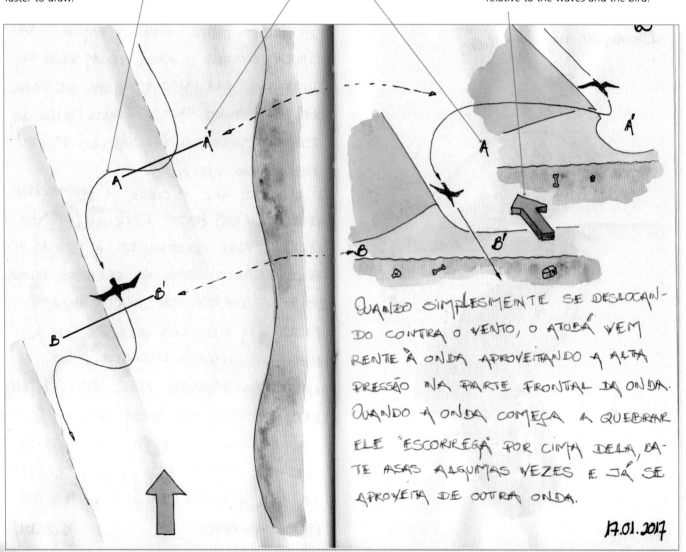

Translation: "When simply flying against the wind, the booby flies very close to the wave, using the pressure in front of it. When the wave starts to break, it 'slips' above it, flap its wings a couple of times and starts to ride another one."

BALANCING FLEXIBLILTY AND STRUCTURE

Striking a balance between planning a page layout that suits the journaling subject and flowing with the moment encourages flexibility and responsiveness, and is a skill students can develop over time.

Playing with Layout

There is no one right way to structure a journal page. Encourage students to play around with integrating words and pictures and how they lay out the information, and to work toward a balance between planning their page structure and being responsive in the moment to what they notice and observe.

Different Kinds of Drawings

Some drawings are meant to show the full nuance and detail of the object; others are diagrams, and focus on specific features. Including both intentionally on the same page is a strategy we can encourage students to adopt.

Laura Cunningham

"Art can be the best way to summarize science. In my nature journal, I can experiment with field sketches that encapsulate my observations of how shadows and light interact with geology. A lot of information can be quickly captured by nature journaling, and this is useful as data to compare the present to the past, as well as the future. Firsthand observations of nature—including art, writing, and measurement—are what science is all about."

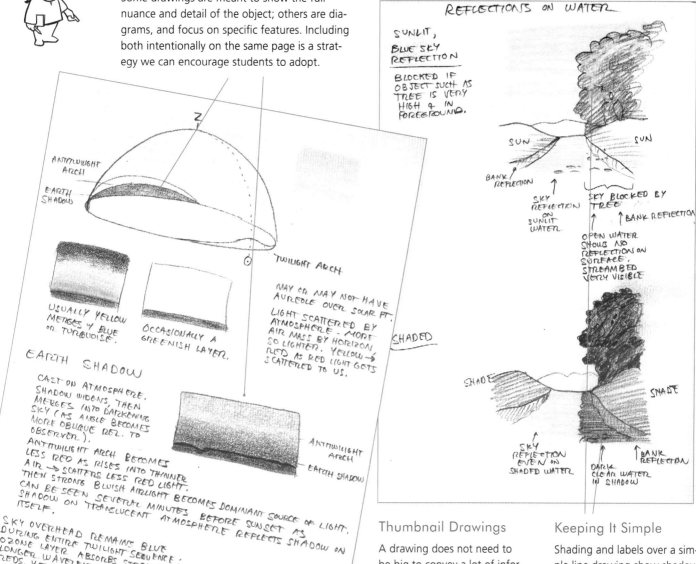

Thumbnail Drawings

A drawing does not need to be big to convey a lot of information. Students should ask, What information am I trying to show with this sketch? What do I need to include to get that across?

Keeping It Simple

Shading and labels over a simple line drawing show shadows and reflections at two times of day. Together, they highlight patterns that would be difficult to show in a drawing or written account alone.

LOOK, AND LOOK AGAIN

Part of the power of drawing is that it slows us down. When students draw, they will notice more and see the moments that a person with only a camera would miss.

Jonathan Kingdon

"It is hardly possible to compare animals without asking questions, and drawing is an exercise in comparisons….The probing pencil is like the dissecting scalpel, seeking to expose relevant structures that may not be immediately obvious…."[32]

Choosing Words Carefully

Written notes can effectively show what is difficult or impossible to capture in a drawing. We can remind students that using words to elaborate on pictures is a key part of visual communication.

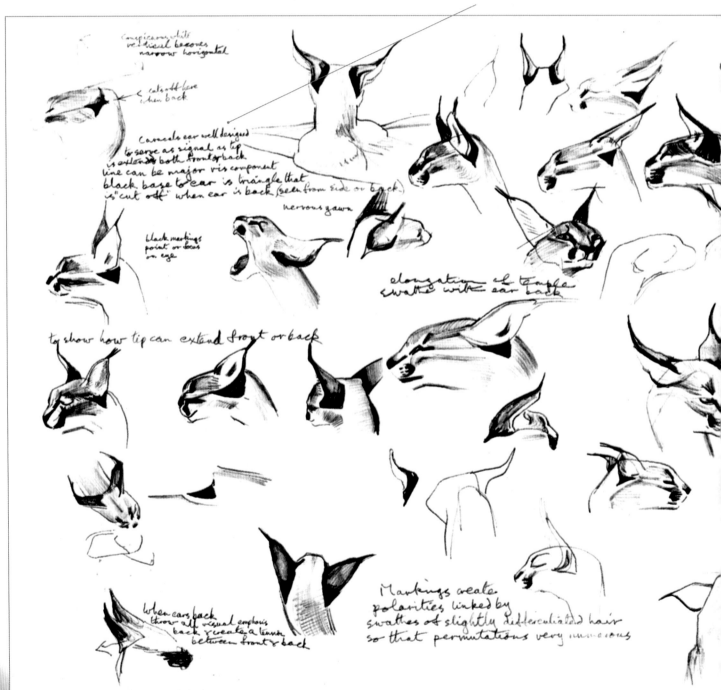

Visual Thinking and Diagrams

We have long considered this little sketch a masterpiece. Teaching students diagramming skills is teaching them to think and express ideas on paper.

Sketching Animals

Animal encounters are rich learning opportunities. Animals also move. Students can use this as a practice opportunity by making many drawings showing different poses. As soon as the animal moves, they can shift to work on a different drawing. Each drawing will capture a different view and level of detail of the subject. This is OK. Drawings from multiple perspectives will lead to deeper, more thorough observations.

179

INSIDE OUT

Students observe a mushroom and draw internal and external views to describe its shape. Then they learn how these views correspond to engineering and architectural plans, and think about how to apply these strategies to future journal entries.

Time

Introduction: 10 minutes
Activity: 30–50 minutes
Discussion: 10–15 minutes

Materials

☐ Journals and pencils

☐ One button mushroom per student

☐ Cutting board or paper towel and a knife (for cutting the apple)

☐ Plastic serrated knives (one per student)

For Extension

☐ One apple

☐ Engineering or architectural blueprints showing plan, elevation, and section of the same object (download from the internet)

Teaching Notes

Being able to use images to clearly display data and ideas is an important part of visual and scientific literacy. The discussion questions at the end of the activity are key to students' process of making sense of the experience and identifying ways to incorporate these strategies into their future journal entries.

Architects and engineers have developed a clear and adaptable system for describing objects with pictures. Instead of drawing a single portrait, they make measured drawings of the top or bottom view (plan) and the side view (elevation). Section views show internal structures by slicing the object from different angles and drawing the view perpendicular to the cut. Architects and engineers can also construct a 3-D drawing of the object as seen from another angle. This powerful approach is easily adapted to nature journaling. This strategy for diagramming can become a part of students' nature journal tool kit, offering a simple way to capture visual information.

Definitions

Plan view: the view from directly overhead, or the traditional map view. The plan view may show the outside of the object (top or bottom) or the internal structure.

Elevation view: the side view, at 90° to the plan view, as seen from the outside. The elevation may show the long axis (side) or short axis (end).

Section view: the cut view, showing the internal structure. A longitudinal section cuts through the long axis of the object. A transverse, or cross section, cuts through the short axis of the object. Reference lines drawn on the plan view show how sections correspond to the other drawings.

NATURAL PHENOMENA

We use apples and mushrooms for this activity because they have a distinct top and bottom and show different patterns in longitudinal and transverse sections. They are also relatively easy to cut with a serrated plastic butter knife. You can also lead this activity in the field with wild mushrooms. Remind students not to taste or eat unknown wild mushrooms.

PROCEDURE SUMMARY

1. Use the smallest possible number of drawings to fully describe the shape and structure of a mushroom.

2. You will need to show different views, and you may cut your mushroom to see what these different views look like.

DEMONSTRATION

When the whiteboard icon appears in the procedure description: Draw a side view of an apple or mushroom, then add other views, including cross and longitudinal sections. *Note:* Unlike in other activities, this demonstration is done during the extension, not the introduction.

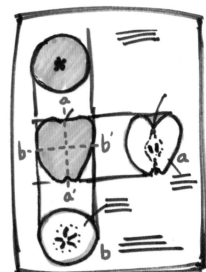

PROCEDURE STEP BY STEP

1. **Pass out one mushroom to each student, then challenge them to figure out the minimum number of diagrams they need to fully describe it, asking them to briefly discuss their ideas with a partner.**

   a. "In a moment, it will be your challenge (should you choose to accept it) to use diagrams to describe the shape and structure of the mushroom."

   b. "What is the minimum number of drawings you would need to fully describe a mushroom's shape and structure? What angles or views would you show? Discuss this with the people next to you."

   c. "You will have fifteen minutes to develop this set of drawings, and after that we will discuss everyone's ideas and approaches."

2. **Pass out plastic knives and tell students that they may use the knives to cut the mushrooms to reveal the inside, but that, before they cut, they should think carefully about which angles would be the most useful.**

3. **As students work, take time to circulate, troubleshoot, and engage students in dialogue about how they are choosing to make their diagrams.**

   a. As you circulate, ask students questions such as "How does the angle of the cut change the view that you see?" or "Why did you choose to show the mushroom from that angle?"

DISCUSSION

Lead a discussion using the general discussion questions and questions from one of the Crosscutting Concept categories. Intersperse pair talk with group discussion.

General Discussion

Call the group back together and ask students to discuss their approaches using the questions here, first in small groups, then with the whole group.

   a. "Please gather with a small group, then each present your drawings and explain why you chose the views you did, then notice differences and similarities in your diagrams."

   b. (With the whole class) "Let's look at how all of us as a group chose to draw our mushrooms. Discuss with a partner: Are there any views that most or all of the groups chose? Why might those views have been so useful? Can you think of any descriptive group names that we could use to categorize these views?"

Patterns

   a. "Were there any patterns in the individual repeating parts of the mushroom?"

   b. "Were there any exceptions to these patterns? Where were they? Why do think these exceptions may have occurred?"

   c. "What are some possible explanations for the patterns of growth we saw?"

   d. "Was there a pattern to the way the individual parts fit into the larger structure?"

Structure and Function

   a. "Look at one of the structures you drew in your diagram, such as a seed or a stem. How do you think this structure works? Why might it be shaped the way it is? How might its form, texture, or color affect how it functions?"

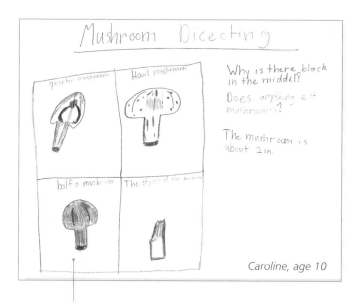

Caroline, age 10

This page includes writing, drawing, measurement, and questions. These were introduced in a previous exercise. Students build skills with each new journal project.

Words help reinforce and explain the drawing. On its own, the meaning of the gray mark is not clear. With the written note, it is clear that it represents a hollow space.

This pale smudge is from a mushroom stain!

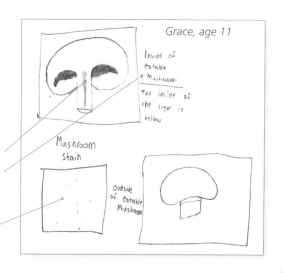

Grace, age 11

EXTENSION: PLAN, ELEVATIONS, AND SECTIONS

1. **Describe how architects and engineers use plan (top view), elevations (side views), and sections (cut views) to show objects. Cut an apple to demonstrate and label different views, and then pass out examples of architectural or engineering plans for students to observe.**

   a. "If we draw this apple from the side and the top, we see different shapes and structure. If we cut the apple in half from top to bottom (so that you get two similar pieces), we see a view of the inside and core that looks like this. This is called a longitudinal section."

   b. "If we cut the apple across the middle (so that the stem and the circle of the sepals are on separate pieces), we get a view that looks like this. This is called a cross section. All of these views and sections show you different aspects of the shape and structure. You can use these views and sections to describe any object."

   c. "Here is an example of [architectural or engineering] plans. Architects and engineers use plan (top view), elevations (side views), and sections (cut views) to describe objects and organize data efficiently on paper."

   d. "Some of these views are similar to those you all used to describe your mushroom."

   e. "With your group, observe the plans in front of you and discuss what you notice. What different kinds of features are revealed in each view?"

2. **Point out how students can now use this approach to diagramming as a way to sketch objects in future journal**

Tracing is a great way to draw something at its actual size (or slightly larger). Indicate scale by writing "life size" or 1/1 next to the drawing.

entries, and ask them to discuss how they might use drawings to describe a nearby object.

   a. "When we study other natural objects, such as shells or flowers, or any human-made objects around us, such as tea cups or pencils, we can use some of these views to show our observations. Sometimes you might choose to use all these different views; sometimes you might just use one or two views."

   b. "With a partner, look at an object nearby and discuss what views would be the most helpful in describing the object, and how you could show and organize that on your paper."

FOLLOW-UP ACTIVITY

Bite by Bite

A fun way to extend this activity and practice drawing is to give each student a whole apple, let them draw it, then tell them to make successive drawings as they eat the apple down to the core.

A cutaway three-quarter view is challenging to construct, but shows structure and texture of the inside and outside more than any other view. The ability to visualize and draw such views is evidence of increasing visual literacy, and develops with practice.

NATURE BLUEPRINTS

Students observe a complex natural object, break it down into component parts, and create a labeled diagram that reveals its symmetry and structure.

It can be overwhelming to draw complex subjects like pinecones or a tower of lupine flowers. Diagrams are tools for exploring and quickly describing structures in a way that is clear and easy to understand. Giving students a basic approach to diagramming is giving them a process they can fall back on when they encounter a complex subject. In the field, this is both a time-saving device and a way to achieve clarity and precision in drawings. Think of the blueprints that an architect makes to build a house. They are comprehensive and detailed, but there is no need to draw the details of every joint if they are repeated throughout the structure; the architect can just show it once. Exploded views show the parts of an object and how they connect, and detail insets highlight important features of a subject. Once your students have this formula for diagramming in their back pocket, they can focus on observation instead of monotonously drawing repeated shapes.

Definitions

Detail: a separate drawing that is an enlargement or elaboration of a part of the structure. This is often used on small, repeated parts such as a flower stamen.

Exploded view: a diagram that shows the relationship or order of assembly of the parts of an object.

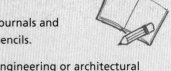

NATURAL PHENOMENA

For this activity, you will need a collection of natural objects with repeating patterns or structure, such as a patch of flowers or a cluster of cones under a tree. This activity is best led once students have completed *Inside Out* or have some experience with diagramming. Once students understand the principles of diagramming, they can advance to more complex objects with repeating parts, as modeled in this activity (e.g., pinecones, fern fronds, lupine flowers).

PROCEDURE SUMMARY

1. To draw a diagram of a complex object with repeating parts: Make a few drawings to show different views of each repeating part, look at how the parts fit in to make the object as a whole, then use diagrams and text to describe how the parts fit together.

2. Make a diagram using the method just described, then record additional observations.

DEMONSTRATION

When the whiteboard icon appears in the procedure description: Create a diagram of a pinecone that shows the details of one scale and how the scales fit into the larger structure of the cone. If students have already practiced making side-, end-, top-, and section-view drawings of an object, use those

Time

Introduction: **10 minutes**
Activity: **30–50 minutes**
Discussion: **10 minutes**

Materials

- ☐ Journals and pencils.

- ☐ Engineering or architectural blueprints showing detail insets and exploded views. You may also collect examples of instructions for Lego kits or furniture assembly.

Teaching Notes

This activity is a great follow-up to *Inside Out.* Students can build on their learning about different views of an object by applying it in the context of drawing a more challenging subject.

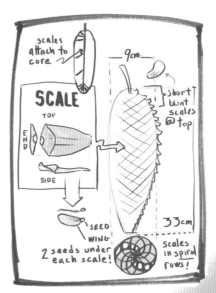

ideas in this activity. Point out to students how you are avoiding drawing the entire cone and instead using a couple of drawings of the repeating parts as a shortcut.

FAIRIES AND TROLLS

A playful way to introduce this activity is to talk about creating blueprints or design plans for fairies. Tell the students that there is an emergency in the forest. Every evening, the forest fairies get out their tools and make new flowers for the next day. But last night the trolls stole all the plans to make the flowers. The forest fairies need your help. Before these flowers wilt, we need to create new design blueprints for the flowers. The goal is to make plans that are as clear and detailed as possible. To this end, students should consider how to use plan, elevation, section, and exploded views, and detail insets.

Most students will understand that this is an imaginary scenario. We do not want to leave the children thinking that there really are fairies and trolls in the woods. If students are very young or seem confused, make it clear that this is pretend.

PROCEDURE STEP-BY-STEP

1. **Explain that scientists and engineers use diagramming to efficiently draw complex objects, such as pinecones [or your chosen natural object].**

   a. "Drawing a pinecone like this one can be difficult. There are so many scales that your hand may cramp up before you get halfway through, or you might just get sick of drawing them. You might get so wrapped up in drawing the same thing again and again that you do not have time to make other observations."

   b. "Scientists and engineers have a great trick to help them handle this problem. It is called diagramming. Instead of drawing to make a likeness of an object, we draw to record as much information as we can in a way that is fast and clear. First, we need to break the object down into parts. Then we look at how the parts fit together."

2. **Show students examples of exploded views and inset details in blueprints or engineering diagrams.**

3. **Pass out one pinecone [or another complex natural object with repeating parts] to each student, and instruct them to examine it, looking for repeating parts.**

   a. "First, let's identify the repeating structures in this object. Are there parts of this object that repeat, or show up more than once or twice? What are they?" (Students might say: seeds, scale things, or spines.)

4. **Demonstrate how to diagram one example of each repeating structure.**

   a. "Let's start by using a couple of drawings to describe a scale [or other repeating part]. What views would help us show its shape and structure?" Make quick sketches representing student suggestions, being sure to include at least the top, end, and side view of the scale.

   b. Repeat step "a" with any other repeating structures (seeds, etc.).

5. **Ask students which views can be used to show the structure of the full cone or object, referring back to the activity *Inside Out* (if you have done it previously) and adding students' suggestions to the whiteboard demonstration.**

   a. "What views will help us show the structure of the cone or object as a whole?"

   b. Make quick sketches representing student suggestions, showing at least the side and end views.

6. **Explain and demonstrate how to use diagrams to show how the repeating parts of an object fit together.**

Top and side views of the same subject help depict the shape and structure.

Green and *brown* are inadequate words to describe color. How can we coach students to modify color words to be more specific?

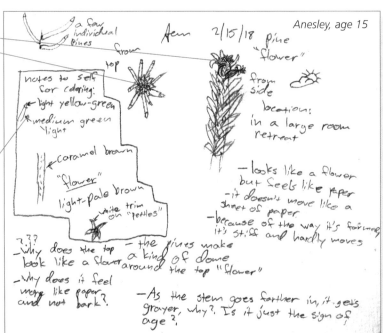

Making a structural diagram helps put the emphasis on recording information rather than making a pretty picture. Show side, end, and cross-section views just as is done in architectural and engineering diagrams.

a. "Now that we have made diagrams showing the scales and the pinecone as a whole, we will use some simple drawings to show how the parts fit together."

b. "What do you see? How do the repeating parts like scales and seeds fit in with the structure of the cone as a whole? Where are they in relation to each other?" (Students might say: scales in diagonal rows, two seeds under each scale, scales attached to a central core.)

c. Make a couple of quick diagrams based on how students describe the repeating parts fitting in to the object as a whole. For example: "Where the cone has opened, I can see how the scales attach to a central core. We can use a cross section to show this view."

7. Summarize the general steps you took to make a diagram of a complex subject: First, identify and draw repeating structures from different views, then use diagrams to show how the repeating parts fit together to form a whole.

8. Tell students that they will now make their own set of diagrams, using labels, descriptive writing, and measurements to describe the object.

a. "You will describe an object with a set of labeled diagrams in your own journal. Look for the easiest ways to describe the most detail and structure without having to draw everything."

b. "Remember, first we identified repeating structures and made a couple of simple drawings of different views of each one. Then we observed how the repeating parts fit together in the object as a whole, and used diagrams to show this."

c. "Your goal is not to draw a perfect representation of the object, or a pretty picture. A diagram is meant to show information efficiently. If many parts have the same structure, you only need to draw one. This is not about art but about seeing details and finding simple and effective strategies to get them on paper. Think, What are the structures, and how do they fit together?"

d. "Add in labels, descriptive writing, and measurements. Once you finish making your diagrams, see how many more observations you can add. You want to create a page that is dense with information and is as clear as you can make it. You will have nineteen minutes to create your diagram."

9. About halfway through, bring the group together to discuss diagramming strategies and ideas in small groups.

a. "Let's pause for a moment to get some good diagramming ideas. At my signal, in groups of ten, lay your journals open on a table and share your diagrams with each other."

b. "Look for effective diagramming ideas: Maybe someone summarized a lot of information with a minimum of effort, or included useful views or ways of showing relationships between drawings or notes."

c. "We don't care if a diagram looks pretty or not. We are interested in the data or information and how clearly it is displayed."

d. "Look for ideas that might improve your own diagram. Once you have seen one another's work, you will have a few more minutes to improve your own diagram using these ideas. "

10. Give students time to use ideas from their classmates to improve their work.

a. "Now take a few minutes to improve your diagram using some of these ideas from your classmates. How can you make it more clear, or show more information?"

11. Call the group together, and ask them how they could use diagramming strategies in future journal entries.

DISCUSSION

Lead a discussion using questions from one Crosscutting Concept category. Intersperse pair talk with group discussion.

Patterns

a. "Were there any patterns in the individual repeating parts of the subject of your journal entry?"

b. "Were there any exceptions to these patterns? Where were they? Why do you think these exceptions may have occurred?"

c. "What are some possible explanations for the patterns of growth we saw?"

d. "Was there a pattern to the way the individual parts fit into the larger structure?"

Structure and Function

a. "Look at one of the structures you drew in your diagram, such as a scale or a stem. How do you think this structure works? Why might it be shaped the way it is? How might its form, texture, and color affect how it functions?"

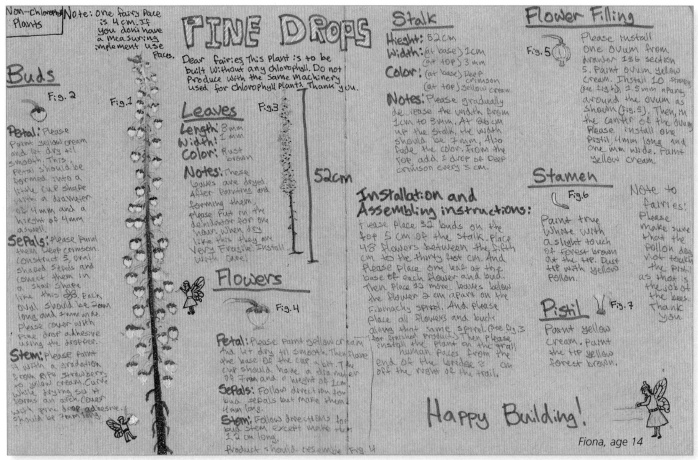

This student created an instruction manual for flower-making fairies. This was a playful angle that can inspire follow-up creative writing projects.

INFOGRAPHIC

The best way to learn anything is to teach it. Infographics combine writing, drawing, and other visuals to explain an idea or synthesize information. Students can start an infographic in the field, then research the topic and add to their journal entry and their understanding.

An infographic is a chart, poster, drawing, or diagram that uses words, pictures, icons, and data to explain an idea or offer information. Making a clear infographic takes as much thought as putting together a good lesson plan. The process guides students through studying a topic and coming up with words and images to explain and describe the phenomenon, leading them to synthesize information. Creating an infographic requires two steps. In the field, students observe and describe the phenomenon, leaving space to add more material later. Back in the classroom, they research the topic and add more information and background to their journal page. This is an opportunity to build visual literacy and to focus on layout skills, such as integrating drawings with other forms of note taking, and thinking about page structure and how to communicate information.

NATURAL PHENOMENA

Students can make an infographic about any interesting discovery in nature, such as a found object, plant, animal, or phenomenon. This activity can be a great way for your class to build common understanding of a part of nature and set the foundation for learning related science ideas in a lesson or unit of study. If you want to use the activity in this way, make sure that students all focus on the same type of organism or phenomenon. Alternatively, students can decide what they want to focus on and search for a part of nature they find interesting.

PROCEDURE SUMMARY

1. Record observations and a description of a species or phenomenon in the field.

2. Research the topic using books, journal articles, or internet resources.

3. Add at least three things learned from your research.

4. Add supporting elements (decorative borders, arrows, frames, titles, subtitles, etc.) to the page.

DEMONSTRATION

When the whiteboard icon appears in the procedure description: Block out areas to place the title and three things learned in research. As you discuss making observations with students, sketch a plant in the center with lines suggesting writing or numbers. Then add some supportive elements, such as frames, titles, subtitles, and arrows to represent connecting ideas on the page.

Time

Introduction: 10 minutes
Activity: 30–50 minutes
in the field and another 60 minutes
for research and completing the
infographic
Discussion: 10–15 minutes

Materials

☐ Journals and pencils

☐ Examples of info-
graphics (clip from magazines,
or download from the internet)

Teaching Notes

You can find great examples of infographics in *National Geographic*, popular science and engineering magazines, newspaper articles (often in the science section), and online. Start a clip file or folder of good infographics. Having a lot of examples will inspire your students and give them many ideas about ways to use images to explain concepts.

Students will need to identify plants or animals for this activity. If students cannot identify species in the field, make sure they collect enough information in their notes to be able to identify the species later from reference material.

This activity can be the basis of a longer project for which students do more extensive, long-term research on a phenomenon or part of nature.

PROCEDURE STEP-BY-STEP

1. **Show students examples of infographics, then ask pairs to discuss how the infographics were designed and how they communicate information.**

   a. "These are examples of infographics. They are carefully planned drawings and diagrams that explain ideas."

   b. "In groups of four, analyze two infographics. Try to figure out what the designer is attempting to communicate. Are there any techniques or methods that they use to help explain things clearly? How do they help you see the most important ideas quickly?"

2. **Explain that students will create their own infographics based on observations in the field and research back at school.**

   a. "You are going to make your own infographics about something you find in nature. Take your journal and find and describe an interesting [plant, mushroom, animal, or phenomenon].

   b. "In the center of your page, use words, pictures, and numbers to richly describe what you see. It is OK if you don't know what species you have found. You will be using your notes later in class to try to identify what you have seen, so be precise and careful in your note taking."

   c. "Later, you will add information to this page based on research you conduct. Leave some blank space at the top and along the bottom of the page for titles and information you will uncover."

3. **Give students time to make and record observations in their journals, circulating to offer support to any who are struggling.**

4. **Return to class, library, or home and offer opportunities to research the subject. Start by trying to identify the observed object based on field notes. If students cannot identify the species they have described, they should make a checklist of the kinds of details they should look for next time to make a more complete description. Then they can add questions, "It reminds me ofs," and titles to their page.**

5. **Encourage research in books, journals, magazines, and online sources, and through expert interviews.**

   a. "Now let's do some research to find out as much as we can about the subjects of your infographics." Methods will vary depending on resources available and students' abilities.

   "When you find something really interesting, write it down and note the book or web page you learned it from."

6. **Tell students to use words, pictures, and numbers to describe the most interesting or relevant facts or ideas they learned through their research.**

   a. "Now use words, pictures, and numbers to show the most interesting or relevant things you learned about this species. Your challenge is to show this information in the most clear and memorable way you can."

   b. "You can use icons or graphics and get creative if you want."

7. **Tell students to cite the sources of the new information in the journal.**

   a. "Write the name of the book or other resource you got the information from next to the fact on the page."

8. **Tell students to add borders, titles, names, scientific names, and other features to make the page easier to scan and understand.**

   a. "Now add some elements to the page to make it easier to scan, and to make the information easier to understand."

Angelica, age 14

Make a clear distinction between information that is directly observed and information from a secondary source.

b. "These could be titles, underlines, arrows, boxes, decorative borders, or icons and graphics."

c. "Think about the infographics we looked at earlier. Are there any design elements from those examples that you could include here?"

DISCUSSION

Lead a discussion using the general discussion questions. Intersperse pair talk with group discussion.

GENERAL DISCUSSION

a. "What are some of the best ways to make an infographic clear and understandable?"

b. "How did you use drawing, writing, and layout elements on the page to show your thinking and ideas?"

c. "If you were to do another draft of this infographic, what might be some approaches to layout or other elements you would want to change?"

d. "Infographics are a fun and interesting way to show what you've learned about something. What is a topic you know a lot about? It doesn't have to be related to school—it could be about a hobby, such as cooking, playing sports, music, and so on. Talk with a partner about what you would include in an infographic about this topic."

e. "When we made our infographics, we were careful to add in the name of the book or article we got information from. Scientists also do this when writing a research paper or communicating about their ideas. Why do you think this is important to do?"

FOLLOW-UP ACTIVITY

Make Your Own Infographic

Give students the opportunity to continue practicing their communication skills by allowing them to make an infographic about a subject that is interesting to them, or a topic they feel they know a lot about.

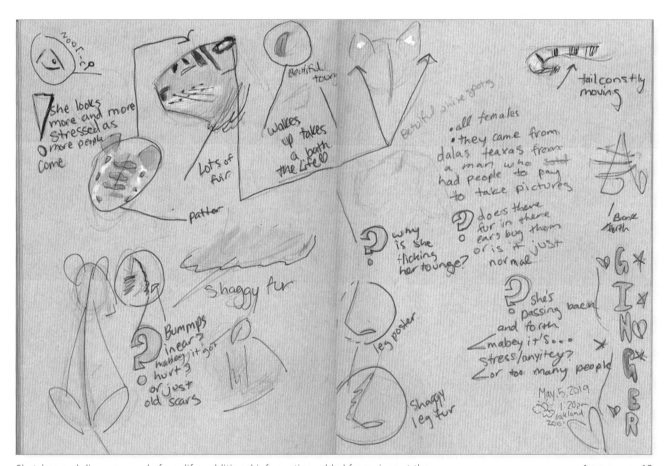

Sketches and diagrams made from life, additional information added from signs at the zoo

Amaya, age 13

PHOTO, PENCIL, AND FOUND-OBJECT COLLAGE

Students make a collage by combining detailed observations (using writing, drawing, and numbers), a photograph, and parts of found natural objects. They then compare drawing and photography as ways of recording data.

Time

Introduction: 10 minutes
Activity: 30 minutes in the field and 30 minutes back in the classroom/ home
Discussion: 15–20 minutes

Materials

- ☐ Journals and pencils
- ☐ Digital or instamatic camera
- ☐ Color printer (back in classroom or home)
- ☐ Glue sticks and tape

Teaching Notes

Using a camera can encourage observation and memory, or shut it down. When people passively take a "snapshot" of an object, they tend to pay it little attention and remember few details. However, if they actively use a photograph to highlight the aspects of an object that they find the most interesting, their memory of the object is improved, even of details they did not focus on.

We can help students learn to take photographs deliberately, using the process as an opportunity for observation. This activity forces students to make decisions about what to show in the photo and how to frame the shot, taking into account the kinds of information that photography is best at capturing. This additional effort will help them remember the phenomenon more deeply and create a more complete record of it.

Photography and drawing are both powerful tools in the natural sciences. Neither one is necessarily better. They are different, each with its own strengths and weaknesses. Drawings are selective; the artist chooses which details to include or exclude. They also take time to produce and require deep observation. Photographs evenly record data across the image without bias, can be produced instantaneously, and do not necessarily require us to observe; however, photography can also become a valuable naturalist's tool and an opportunity for observation.

While waiting patiently to photograph a heron in a certain pose, you can notice its typical postures, preferred locations, and common behaviors. When photographing a streambed, you must decide how far away or close up to take the photograph, excluding or including different details. In the process, you are more likely to notice these details. When we discuss how to approach taking photos thoughtfully with students, cameras can become tools for active observation, used together with drawing and writing to create a powerful record of information.

In this activity, students draw and take a photo of the same phenomenon, then compare the types of information that each method captures. Students engage in discussion about the strengths and weaknesses of each medium, and the types of information each is ideal for capturing.

NATURAL PHENOMENA

In this exercise, groups of four to five students will observe a natural object together. Find some small to mid-sized objects that can be easily drawn, photographed, and sampled by these groups. A shrub with flowers or fruit works well. If collecting is not an option, students could omit this portion of the activity and instead focus on a landscape feature, plant, or other phenomenon.

PROCEDURE SUMMARY

1. Mark a place on the page to add a photo.
2. Use words and pictures to record observations of a subject.
3. Make a decision with your group about how to photograph the object.
4. Paste the photo into your journal.
5. (Optional) Paste a found object into your journal.

DEMONSTRATION

When the whiteboard icon appears in the procedure description: Draw a double-page spread of

Amaya, age 13

Tape photos and interesting found objects into your journal.

journal pages. Mark off a place to paste a photograph on the left side. Trace around a leaf or other found object to reserve a place to add the object to the page. Demonstrate using writing, drawing, and numbers to record information in the remaining space.

PROCEDURE STEP-BY-STEP

1. **Tell students that they will use drawings and photographs to record observations of a natural subject, then ask students to list the kinds of details that are captured through each modality.**

   a. "We are going to explore a phenomenon using drawings and photographs to record our observations."

   b. "Each of these approaches captures a different kind of information. Please discuss with a partner: What kinds of details can you capture in a photograph? What about a drawing with written notes?"

   c. "We are going to compare these two ways of recording information as we explore a natural object."

2. **Give students a moment to block off a rectangle for a photograph in their journals.**

   a. "Open your journal to a new double-page spread. On the left side, block off a four-by-six-inch rectangle. We will paste a photograph into this space when we return from the field."

3. **Tell students to divide into groups of four or five, select an object to focus on, discuss how to record information about it in their journals, then begin using words, pictures, and numbers to document what they notice.**

   a. "In a moment you will divide into groups of four or five. You and your teammates will observe an object and discuss how to best record information about it in your journals."

   b. "Then, each of you should use words, pictures, and numbers to document in your own journals what you notice about the object. Leave the rectangle for the photograph blank."

4. **Explain the procedure for taking turns with the camera, and how groups should thoughtfully make decisions together about the angle, focus, and details of the photograph they will take.**

   a. "As you work, I will come around with a camera."

   b. "So often, we take snapshot photographs without taking the time to slow down and make decisions about what we are trying to capture."

   c. "As a group, you will be able to take one photograph of your object. You will need to figure out what view and angle will be the most useful for describing the object or some significant detail on it. You will also need to decide how close up to take the photograph. These decisions will affect what you are able to capture in the photograph."

   d. "You will need to make these decisions together, so listen to one another's ideas, and share the reasoning behind your ideas."

   e. "For example, you might say, 'I think the holes in the petals are an important detail, so I think we should take a photo at this angle, because it will show the holes well.'"

5. **Tell students to begin observing and recording notes in their journals, and discussing their approach to taking the photograph.**

6. **While students are working, circulate among the groups, asking them about what angle to take the shot from and how zoomed in the shot should be. Then take the photograph with a digital camera and record which group's photo it was.**

7. **(Optional) Once students have completed their observations and field notes, have them glue or tape a collected sample of their natural object into their journals.**

   Note: Wet plant samples should be pressed and dried in the pages of a thick, heavy book (phone books work well) before being glued into a journal.

   a. "Please take a moment to collect one or more parts of the real object that can be taped or glued into the journal. Samples that are relatively dry and flat work the best."

b. "Glue or tape your found object into your journal. Add written notes describing what it is and any key observations about it."

8. **(Back in class or at home) Print out photos and instruct students to cut and paste them into their journals.**

9. **Point out how field scientists use both photographs and written notes intentionally to capture different kinds of information, and encourage students to do the same.**

   a. "Field scientists sometimes use both written notes and drawings, as well as photographs in their journals."

   b. "Using both together, and making thoughtful decisions based on what each medium is best at capturing, can lead to a more complete record than using one alone."

   c. "We can also use both of these approaches in our journals, adding photographs from time to time when they will help us capture significant information."

   d. "These two approaches do not replace each other; they are just different ways of capturing information."

10. **Give students a moment to integrate their photographs into their written and drawn observations by adding arrows, circles, and more written notes to clarify observations and connect them to ideas.**

   a. "Let's take a moment to integrate our written notes, drawing, and photograph."

   b. "This will help clarify our observations and make for an even more in-depth record of this phenomenon."

   c. "Draw circles, arrows, and written notes directly on top of the photograph to highlight the most important features of the object."

   d. "You can also use arrows, labels, written notes, or icons to connect what you have already recorded through writing and drawing to what is shown in the photograph."

   e. "For example, you might use a line to connect a feature you recorded through drawing to where it appears in the photograph, or use writing to describe the difference in the angle of the drawing vs. the photograph."

DISCUSSION

Lead a discussion using the general discussion questions and (optional) one of the Science and Engineering Practices questions. Intersperse pair talk with group discussion.

Note: We recommend discussing the whole series of "General Discussion" questions to fully conclude the activity.

General Discussion

With students in small groups, prompt them to discuss the questions here and think about the types of information each medium is best at capturing:

a. "What kinds of information did you capture in your photograph? What kinds of information or observations were you not able to record?"

b. "What did you learn or notice about the object through photographing it?"

c. "What types of information and observations did you capture through drawing and writing? What kinds of information or observations were you not able to record?"

d. "What did you learn or notice about the object through drawing and writing to describe it?"

Lead a group discussion about the strengths and weaknesses of written notes vs. photographs.

e. "For scientific note taking, what are the strengths and weaknesses of written notes and drawings, in comparison to photographs?"

f. "What kinds of information can be captured in a photograph, but not in written notes or a drawing?"

g. "What kinds of information can be captured through written notes and drawings, but not a photograph?"

Obtaining, Evaluating, and Communicating Information

a. "When do you think scientists might choose to record information in a photograph during their studies? Why?"

b. "When do you think scientists might choose to use words, sketches, and numbers to record information?"

c. "When would it be difficult or inappropriate to collect samples of things you observe?"

d. "What forms of note taking or recording information are more useful in communicating or explaining ideas to someone else?"

e. "How could we combine different approaches for recording information to communicate ideas?"

f. "What are some ways we might thoughtfully include photography in future nature explorations or journal entries?"

FOLLOW-UP ACTIVITIES

Finding Opportunities for Collages

Encourage students to look for other opportunities to use both photography and collected samples in their journal. They may find inspiration from looking at collage journals or scrapbooking resources online.

Looking at Examples

Guide students to look at journals of scientists and naturalists (the Museum of Vertebrate Zoology from the University of California archives is a good source for this) to see examples of journal pages that integrate photographs with text and drawings to record information. Engage students in conversation about why the authors chose to use these different ways of recording information. Many scientists use photographs to take a snapshot of an exact location of an observation they made, or to record a specific individual's markings along with a drawing, or to capture a landform feature they will study over time. Giving students the opportunity to see how scientists use photographs in this way will offer them ideas about their own journaling.

Working in Photography

In the future, give students the opportunity to occasionally use photography in their journaling. This could take place during an activity such as *Mapping,* where students snap a photograph of the site they map afterwards; this could also work with *Landscape Cross Section.* Or, in *Change over Time,* students can use photos to capture details along with the drawing. The use of photographs should not replace drawing in the journal; photos should be included thoughtfully as a way to augment what is recorded through drawing and writing.

NUMBERS

QUANTIFICATION AND

MATHEMATICAL THINKING

NUMBERS: QUANTIFICATION AND MATHEMATICAL THINKING

FINDING THE NUMBERS, FINDING THE PATTERNS

Numbers reveal patterns and significant details. Students can learn to find the numbers behind their observations through participating in the activities in this chapter. Quantifying distances, numbers of species, or processes of change will add another layer of understanding and intrigue to students' journal pages. Approaches for organizing and analyzing data will prepare students to use numbers to construct explanations and gather evidence. The more comfortable that students feel manipulating numbers, the more they will use them as a tool in their journals and beyond.

ACTIVITIES IN THIS CHAPTER

NUMBERS AND QUANTIFICATION

Students can learn to "find the numbers" within their observations, and practice counting, measuring, timing, and estimating. Once students are exposed to basic techniques, they can continue to build their skills as they use numbers in their journals and creatively adapt their approach.

Math and quantitative thinking are often taught out of context. Many of us learned math isolated from other subjects, with fabricated problems as the main form of practice. What if, instead, math provided a new lens for seeing the world, and was a tool for figuring out how things work?

Nature is an authentic and dynamic context in which students can develop and apply mathematical thinking skills. We can give students physical and mental tools that enable them to count, measure, time, and estimate their way through the world. This can result in an explosion of inquiry during their nature explorations and a proliferation of data in their journal entries.

Adding numbers to a journal page is one of the quickest ways to spark questions and intrigue while studying a phenomenon. Once a student starts measuring anemones in a tide pool, the questions start flowing: "What's the average size of anemone in the area? What are the largest and smallest ones we can find?" "Why are there so many smaller anemones over there, and larger ones over there?" "What causes the variations in size?" "Could it be related to their age, their proximity to food, some environmental factor?" When we count, a whole other set of questions emerges: "Is there a carrying capacity for anemones in a tide pool, or are their numbers more affected by reproduction and dispersal?" Quantification leads to deep and interesting questions, reveals patterns, and sparks intrigue. It is also a skill needed for answering questions and collecting data.

Students can integrate numbers into their entries alongside words and pictures, using numbers to build on observations and spark inquiry. They can discuss numbers nature in three ways: counting, measuring, and timing. Students can record their data precisely or, if this is not possible, make estimates.

BEST PRACTICES

In their journals, students

- Use numbers to show observations (counting, measuring, timing, estimating), discover or describe patterns, generate questions, and answer questions.
- Use numbers to quantify features of drawings or written descriptions.

- Collect samples of data (i.e., make more than one measurement) and record sample size when describing variation.
- Standardize sampling methods (i.e., use the same methods for measurement each time).
- Randomize samples to avoid bias.
- Display data using histograms, bar graphs, stem-and-leaf plots, or scatter plots.
- Manipulate data (calculate ratios or averages, simplify fractions, etc.).
- Use numbers or symbols to create formulas to explain or describe a phenomenon. (*Note*: This is appropriate only for older students who have learned how to use formulas to describe real-world phenomena.)

The activity *Hidden Figures* is a great way to introduce students to different approaches to quantification in the outdoors. Offer students time before this activity to write down their biometrics and familiarize themselves with how to use rulers, meter sticks, stopwatches, and other measuring tools, as described in the rest of this chapter.

> "If you want to ask questions, start counting."
> —Todd Newberry

COUNTING

The easiest way to start adding numbers to our journals is to start counting. We can count numbers of individual organisms (e.g., ducks or newts). We can also count parts of organisms (e.g., petals, leaves, or spines). We can count clouds, numbers of rough rocks vs. smooth rocks in a part of a stream, or repeating landscape features (e.g., ridges on a mountain). Students should record the time and location of their data collection. If there is a note that says "25 ladybugs," that could mean the number of insects seen that day, on a single branch, or in a half hour of observation under an oak tree. The only way to prevent confusion is to be explicit.

MEASURING

Recording the exact size of an object enables us to compare it to other similar things, helps us track change over time, and brings a degree of precision to our observations and journaling. Encourage students to carry a ruler or a measuring tape with their journaling supplies. You can also print out rulers and paste them

into student journals. Relative measurements are useful when there is no ruler or yardstick. For example, a
ranger in the Sierra Nevada observed a wolverine-like
animal across a lake. In his notes, he recorded the size
of the animal relative to boulders it was near. Later,
when the animal had gone, the ranger measured the
boulders, enabling him to more accurately gauge the
animal's size.

Metric vs. Standard

Most of the world has adopted the metric system of
measurement. It is consistent, easy to understand,
and easy to use in calculations or conversions. This
is the preferred system for scientific investigations.
The US uses standard units of measure. One difficulty
with the system is that there is no consistency among
the units of measurement, making calculations confusing. Inches are divided into sixteenths; there are
twelve inches in a foot, three feet in yard, and 1,760
yards in a mile. This is one reason why the US mili-
tary abandoned standard measurements in favor of
the metric system, in spite of native familiarity with
standard units. Metric units can be confusing at first
to those who do not have an intuitive sense for them. With a
little practice, you and your students can develop comfort with
metric measurement. It is worth the effort.

Teaching Students to Measure with Biometrics

Teach students to use the length of their body parts to help measure objects when they do not have a ruler. If they measure and
record the lengths of their stride, they can measure distances
in nature. See more about setting up a biometrics tool kit on
page 203.

Measuring with Degrees of Arc

We can estimate the spacing of objects in the sky, at sea, or on
the landscape using our outstretched hands. What is the distance
between the sun and the horizon? Where is the bird on the fence
post, relative to the barn? We make these measurements in terms
of the number of degrees of a circle that separates them (for
example, "Raptor on the fence post 25° to the left of the barn.")

TIMING

Anything that moves or shows a periodic behavior can be timed.
Time how long a loon stays underwater after it dives, or the times
between dives (does it tend to come up with fish after longer or
shorter dives?). Time how long it takes for clouds to move past
a landmark, such as a mountain or building. Time the interval
between owl calls. Time how long a squirrel spends doing one
behavior before it changes. Place a small branch near the head
of a slug moving across the forest floor; 1 minute later, measure
how far it has gone to calculate the slug's speed in centimeters
per minute.

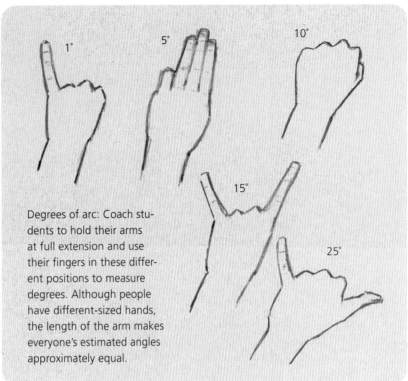

Degrees of arc: Coach students to hold their arms
at full extension and use
their fingers in these different positions to measure
degrees. Although people
have different-sized hands,
the length of the arm makes
everyone's estimated angles
approximately equal.

Practice estimating time by having students count seconds out
loud while watching a digital clock for 1 minute. Instruct them
to keep time with hand motions as they do, feeling the beat of
the seconds. Then test the students' sense of time by having them
sit at their desks, close their eyes, and estimate 1 minute, silently
counting but keeping time with the rhythm of their hands. When
they guess a minute has passed, they raise a hand and open their
eyes to check the time. After a couple of rounds, most students
will be able to accurately estimate seconds without looking at a
clock.

ESTIMATING

Sometimes it's not possible to make an exact measurement or
count. We cannot approach a whale with a measuring tape,
or count a flock of a thousand birds. In this case, an educated
guess is better than nothing at all. Teach students to round their
numbers to the level of precision of their estimation. If they
see approximately 500 geese in one field by counting by clusters of 50, then see 25 more, they would not report 525 geese,
as this would imply an exact count. Instead, they would keep
their estimate rounded to the nearest 50. Another way to handle uncertainty is to give a range instead of an exact measurement (e.g., "There were 500–550 geese"; "The whale was 15–20
meters long").

Look for natural opportunities to practice estimating numbers.
Students could, for example, guess the number of people in a
room or at a bus stop, then estimate the number by counting by
groups, then do a precise head count.

Estimating Percent Cover

Estimating how much of the sky is cloudy or how much of a hillside is covered with maples is difficult, but this can be useful information to collect. Just as we have to practice estimating large numbers, we lack experience in estimating percent cover. Students can look at percent cover circles to improve the accuracy of their cover estimates. If they are not sure of the percent cover, they can give a range.

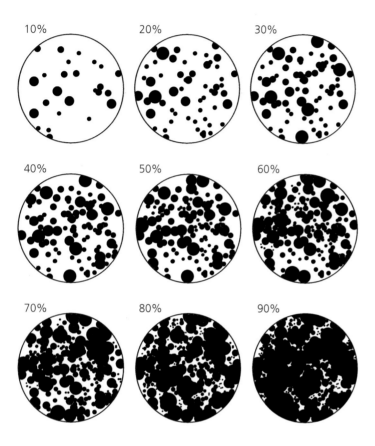

Estimating Numbers

Counting by groups is a good way to estimate numbers. Coach students to count a group of 10 individuals, visualize what 10 looks like, then count by 10s. Count larger groups by grouping 50s, 100s, or any other increment.

Looking at a card that shows 50, 100, and 500 dots helps us recalibrate our estimations. Students can frequently refer to these charts when estimating large numbers. Offer copies of the Cut-and-Paste Quantification Tool Kit for students to paste in their journals and use in the field.

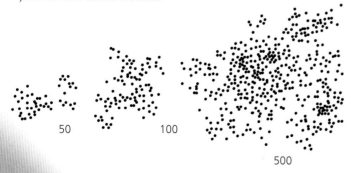

VISUALIZING DATA

Visualizing the spread and central tendency of data is a helpful tool for understanding observations and noticing patterns in data, but calculating the mean and standard deviation is time consuming in the field, and it requires background knowledge in statistics to understand the results. Scatter plots, histograms and bar graphs, tally bar charts and tally histograms, and stem-and-leaf plots are easy ways to collect and display the shape of data. Teach the following data visualization tools to students in the classroom, then encourage them to look for opportunities to use the approach in the field.

Scatter Plots

Scatter plots are a useful tool to help scientists and students see the relationship between two related variables, where one factor may affect or influence the other. For example: time of day and number of birds singing (counted every 10 minutes starting 1 hour before sunrise to 1 hour after sunrise); air temperature and the number of cricket chips per second (counted over a 30-second interval); or heights of goldenrod plants and estimated numbers of aphids on each plant. The variables (sets of data) are plotted on an x-y coordinate graph. Plot the variable that is thought to change because of the other on the vertical (y) axis. A ball of random points means no relationship between x and y; a cloud of points that forms a line sloping up (from the lower left to the upper right) is a positive correlation (the higher the x, the higher the y); a cloud of points that points down is a negative correlation (the higher the x, the lower the y).

Histograms and Bar Graphs

If you collect measurements, speeds, or any other variable in a data table, you can convert the data to a histogram or bar graph by stacking each observation from the baseline. You can do this most easily by collecting the data in round numbers, such as to the nearest centimeter. Once you collect your data, create a number line from your lowest value to your highest and stack the observations out from the line, using one square for each measurement. For best results, students should draw same-size boxes for each observation, with no space between them.

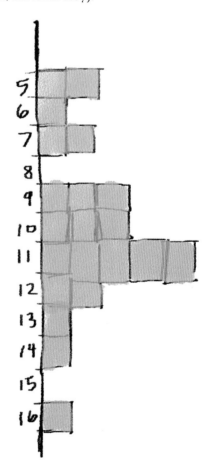

200

Tally Bar Charts and Tally Histograms

If students will be counting different types of objects (birds, types of flowers, etc.), instruct them to make a vertical list of the objects. Next, have them draw a vertical line down the right side of the list. As they count, students add tally marks to each category. Make sure you clarify that students will need to keep the tally marks an even distance from one another, aligning each tally mark or group of five vertically, as illustrated here. The result is a set of numbers that also visually shows the spread and any clusters of data.

Fiona is a young naturalist in California—some pages from her nature journal are featured on pages 59, 62, 64, 98, 186, 210, and 221 of this book; when she was 13, she innovated a way to use tally bar charts to record color gradients. Make a vertical column of color swatches and, to the right of the column, tally the number of individuals that show each color. Encourage your students to play around with collecting other types of data in tally bar charts.

Stem-and-Leaf Plots

Stem-and-leaf plots preserve original numerical data and visualize it at the same time. They are appropriate for sets of numbers that students might collect—for example, the height of all the wildflowers in a patch of the yard, the diameter of spider webs, or the number of seconds between owl calls. As with the tally bar chart, the length of the data rows form a graph, showing the shape and overall pattern of the data. Use the example shown in the box "How to Make a Stem-and-Leaf Plot" to teach the approach to your students, or find an opportunity for students to create a data set using their own measurements. Younger students may need more scaffolding and support to develop, understand, and integrate stem-and-leaf plots into their journaling.

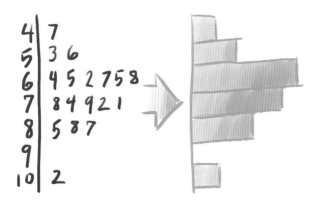

Random sets of measurements often create a normal (bell-shaped) curve. This is not always the case. What is the shape of your data? If several students sample the same kind of objects, they will get similar but different distributions. This is a great opportunity to discuss variation and sampling.

HOW TO MAKE A STEM-AND-LEAF PLOT

Students do not need to calculate the mean and the standard deviation to understand their data. Use this process to teach students to plot their numbers as they record them, creating a picture in the form of a bar chart.

We are going to learn a strategy for recording data that creates a picture of it, because it can be hard to make sense of a long stream of numbers, and it takes time to calculate averages.

Let's imagine that we observed plants in two locations—on a ridge and in the forest—and noticed that the plants on the ridge tended to be shorter than those in the forest. You could stop there with this general observation, but by measuring the heights of one kind of plant at both locations, you may be able to describe this pattern more clearly. If you just measured the plants and recorded your findings as a list, you would have a jumbled mess like this:

RIDGE: 24 36 41 22 16 42 37 35
30 4 16 54 7 66 42 34
54 23 21 44 32 48 43 31
23 18 10 4 54 72 33 24
34

FOREST: 25 38 63 54 49 43 36
41 89 41 62 51 94 71
77 82 64 58 104 66 57
51 40 42 56 32 13 53
92 74 57

If you are good at statistics, you might calculate an average, but that is hard to do in the field, and you end up with a number that is meaningless to many people. There is a better way. Instead of collecting your data and then analyzing it, enter the data directly into a stem-and-leaf plot as you record it. Here's how.

1 Construct the frame of the stem-and-leaf plot. The stem is the column of numbers on the left side. The stem represents the digits that are in the tens and hundreds place.

FOREST
0
1
2
3
4
5
6
7
8
9
10
11

2 Now we begin to add "leaves" as we measure each plant. When you take a measurement, add it to the ones place on the right side of the graph. If our first number is 25, the 2 in the stem represents the twenty in the tens place. The 5 represents the 5 in the ones place.

FOREST
0
1
2 5
3
4
5
6
7
8
9
10
11

3 Continue measuring and adding to the list. The next measurements are 38, 63, 54, 49, and 43. Observe how each is represented in the chart on plot. The 3 of 43 is added next to the 9 of the previous 49. As you add numbers, try to keep your writing a consistent size, and the numbers a consistent distance away from each other horizontally.

FOREST
0
1
2 5
3 8
4 9 3
5 4
6 3
7
8
9
10
11

4 When you are done, you have made a graph! The plot is a histogram of the data. A quick glance reveals that the data centers somewhere around the 50s, with a spread from 13 to 104. The plot is clear and intuitive to read, and while you were measuring, you also identified a general pattern.

FOREST
0
1 3
2 5
3 8 2
4 9 3 1 1 0 2
5 4 1 8 7 1 6 3 7
6 3 6 2 4 6
7 1 7 4
8 9 2
9 4 2
10 4
11

5 To compare the ridge with the forest, simply create another plot on the other side of the stem. You can now see that the plants not only are shorter on the ridge but also have a narrower range of heights.

RIDGE		FOREST
	0	
4 7 4	0	
0 8 6 6	1	3
4 3 1 3 2 4	2	5
4 3 1 2 4 0 5 7 6	3	8 2
3 8 4 2 2 1	4	9 3 1 1 0 2
4 4	5	4 1 8 7 1 6 3 7
6	6	3 6 2 4 6
2	7	1 7 4
	8	9 2
	9	4 2
	10	4
	11	

MAKING QUANTIFICATION TOOL KITS

Bring measuring tools outside. With the right tools, students can find numbers behind every observation.

MORE THAN A HAMMER

If the only tool you have is a hammer, every job looks like a nail. The more tools your students have in their quantification tool kits, the more they can do. If they have a measuring tape, they will find opportunities to measure. With a watch, they will time things.

QUANTIFICATION TOOLS

Portable, practical measuring tools are a great addition to any journaling kit. Here are a few favorites:

- Short ruler

- Retractable measuring tape: Often sold in sewing stores, these tapes show centimeters on one side and inches on the other. (Use centimeters whenever possible.)

- Goniometer: Used by physical therapists to measure joint movement, these tools are easily adapted to field use to measure angles more easily than with a protractor. What is the angle of a slope, a leaning tree, the branches of a plant, or cracks in the mud?

- A watch with a second hand: If you can time a phenomenon, you can time observations and convert counts to rates (e.g., counting the number of seconds a slug takes to go 10 centimeters, and dividing to get cm/second.

- Vernier calipers: This tool is used for more accurate measurements of small objects.

BIOMETRICS

Capturing students' biometrics is a part of setting up their quantification tool kit. Set up a biometrics workshop and help your students record their personal measurements. Remeasure biometrics at least once a year while children are growing. Decide whether to use metric units (preferred) or inches and feet. Then have students measure the following:

- Height: total height and navel height.

- Length: span of outstretched arms; length of the finger and hand measurements in the same positions as used for degrees of arc; boot length (for toe-to-toe measurements), and personal cubit (distance from tip of middle finger to elbow).

- Distance: the distance an individual student travels in one step—their pace or stride. This is best measured over a long distance (100 feet or 50 meters) and divided by the number of normal (not exaggerated) steps. This will give you an accurate average step length. Multiply the pace by 10 to get distance in 10 paces. Challenge students to figure out how to calculate

how many steps it would take them to walk 10, 50, or 100 feet if they are using standard units or 1, 5, and 10 meters if they are using the metric system. Then students should write these measurements into their journal so that they can use them again and again.

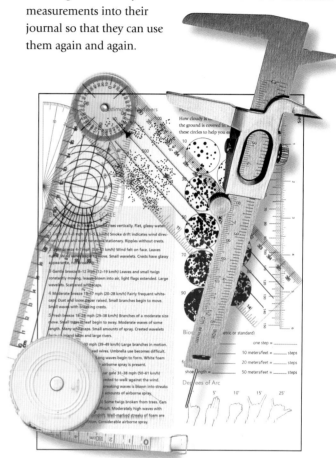

Cut-and-Paste Quantification Tool Kit

Photocopy and paste the quantification tool kit (appendix F) into the back of every student's journal. The tool kit is sized to fit in a standard composition book. Do not enlarge or reduce the photocopy, or the scale of the ruler will be distorted.

The first time you use any new tool with students, it will be a distraction. Before bringing tools into the field, spend some time in class familiarizing the students with how to use each tool and having them practice with real objects.

ADDITIONAL RESOURCES

Mathematical Mindsets: Unleashing Students' Potential through Creative Math, Inspiring Messages, and Innovative Teaching, by Jo Boaler

The Impact of Identity in K–8 Mathematics Learning and Teaching: Rethinking Equity-based Practices, by Julia Aguirre, Karen Mayfield-Ingram, and Danny Martin

SHOW WHERE THE NUMBERS COME FROM

Numbers are a way of describing the world. Students should show where the numbers come from, so that the data can be used to deepen their understanding.

Chloé Fandel

"These two pages record a variety of field measurements, all referenced to the map on the upper left. There are several other pages with columns of data for the other locations on the map. The map is extremely important, because it shows where each data column comes from, and which bodies of water are connected to each other. For the measurements on the right-hand page, instead of writing down "length of tube 134 cm, depth of hole 38.5 cm," etc., I drew a diagram [of the instrument], because it's more specific and less likely to cause confusion later. I also calculated some results quickly in the field, which we checked later using Excel. These [field calculations] are to see if the results are reasonable. Sometimes they are not, and then I know that I made a mistake and need to re-take measurements while I am still in the field."

Maps and Context

A map puts data in context. The location of each sample is essential to seeing and understanding patterns.

Recording Methods

In a study, it's important to record how data is collected, not just the numbers. Using a diagram to show the tool used to capture data is one way to show the procedure.

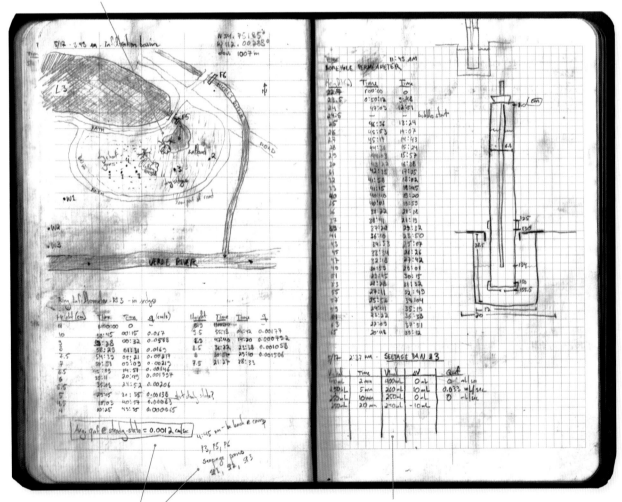

Quick Calculations

An average, calculated in the field, can spark questions, reveal patterns, and help catch measurement errors.

Tables and Columns

Data doesn't need to be fancy. Simple graphic organizers, such as columns and tables, can help students structure their data and scan it easily when they review it.

USE NUMBERS TO FIND THE PATTERN

Numbers are another language with which to explore patterns and describe what is seen. Using diagrams and consistent measurement procedures sets students up to use their data to deepen their understandings of nature.

Nalini Nadkarni

"These notes are from a project that involved estimation of the total biomass of canopy-dwelling communities, with a focus on canopy-dwelling plants (epiphytes) and the arboreal soils they create in the canopy. To keep track of which branches we inventoried, we drew an image (stick figure) of each tree, and visually numbered each of the branches, and estimated the length of each. We multiplied the number of branches of a given length (1 m, 2 m, 3 m…). Then we added these up per tree (the columns on the right page). We entered the data into our files. We used a red marker to note which numbers had been entered."

Using Numbers to Understand Processes

Making measurements and collecting numerical data lead to a more precise record of patterns that can reveal information about underlying processes and phenomena. We can coach students to use numbers along with words and pictures to learn about their surroundings.

Metadata

Without the context of the specific tree visited, the date, and the time, the entry loses context essential to the analysis that will come later.

Tables for Organizing Data

Tables structure and systematize data collection. If observations are recorded haphazardly, it's easy to forget a key piece of data.

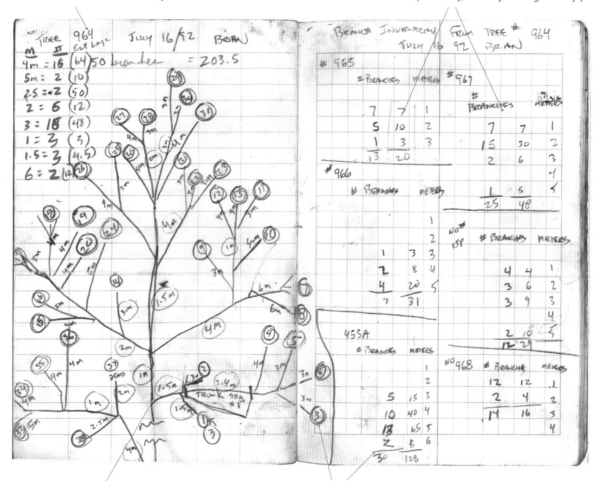

Diagrams vs. Drawings

Diagrams capture the most relevant information. The goal is not to make a pretty picture or a portrait but to record observations. Here the goal is to show the number and relative positions of branches on a tree. Drawing the whole tree in detail would take away from the focus of the data and project.

Contrasting Colors

Red pencil shows data that has been transferred to a permanent file. This makes it easier to keep track of your process.

HIDDEN FIGURES

There is a matrix of mathematics behind everything we see. Students can learn to find the numbers or quantify their observations to reveal and explore unseen patterns.

Time

Introduction: 10 minutes
Activity: 30–50 minutes
Discussion: 10 minutes

Materials

☐ Journals and pencils

optional

☐ Stopwatches or watches with second hands

☐ Measuring tapes/rulers (metric preferred)

☐ Protractors or goniometers

☐ Cut-and-Paste Quantification Tool Kit page

Teaching Notes

Many students have math anxiety. Help them reframe the fear. Numbers are just another language to describe the world. The goal at this introductory stage is to practice different approaches to quantification, not to follow a strict process in order to reproduce a calculation; do not penalize students for a wrong calculation in this context.

Stem-and-leaf plots and tally histograms (see p. 202) are great approaches for visualizing data. Teach students about these strategies in math class, so that they can apply them in the field in this activity and in other journaling experiences.

There are numbers behind our observations. Math is as beautiful as the world it describes. If you have an aversion to math, it is likely because you learned it outside of any meaningful context. If applied thoughtfully, mathematics is a simplified and symbolic language with which you can describe the world. On a plant, for example, the number of petals, its overall height, the distances between nodes, the number of visitations by pollinators, the number of aphids per leaf, and its speed of growth are all quantifiable measurements, and they all reveal something interesting about the plant. Teaching students to find the numbers behind their observations gives them a way to use math to reveal patterns they otherwise might not notice. They will be able to apply these skills in their future explorations and journal entries, adding another language and level of precision to the page.

Make a decision about how far to go with this activity. Introducing counting, measuring, timing, and estimating all at once might be too much for your group. If so, you might choose to focus on only one skill per outing. If your students are new to journaling, consider running this activity without journals, so that students only have to focus on one new skill at a time.

NATURAL PHENOMENA

The goal of this activity is for students to practice quantifying parts of nature. Any event can be timed, any object can be measured, and there are things to count everywhere. Any area with different natural organisms and features will work.

PROCEDURE SUMMARY

1. Count [time, measure, or estimate] things around you, trying to find different ways to quantify your observations.

2. Be creative; there is no penalty for being wrong. Numbers are just another way to show observations.

3. Use your journal to record what you quantify.

DEMONSTRATION

When the whiteboard icon appears in the procedure description: Draw a sample journal page showing use of writing and drawing. Then add measurements, counts, scale, stem-and-leaf plots (if students are familiar with them), and other numbers.

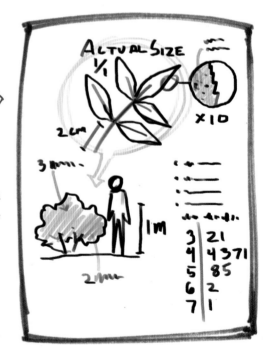

PROCEDURE STEP-BY-STEP

1. **Explain that numbers, like words and pictures, are a way of describing observations; reference a nearby object in nature as an example.**

   a. "We can show observations through words and drawings. We can also record observations through numbers."

   b. "Look—over there, I'm noticing that there is a clump of gopher holes. I could say, 'There is a clump of gopher holes,' but I could also go count the number of gopher holes, or measure how wide each one is. This is a deeper level of observation."

2. **Tell students that they will practice using numbers to record interesting observations of their surroundings, learning simple tricks and strategies.**

   a. "It takes practice to see the numbers behind our observations. We are going to learn some simple approaches to quantifying our observations that you will be able to apply in our future explorations."

   b. "Right now, our only goal is to use numbers to show or describe our surroundings. As long as you are doing that, you are successful."

3. **Explain the value of including numbers in nature study.**

   a. "Using numbers is helpful. It reveals interesting patterns and leads us to questions. For example, if I measure the diameter of all the gopher holes and notice that they all fall within a certain range, I can wonder: What might be causing that pattern? Would I expect to see something different at another time of year, when the population of gophers might be a different size or when the soil is a different texture?"

   b. "These techniques for finding numbers give you another way of exploring and figuring things out in nature."

Note: Offer one or more of the quantification techniques described in the next sections, giving students time to play around with the technique(s). Make a decision about how many of these techniques to introduce in a session. With a younger group, you might focus on just one skill (e.g., measuring) and give students extended practice and support using it. To support groups that need even more structure, give students specific things in nature to quantify (e.g., "Count all the gopher holes you can find. Now measure their diameter"). Give more independent students a few minutes to practice using each quantification approach (measuring, timing, counting, and estimating) and then a longer period of time during which they can play around and choose which approach to use. Estimation can be more challenging for younger students, so consider introducing it later, in a context where you are able to spend time devoted to offering specific approaches and dedicated practice.

Measuring

1. **Explain how students can use measuring to find out the lengths or distances between things, referencing nearby examples and referring to any available tools.**

   a. "First, we are going to look at measuring. We can measure almost anything we find! Getting data on the lengths or distances between things can reveal interesting patterns."

   b. "For example, you could measure the size of leaves on a bush, the distance between leaves, the length of a scar on a tree, the distance between plants, the width of holes in the ground, or the distance from tree to riverbank in different areas."

   c. "Try to choose clear points to start and end your measurements. For example, measuring the distance between branches and starting your measurement in middle of a branch could cause confusion—measuring from where both branches emerge from the tree has a clear start and end point."

   d. Offer any tools you have for measuring, reminding students to be careful and to share materials.

2. **Tell students to pick a nearby natural object and measure as many parts of it as possible, recording their data in their journal and adding sketches, writing, and questions if necessary.**

   a. "Go pick something to focus on and write and draw about it a bit in your journal. Then see how many different parts of it you can measure, and record your measurements in your journal."

   b. "Numbers are great for leading us to questions. Intentionally ask questions based on your measurements as you go. What is the most interesting question that comes to you?"

   c. "If you want to switch topics and measure something else instead, that's fine, too. You have four-and-a-half minutes. Go!"

3. **After students have had some time to measure, but before they lose interest, call them back and debrief measuring, leading a brief discussion using the questions here.**

   a. "How did that go? What kinds of different things did you measure? What was your most creative measurement?"

   b. "Did you notice anything that surprised you? How did measuring help you learn about what you were looking at? Did your measurements lead to any questions?"

Timing

1. **Explain how students can use timing to record anything in motion, referencing nearby examples and appropriate tools and techniques for your setting.**

a. "Let's look at timing. We can time anything in motion."

b. "For example, can use timing to learn about animals. If we see an animal repeating a behavior, we can time the intervals between behaviors. If we see a hawk flying above us, we can time how long it soars before flapping its wings. We can time how long the sparrow's call is. If we see a duck dive underwater, we can time how long it is until it comes up."

c. "If there are interesting weather patterns, we can also time them. For example, we can record the amount of time between gusts of wind, or how long it takes one quickly moving cloud to pass over the horizon."

d. "If you have a stopwatch, you can use that to record timing. If you don't have a stopwatch, you can count the seconds yourself. A second lasts about as long as it takes to say, 'one hippopotamus.'"

2. **Tell students to find things to time nearby, and to record their data in their journal, asking questions and using writing or pictures to show their observations when necessary.**

a. "Find things nearby to time."

b. "Record what you time in your journal. Intentionally ask questions based on your measurements as you go. What is the most interesting question that comes to you?"

c. "If written notes or drawings would be helpful, you can add them. You will have four-and-a-half minutes to time whatever you can. Go!"

3. **(Optional) Call students back and introduce the process of using time and distance to calculate speed; ask students how they might time the speed of a banana slug, then explain that they must calculate distance over time.** *Note:* **This will be most appropriate for older students.**

a. "How might you calculate the speed of something like a banana slug?" (Hear student responses.)

b. "For this type of timing, you must measure the distance that the slug covers over a period of time. You could put a stick next to the slug as a marker, and a second stick after one minute. Then, measure the distance between the sticks."

4. **After students have had some time to measure, but before they lose interest, call them back and debrief timing, leading a brief discussion using the questions here.**

a. "How did that go? What kinds of different things did you time? What was the most creative way you thought to time something?"

b. "Did you notice anything that surprised you? How did timing help you learn about what you were looking at? Did your timing lead to any questions?"

5. **Explain that for faster processes, a stopwatch is a good tool, but for longer processes, such as a leaf changing color, a calendar is better.**

a. "Some things, such as a leaf changing color, occur on a time scale that is too long for us to observe with a stopwatch, but we could time this phenomenon and other longer-term processes over the course of days or weeks, with a calendar instead of a stopwatch."

b. "Consider the units for your timing when you write down numbers in your journal. When might seconds be appropriate? When might we use days or weeks?"

Counting

1. **Explain how counting is a way of observing that reveals interesting patterns, and point out nearby examples of what students could count.**

a. "As soon as we start counting, we reveal patterns that we can then think about and use to better understand our surroundings."

b. "We can count numbers of individual organisms, such as the number of lizards sunning on the rock, the number of anemones in each tide pool, or the number of spider webs on a tree. We can also count the number of parts of organisms, such as the number of pine needles in each bundle, the number of spines on an oak leaf, or the number of spots on different ladybugs."

c. "You can also count nonliving parts of the landscape, such as holes in the ground."

2. **Explain that while counting, it's important to include the context. (For example, is it the number of ladybugs found on one single bush? across a field? over the course of a day?)**

a. "When you record what you have counted, you need to add some context. For example, for a number of ladybugs, you need to specify: Is it ladybugs found on a bush, across a whole field, or seen over an entire day?"

3. **Tell students to find things to count nearby, recording their data in their journal, asking questions, and using writing or pictures to show their observations when necessary.**

a. "Try counting as many different things as you can. Record what you find in your journal and add writing and drawing if it is helpful. Include any questions that come to you."

4. **After students have had some time to count, but before they lose interest, call them back and debrief counting, leading a brief discussion using the questions here:**

a. "How did that go? What kinds of different things did you count? What was the most creative way you thought to use counting?"

b. "Did you notice anything that surprised you? How did counting help you learn about what you were looking at? Did your counting lead to any questions?"

Estimation

1. **Explain when and how students might want to use estimation to gauge large numbers, distances, percent cover, and other values that would be difficult to count or measure.**

   a. "Sometimes it isn't possible to count or measure."

   b. "Maybe there are so many leaves it would be difficult to count them all, or maybe the birds are moving too fast to count each one. Or maybe we want to get a sense of how much of the sky is clouded over, and counting the number of clouds will not reveal this information."

   c. "In these situations, 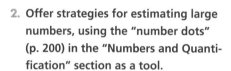 we estimate. Estimation isn't just guessing; it's using a specific approach to make as accurate of a guess as possible."

2. **Offer strategies for estimating large numbers, using the "number dots" (p. 200) in the "Numbers and Quantification" section as a tool.**

   a. "One approach for estimating large numbers of individuals is to count just ten, see what that looks like, then count how many chunks of ten seem to be present."

   b. (Show the "number dots" on p. 200.) "You can also use this chart that shows what ten, fifty, one hundred, and five hundred look like to help you estimate."

3. **Offer strategies for estimating percent cover, using the "percent cover" chart (p. 200) in the "Numbers and Quantification" section as a tool.**

   a. (Show the "percent cover" chart on p. 200.) "To estimate something like the amount of sky covered by clouds or the amount of ground covered by snow, use this chart to make a quick guess."

If you and your students find some insects on a leaf, these are examples of the kinds of quantification questions you can ask:

- How many bugs?
- How many instars (age/size classes)?
- How many bugs in each stage of development or life-cycle phase?
- What proportion of the plants in the area are infested with bugs?
- Can you find insect-damaged leaves?
- Can you quantify levels of leaf damage?
- Is there a correlation between number of bugs and leaf damage?
- Can you group the bugs by color?
- How many bugs are in each color class?
- Does color correlate with size?
- Are there more bugs above or below the leaves, at the top or base of the plant, or in the shade or sun?

Then ask why…

4. **Tell students to go out and quantify nearby parts of nature, counting exactly if possible, estimating to determine large numbers, and recording estimates in their journals.**

   a. "Go out and find some things to count or estimate. If you are able to count something exactly, go for it. If you can't, then use estimation to help you. Notice in what situations it is better to estimate than count."

b. "Record what you find in your journals. Keep your estimate rounded off. If you count around a hundred ducks and then see two more, Say 'about one hundred,' not 'one hundred and two.'"

5. After students have had time to estimate but before they lose interest, call them back and debrief estimation, leading a brief discussion using the questions here.

   a. "When did you choose to estimate something rather than get exact data?"

   b. "Did you notice anything that surprised you? How did estimating help you learn about what you were looking at? Did your counting lead to any questions?"

Quantification Practice

1. Instruct students to explore the nearby area, using their new quantification skills to record data about anything that is interesting to them, and including words or pictures when necessary.

   a. "In a moment, you will get to apply your quantification skills by exploring anything that is interesting to you."

   b. "Record observations in drawing and writing, and look for ways to count, estimate, measure, or time what you are looking at. This takes creativity. See if you can discover interesting or unexpected opportunities for quantification."

c. "You could use all these approaches to quantification or focus on just one that you think is particularly fun or interesting."

d. (Optional: if you have taught students how to visualize data—e.g., graphs, histograms, scatter plots) "Whenever possible, visualize your data."

2. Explain that there is no penalty for being wrong or making a mistake, because the goal is to see how many ways students can use numbers to describe their surroundings.

   a. "The more relevant numbers you find, the better, but there is no penalty for making a mistake. Just try things out. Our goal is to see how many ways we can use numbers to describe our surroundings."

3. Ask if there are any questions, reference any available measuring tools, and send students out to practice quantification.

   a. "When we return, we will see how we found the hidden figures in our observations. You will have ten minutes. Does anyone have any questions before we start?"

   b. (Optional) Set out an array of measuring tools that students may use. You may want to introduce the tools one at a time in class under more controlled conditions. (New tools are often a distraction.) Set clear expectations about bringing the measuring tools back at the end of the study.

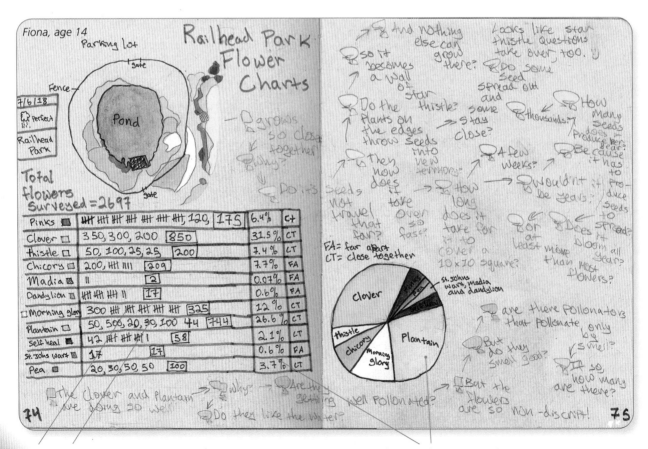

Counts and approximations of numbers of plants in a study area, tied in to a color-coded key and map

Plant numbers converted to percentages and displayed on a pie chart

210

DISCUSSION

Lead a discussion using the general discussion questions and the Science and Engineering Practices (SEP) questions. (For more information about SEPs, see page 239 in the section "Journaling and the NGSS.") Intersperse pair talk with group discussion.

General Discussion

Call the group back and debrief the experience, asking students to share what they noticed through quantification, and any questions that arose.

a. "Look at your numbers. What did you notice? Did quantification help you reveal any interesting patterns? What were they?"

b. "Did you come up with any interesting questions as you were quantifying? What were they?"

c. (If students didn't come up with any questions) "Now use your quantitative data to come up with as many questions as possible. We can use quantification to reveal patterns and ask questions."

Using Mathematics and Computational Thinking

Ask students questions about the kinds of observations they were able to make using numbers, and how they might integrate numbers into their future journal entries.

a. "What did you notice in your explorations? What did you learn by using numbers to describe your observations?"

b. "What could you measure, count, estimate, or time while doing a journal entry on plants? animals? the weather?"

ADVANCE OR FOLLOW-UP ACTIVITIES

Practice Estimation

You can have students practice estimating numbers by throwing beans on a blanket, doing a quick gut estimate, making an approximation roughly counting by 10s, 50s, or 100s, and then making an exact count. This will help students develop the skill of quick estimation and will also be an opportunity to see how accurate your students are. Do they tend to estimate too high or low?

Collect Biometrics

Measure and record students' metric or standard biometrics for measuring without tools: shoe length, distance traveled in one regular step (averaged over 100 feet or 50 meters), distance traveled in ten steps, knee height, navel height, total height, arm span, number of steps needed to travel 10 meters and 100 feet. Then go out and measure and estimate objects near your home or school.

Find the Numbers

Show examples of naturalists' or scientists' journal entries that use numbers along with other modes of recording information. Engage students in discussion about how quantification is used, what types of things are quantified, and how this works with other modes of recording information to show observations and thinking.

Record Data over Time

Engage students in long-term scientific studies that involve mathematical thinking and data collection. Connect field methods for gathering data to this activity.

BIODIVERSITY INVENTORY

Students record the diversity of species in two study areas and use graphs and diversity indices to describe and analyze the data.

Collecting biodiversity data and making species lists reveal patterns. Scientists use these approaches to gather information about ecosystems, assess how the distribution of organisms in an area has changed, and think about possible impacts of different disturbances or changes in environmental conditions. Students can use this activity to think about similar concepts, get a sense of the range of different species in an area, consider factors that influence organisms, and use their data to make predictions about what might happen should conditions change in the future. In this activity, students look for species, count individuals, then create pictures of the patterns. In the process, they develop an intrinsic understanding of the data and of what math can do, and learn a reproducible approach for gathering data.

NATURAL PHENOMENA

Find two natural areas with a diversity of trees, shrubs, and herbs. Make the two study areas the same size, and spend the same amount of time collecting in each. It's helpful if students have had some time to explore the area previously, so that they're familiar with the species there to some extent. Local parks are ideal locations. Vacant lots or unmowed edges of a ball field will also work. Limit the area of study based on the range of species there. The size of the area will depend on the density of plants and the ease with which you can find them. If the area is not very diverse, make the study area larger so that you can work with a bigger range of species. If there are a lot of species, make the area smaller so that students aren't overwhelmed.

Unless it is logistically impossible, plan to take two surveys in different, distinct areas. (The two observation periods could also take place on two different days.) A comparison will offer a wealth of information and provide a starting point for discussing possible causes of the variations. Find plant communities or habitats with contrasting environmental conditions or where one area has received a different "treatment" than another. For example:

- Plant communities in wet vs. dry areas

- Disturbed areas vs. restoration projects and natural areas

- Burned areas and unburned areas

- Areas that were burned in different years

- North- versus south-facing slopes

PROCEDURE SUMMARY

1. Make a biodiversity inventory of the species in the area.

2. Record some simple drawings and text for each species so that you remember it (but you don't need to know the name of it).

3. Compare notes with classmates if you think you are done. They might have found species you missed.

DEMONSTRATION 1

When the whiteboard icon appears in the procedure description: Use writing and drawing to describe different plant species, modeling making a simple diagram of leaves with a few written notes.

PROCEDURE STEP-BY-STEP

1. **Tell students that they will explore the area looking for different kinds of trees and shrubs, cataloguing each one in their journals.**

   a. "We are going to take twenty minutes to record every kind of tree and shrub in this zone."

   b. "You will use your journal to help you remember what you see."

2. **Tell students that they will use drawing and writing to briefly capture key details (leaf shape, fruits or flowers, form of growth) of each plant, but that they do not need to make a detailed, pretty picture of each one.**

   a. "The goal is not to make an exact drawing showing everything about each species you find but to capture enough information so that you or someone looking at your journal could identify the plant later."

   b. "Record important details such as leaf shape, any fruits or flowers, and the form of the plant growth. A fast way to capture information is to trace the leaf, then draw in the veins."

3. **Set up the boundaries of the survey area. As noted earlier, the size of the area will vary depending on the plant diversity, access, and hazards.**

4. **Tell students to begin by working alone, to focus on a few important details of each plant before moving on, and to compare notes with classmates once they think they have found all the species in the area.**

   a. "Start by working alone. You will have lots of time to find different species. Remember not to get lost in making a pretty picture of the plant. Make a quick sketch of the leaves along with some written notes, then move on."

   b. "When you think you have found all the species, compare your notes with a few of your classmates and see whether they can direct you to a species you may have missed."

5. **Send students out to do their survey and circulate to offer support as needed.**

6. **Repeat steps 2–5 to measure the biodiversity in a second area (either right then or at a later time).**

DISCUSSION

Lead a discussion using the general discussion questions and questions from one of the Crosscutting Concept categories. Intersperse pair talk with group discussion.

General Discussion

Call the group back together to discuss their notes, using the questions here.

   a. "How many species did you find in each place?"

   b. "What differences were the most striking to you when you saw the study areas?"

   c. "What are some possible explanations for the differences in species and numbers of species in these two areas?"

   d. "How might you expect the biodiversity of these two areas to change over time?"

Ask students how making a biodiversity inventory helped them learn about the area, and point out how students can catalogue the plants they find in other settings.

   e. "Looking at the biodiversity of an area can be an interesting way to study any place you go and to begin to understand it better."

   f. "How did making a biodiversity inventory of this place help you learn about it?"

Patterns

   a. "What differences did you notice between the two study areas?"

   b. "What kinds of differences do not show up when we use this way of collecting our data?"

   c. "What unseen forces may be behind some of the patterns we observed?"

d. "Look at the range and shapes of the leaves and plants you recorded in your journal pages. How are they similar to or different from one another? How do the leaf shapes and plants compare from one site to another?"

e. (*Note:* This question would require that students engage with a field guide or other resource.) "A species can only survive where its needs are met, and different types of plants have different needs. Knowing some things about the preferences of different plants can help us learn about the trends in environmental conditions in an area. Refer to this field guide [or other resource] to read about the plants you found in each area, and use this information to make some general statements about what you think the environmental conditions are like there."

Stability and Change

a. "Ecologists have found that ecosystems with more diversity tend to be more stable. Why might this be the case?"

b. "Which system do you think will change the most over the next five years? Why? What would cause the changes?"

Cause and Effect

a. "What were some of the differences between the two study areas?"

b. "What are some possible explanations for those differences?"

FOLLOW-UP ACTIVITY

Field Guide Identification

Students do not need to know the names of plants in order to have a general conversation about the biodiversity of an area, but identifying the plants that they have taken care to sketch can be rewarding and fun. Offer regional field guides and time for students to identify the plants they found.

EXTENSION

A Deeper Dive into Biodiversity: Richness and Evenness

Note: This extension is intended for seventh grade and higher.

Diverse ecosystems are more stable, productive, and better able to respond and adapt to environmental changes. Species richness and evenness are two ways of describing and quantifying biodiversity that offer an understanding of ecosystems. Measuring the number of species (the *richness*) in an area is a useful exercise, and is appropriate for younger students.

Richness leaves out an important part of the picture, however. Having an idea of the relative numbers of individuals within

each species gives you a more nuanced understanding of diversity. Two areas might have the same total number of species, but one might have high numbers of all species, whereas the other is dominated by one or two species. This second pattern often occurs after a major disturbance or the introduction of an invasive species. The measure of the abundance of individuals within each species is called *evenness*.

Species richness and evenness are two diversity indices that scientists use to describe natural systems. Students can visualize this data with graphs or, at higher grades, algebra and logarithms.

DEMONSTRATION 2

When the whiteboard icon appears in the procedure description: Add tally marks next to the notes about each species to record the number of individuals in the study area.

PROCEDURE STEP-BY-STEP

1. **Define species richness as the number of species found in an area, referring back to the systematic observation students just did.**

   a. "Species richness refers to the number of species you find in an area. We made a species richness inventory just now when we recorded the plants, trees, and shrubs in this area."

   b. "In your journal, write 'species richness' at the bottom of the page with the number of species you found."

2. **Explain how to collect species evenness or abundance data: Walk through the study site making tally marks next to each species for every individual observed.**

a. "Now we are going to collect abundance data about each of these species. You will walk through the study site and count how many individuals there are of each species."

b. "Use tally marks (or approximations if there are too many) to record the number of trees and shrubs of each species. Keep track of the number next to your sketches of the species." (Demonstrate on your whiteboard.)

3. **Ask students to share some ideas about how to be systematic in their counts to avoid marking the same plant twice or missing a part of the study area.**

   a. "You are going to need to come up with a way of being systematic about how you count the plants so that you do not count the same one twice or miss a section of the study area. Turn and talk with a partner about how you might do this."

   b. "What are ideas for how to make our data collection systematic so that you do not cover the same area twice or miss big areas?"

4. **Send students out to collect evenness data, circulating to provide support.**

   a. "You will have fifteen minutes to collect this data. Are there any questions?"

5. **When the time is up, call the group back together and explain how to analyze diversity richness and evenness by creating graphs: Give each species a letter code (A for the highest number of plants, B for the next highest, etc.), listing those codes on the x-axis of the graph, making the y-axis as high as the most abundant species, then creating vertical bars to show numbers of individuals.**

   a. "Give each species in your notes a letter code, starting with A for the species with the highest count, then B for the next highest, and so on."

   b. "Now let's use your diversity data to make a bar graph. On the horizontal (*x*) axis, list all the species you found by their letter code."

   c. "On the vertical (*y*) axis, make a scale that goes as high as the species that was the most abundant."

   d. "Now draw bars to show the number of individual plants you observed for each species."

6. **Give students time to make their graphs, then call for the group's attention and discuss patterns in their data.**

   a. "Look at your graphs. What statements can you make about the richness and evenness in each site?"

   b. "Based on your data, which ecosystem is more likely to be resilient in the face of disturbances? Explain your thinking."

   c. "How might the plants change in this ecosystem if there were some kind of disturbance or shift in environmental factors, such as a change in weather, the amount of water, and so on?"

   d. "Ecologists have found that systems which have higher evenness tend to be more stable and resilient when there is a disturbance. What are some possible reasons for this?"

7. **Use the Crosscutting Concept Scale, Proportion, and Quantity to examine the relationship of size and abundance.**

   a. "We did not directly take notes on this, but see whether you can pull any ideas from your notes or observations: Is there a relationship between the size of the species and its abundance? What are some possible explanations for this?"

   b. "Were there any species that may be important parts of this ecosystem even though there were not very many individual members of them? What is your evidence?"

Going Further: Constructing Rank-Abundance Curves

1. **Rank-abundance curves enable students to visualize richness and evenness. The calculations are straightforward and only require division and an understanding of proportions. The graphs are useful if you plan to ask your students to compare two or more communities.**

   a. "To begin, create a table. [You can refer to the one on this page to demonstrate what this table should look like, or reproduce this table on a whiteboard or poster.] In the first, left-most column, list all the species you found. You do not need to know the names of the species; you can use letters or your own descriptive names."

Species	Number in Sample	Number in Sample/Total	Rank
A	32	32/280 = 0.11	4
B	12	12/280 = 0.04	6
C	8	8/280 = 0.03	9
D	55	55/280 = 0.20	2
E	46	46/280 = 0.16	3
F	31	31/280 = 0.11	5
G	10	10/280 = 0.04	7
H	74	74/280 = 0.26	1
I	3	3/280 = 0.01	10
J	9	9/280 = 0.03	8
Total	280		

   b. "In the second column, write the number of individuals you counted in the sample. Add up the numbers in this column to get the total number of individuals you counted. Write this total at the bottom of the column."

c. "In the third column, calculate the proportional abundance of species by dividing the number of individuals by the total number of individuals in all the species combined."

d. "Assign a rank to each species starting with 1 for the most abundant species. If two species have the same number of individuals, choose one to be a higher rank, the other to be the next highest."

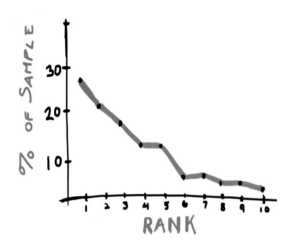

e. "Repeat this process for the second community or system."

f. "Graph the results with the rank on the horizontal axis and the percent of the sample on the vertical axis. To convert proportional abundance (number in sample divided by total) to percent, move the decimal point two places to the right."

g. "Now graph the second community on the same chart and compare the shapes of the two lines. How do the lines describe richness and evenness? What differences do you see between the two communities?"

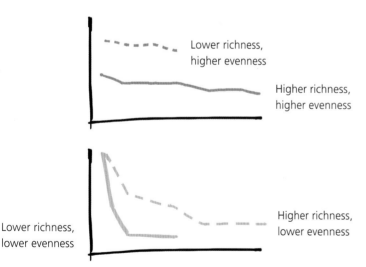

If graphed lines are roughly horizontal, the community they represent has high evenness, whereas lines that start high and then dip sharply indicate low evenness, as you might see in a highly disturbed area or a habitat being overrun by an invasive species. Short lines indicate low richness (fewer species). Longer lines show higher richness.

Going Further: Advanced Math—Calculating Diversity Indices Using Algebra and Logarithms

Species richness, the Shannon Diversity Index, and evenness are useful ways of quantifying biodiversity.

- Species richness, denoted by R, is simply the number of species found.

- The Shannon Diversity Index, denoted by H, takes into account the number of individuals found in each species. To calculate this index requires the use of algebra and logarithms.

- Evenness, denoted by E_H, converts the value of the Shannon Diversity Index to a number between 0 and 1, where 1 is total evenness, with the same number of individuals found in each species.

If you are not familiar with these indices, there are many good free online tutorials, videos, and step-by-step walkthroughs of how to calculate them. You may choose to calculate these by hand or to help your students format spreadsheets to calculate the indices for their data sets. Learning how to format spreadsheets is a skill equally as valuable as solving problems by hand.

TIMED OBSERVATIONS

Students observe the behavior of a group of animals and use a sample protocol to quantify what they see.

Many people think of math as calculations. But numbers are just another way of visualizing and modeling things we observe. The more that students are exposed to math as a tool for visualization and problem solving, the more they will be able to creatively employ it in their own learning. Conducting timed behavioral observations of animals is one way of using numbers to help us learn about an organism. This turns general observations into more precise data that can be used for deeper analysis and understanding of the animal. Quantitative data on animals' behavior offers a window into patterns that we otherwise might not be able to see. Making timed observations at intervals is also a strategy students can use in other settings to gather numerical data.

NATURAL PHENOMENA

This activity requires a group of animals that can be easily observed. This could be a flock of birds (ducks, pigeons, blackbirds, robins on a lawn), ground squirrels, deer, antelope, lizards, or other cooperative species. Animals that exhibit repeated behaviors and that are less likely to run away, hide, or fly off are ideal. The animal group should be close enough that you will not need binoculars or other magnification to observe their behavior, but not so close that they will be disturbed by students talking. This activity could occur spontaneously when you see a group of animals. You can also try to find local areas where animals tend to congregate so that you can plan the activity. Keep your eye out. Lizards often cluster in the morning sun, ducks and waterfowl congregate at flat bodies of water, and larger birds such as crows, ravens, or seagulls frequent many urban areas.

PROCEDURE SUMMARY

1. Record the behaviors of five different individuals every 20 seconds.
2. At every 20-second mark, make a tally next to each type of behavior one of the five animals is doing.

DEMONSTRATION

When the whiteboard icon appears in the procedure description: Build your demonstration page in four stages. First identify the behaviors you observe. Then create sample data (tally marks) as you describe making timed observations. When the data is in, show students how to create a graph from the data. Finally, use the data you collected to record questions or describe patterns you see.

Time

Introduction: 10 minutes
Activity: 30 minutes
Discussion: 20 minutes

Materials

☐ Journals and pencils

☐ Stopwatch(es) with countdown function or watch(es) with second hand

Teaching Notes

If you have many stopwatches, give one watch/stopwatch to each group. (Wristwatches with second hands or phones with stopwatch features work too.) If you have only one stopwatch, you or an assistant can call out a signal at the appropriate intervals. It's best to give students time to play around with a stopwatch beforehand so that they are familiar with using the tool when they begin this activity.

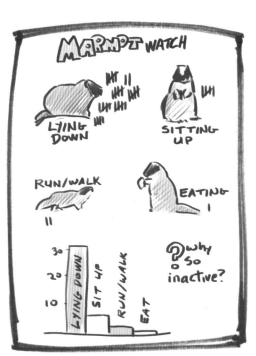

PROCEDURE STEP-BY-STEP

1. **Observe the animal species for 5 minutes, asking students to make a list of the kinds of behaviors they see.**

   a. "We are about to observe these animals for five minutes. As we do, focus on the animals' behaviors. What are they doing?"

   b. "In your journal, draw or write to make a list of the kinds of behaviors you see. What are the behaviors that you see the most often or poses or positions that the animals hold for longer periods?"

2. **Tell students to focus on making observations of the animals' behaviors, not trying to explain the behaviors.**

   a. "Try to avoid making assumptions about the reasons for behaviors; instead, describe what you see."

   b. "For example, instead of saying 'looking for predators' say 'standing, head and ears up.' Any questions? Begin."

3. **After 5 minutes, ask students to share some of the behaviors they observed, then put the common behaviors into categories (feeding, lying down, vocalizing, etc.).**

   a. "What were some of the most interesting or common behaviors that you saw? What poses did they do frequently or for long periods of time?"

   b. "Let's sort these into major categories, such as feeding, being still, moving, calling, and so on."

4. **Compile what your students share into a short list of clear, unambiguous categories on a whiteboard. (Four to ten categories is a good range, fewer for younger students, more with older observers.)**

5. **Tell students they will record their observations every 20 seconds.**

   a. "In just a moment we will use a system to quantify and record our behavioral observations. Our goal is to find out what these animals do and how they spend their time."

   b. "To do that, we will look every twenty seconds and tally the behaviors we see."

6. **Divide students into groups of three, then explain the procedure.**

   a. "In each group, there will be an observer, a recorder, and a timekeeper."

   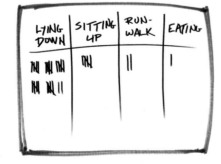

   b. "Every twenty seconds, the timekeeper will say 'Now!' Then

the observer will look up and say the behaviors they see each individual animal doing at that moment. Then the recorder will put down tally marks on the data table (one for each animal) each round.

7. **If the animals are relatively still, students can observe the behaviors of the same individuals in each observation. If this is difficult to track, they can choose individuals at random with each observation.**

   a. "Try to record the behavior of the same five to ten animals with each observation. If they move too much to track, choose random animals each time."

   b. "Keep track of observations with tally marks next to your behavior sketches, or make a table of the behavior categories like this."

   c. "We are going to observe for fifteen minutes. We will change roles every five minutes. Each of you will get a chance to take all of the roles."

8. **Give students a moment to copy the template data table (as shown on your whiteboard) and record the date and time in their journals.**

   a. "Take a moment to copy the template for the data table into your journal."

   b. "Before you begin, write the time, location, and weather in the corner of your page."

9. **Distribute stopwatches and explain how to use them. If these are a new tool, students will need a little time to play with them.**

10. **Tell students to begin their observations, then pay attention to how they are doing, offering support or redirection as needed.**

11. **After the animals have left, or after students have completed 10–15 minutes of observations, call for their attention.**

    a. "Please stop your observations, and write down the time that you stopped recording behavioral observations."

12. **Explain how to graph the data: Make columns for each behavior on the x-axis, and the frequency of each behavior on the y-axis. Demonstrate on a whiteboard.**

    a. "Make a simple bar graph to compare the frequencies of each behavior type. On the horizontal (x) axis, you will create one column for each behavior type."

    b. "On the vertical (y) axis, you will show the frequency of observations for each category."

    c. "Figure out how high the longest bar will go first to make sure you have enough room for your graphs, then mark the units of measurements on the horizontal (x) axis."

DISCUSSION

Lead a discussion using the general discussion questions and questions from one of the Crosscutting Concept categories. Intersperse pair talk with group discussion.

General Discussion

Use the graph and students' initial observations to discuss trends and patterns in the data and to build group understanding of the organisms' behaviors.

a. "Let's use our graphs to discuss what we learned about these animals. What behaviors did you see the most?"

b. "What behaviors were uncommon?"

c. "What general statements can you make about the behavior of these animals at this time?"

d. "What are some possible explanations of these patterns?"'

e. "Do you think this pattern in behavior is consistent for this species in other geographic locations? Why or why not?"

f. "How do you think the patterns of behavior might change if we repeated the activity at different times of the day or year?"

g. "How might this change if there were predators present?"

h. "What could you do to get an even better understanding of the way these animals behave? How would you design your experiment or structure your observations?"

Cause and Effect

a. "Did you see any behavior that seemed to have been triggered by some other event or stimulus? What is your evidence for this?"

b. "Some of the behaviors we saw may have been a response to forces or threats that we cannot see. Can you think of anything that we can't see that might influence the animals' behavior?"

Structure and Function

a. "What were some of the structures you noticed while observing these animals? Describe them in detail."

b. "Pick a structure and think about how it might function to help this organism survive. Connect your explanation of a specific structure to the environment. For example, don't just say 'Its legs help it run.' Think about the specific shape of the animal's legs, and what you observed when it was running. Do the legs help it run quickly over a specific kind of surface? How do they work?"

Stability and Change

a. "Did the level of animal activity stay the same, or were there times when the animals were more still and more active? Why?"

b. "After a disturbance, how long did things take to get back to a less active state or for the animals to return to the most common behaviors you observed?"

c. "After the disturbance, did things go back to the way they were before, or did there seem to be some new level of activity or behavior?"

d. "How might these patterns change at a different time of year?"

FOLLOW-UP ACTIVITIES

Practicing Percentages

1. **Convert the raw observations to percentages. This is good way to integrate math skills with a science exploration.**

   a. "How can we convert the number of observed behaviors to a percentage? [Divide the observation count for one behavior by the total number of recorded behaviors.] Let's calculate the percentage of time these animals spent doing each behavior."

   b. "Add the percentages to the graph below each column."

Responding to Disturbances

Repeat the activity after a mild disturbance (e.g., a student walking near the animal group). This time, keep separate logs for each minute after the disturbance. How long does it take for the group behavior to return to the predisturbance baseline?

Engaging in Further Research

Do further research and reading on the species students observed. Students' baseline observations will set them up to be curious to know more. They could read scientific papers, internet articles, field guides, or other sources to learn more about the characteristic behaviors and life histories of this animal. Or they could read about animal behavior in general, looking up information on how groups of animals tend to interact, behave in the face of disturbance, and the like.

CHANGE OVER TIME

Students describe a developing organism (e.g., a bean plant in a cup, a marked wildflower); a changing object (e.g., a decomposing orange); or a landscape feature (e.g., a sandy stream bank) as it changes over multiple observation sessions.

Time

Introduction: 10 minutes

Activity: 30-minute initial observation, with 10–20-minute follow-up sessions over the period of change (which may be months, with more frequent sessions during periods of intense change)

Discussion: 10–15 minutes

Materials

☐ Journals and pencils

☐ Twist ties to loosely mark plants or other objects in the field

☐ Rulers or measuring tapes

Teaching Notes

Students will be most successful with this activity if they have an approach or plan for structuring the page to facilitate data collection over time. If necessary, guide students through the process of planning how to arrange sequential observations on the same page or a spread of pages, and how to add symbols (such as arrows) to show the progression of observations over time.

Most times when we go outside to look at nature or record observations in our journals, we just see snapshots of longer processes. Yet the fruit on the trees, the spit of sand by the river, and the leaves just fallen are all in states of change. This activity gives students the experience of observing and recording changes in a phenomenon over time, leading to deeper understanding of the subject, the forces that cause it to change, and change as a general process.

NATURAL PHENOMENA

Look for objects that change over time at a rate that can be regularly observed and documented. This might include developing plants in a nearby garden or natural area, a potted plant or sprout in a cup; butterfly caterpillars (or eggs if you can find them; separate and place individuals in labeled containers to keep track), frog eggs in an aquarium (avoid releasing invasive species into the wild), a section of a branch of a deciduous tree (beginning when the tree is in bud), a decomposing vegetable in a compost bin (mark with a nondegradable tab so that you can find it as it rots), or a dead animal.

PROCEDURE SUMMARY

1. Use writing, drawing, and numbers to describe the subject.
2. Take detailed notes, especially about parts of the object that you predict will change the most.
3. Measure the object and create a table to record future measurements.
4. Leave space on the page for future observations, and label this 1.

DEMONSTRATION

When the whiteboard icon appears in the procedure description: Draw a bud of a plant (not the species that the students will be working with). Draw sets of horizontal lines to suggest written notes. Number the drawing 1. Then add and number subsequent stages until you fill the page. Then explain that you would continue your notes on other pages in your journal as the plant continued to develop.

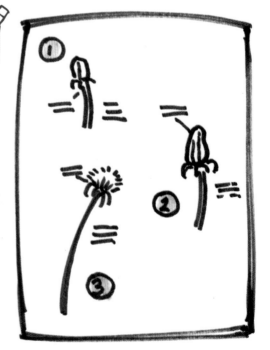

PROCEDURE STEP-BY-STEP

1. **Explain that students will use their journals to record the changes in an object (or phenomenon) over time, using words, pictures, and numbers to show observations.**

   a. "Many changes are too slow for us to see. We can use our journals to freeze one moment in time and then compare that with what we see in the future."

   b. "You will use words, pictures, and numbers to describe the object [this orange, this leaf, this section of stream bank] as accurately as you can."

   c. "Then we will come back and observe it every day [or once a week, once a month, etc.], making a new journal entry and looking for things that have changed."

2. **Tell students to begin by making a drawing on one half of a journal page and to focus on parts of the object that they think will change the most, then ask them to share which details it will be important to record.**

   a. "To begin, you might draw the subject on the top half of your paper with written observations next to it, leaving space below for your next observation."

   b. "As you describe your subject, try to predict aspects that you think will change the most, and include notes about them so you will be able to compare future states."

   c. "What kinds of details might be important to record?"

3. **Talk students through the process of measuring a few "landmark" aspects of the object they will be able to find again, using consistent units.**

   a. "We are also going to measure one part of this object and remeasure it each time we make observations. This will help us be more accurate in noticing changes."

   b. "Choose clear landmarks [e.g., a stem from the attachment point of one leaf to the base of a bud; one side of a decomposing orange to another] at which to start and stop measuring. You need to be able to find them again." (For a stream or other landscape feature, make sure one

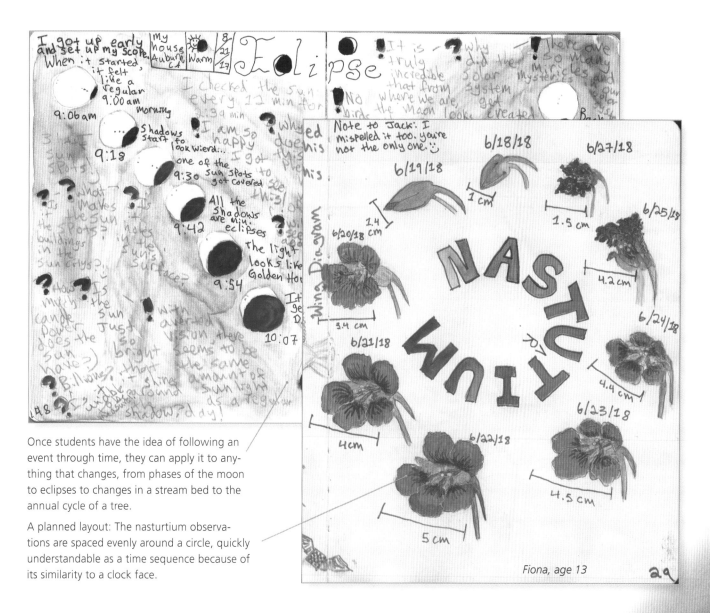

Once students have the idea of following an event through time, they can apply it to anything that changes, from phases of the moon to eclipses to changes in a stream bed to the annual cycle of a tree.

A planned layout: The nasturtium observations are spaced evenly around a circle, quickly understandable as a time sequence because of its similarity to a clock face.

Fiona, age 13

landmark is something that is unlikely to wash away, such as a tree.)

   c. "We also need to decide what units we will use to record our measurements. If the units are too big, we will not be able to notice change over time. Please talk with the people near you about whether you think it would be better to use millimeters, centimeters, or some other unit of measurement."

4. **Set boundaries, ask students if they have any questions about what is expected of them, and send them out to journal.**

   a. "Are there any questions? You'll have ten minutes to do this initial observation. Go for it!"

5. **As students journal, take time to circulate and support them, offering reminders about time and the goal of the journal entry.**

   a. (When there are about 5 minutes left) "Remember to use words, pictures, and numbers to record what you see."

   b. "This will be the only opportunity you have to see your subject in this state. It will be different when we return. Are there any other important details to include? Are there parts of the subject that you want to be sure to record now so that you can compare possible changes later?"

6. **Call the group back together and explain how to set up a table to record measurements over time by making a column for the date and a row for the length, demonstrating this on a whiteboard.**

   a. "We will create a simple table to record your measurements over time. You will need a column for the date (and the time if what you are observing will grow quickly) and the length, distance, or other measurement. Be sure to specify the units of measurement."

7. **Explain how to create a graph to record measurements, showing time on the *x*-axis and length on the *y*-axis.**

   a. "We'll also create a graph that has time on the horizontal (*x*) axis and length on the vertical (*y*) axis. Leave room on the right side of the graph for future observations."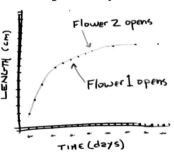

8. **Ask students to discuss some predictions of how the object might change, and the rate of change they expect to see.**

   a. "Discuss with a partner: When should we next check in with the object to see whether changes have begun?"

   b. "What changes do you expect to see first? Why?"

   c. "Do you think that this object will change at the same rate during the whole period we are observing it, or do you expect the rate of change to vary? Why?"

   d. "It is hard to predict how fast this object will change without previous experience. To be on the safe side (so we do not miss a state of change), we will check again [state the time until the next observation period]."

9. **During each follow-up observation, students should do the following:**

   a. Repeat the measurement, record it in the table, and add a new point on the graph.

   b. Redraw the object, adding written notes detailing changes or newly observed details. (Depending on the size of their journals, this could occur on the original page or on subsequent pages.)

   c. Make diagrams of any interesting features.

   d. Record any questions that have been stimulated by their observations, and discuss their questions with their peers.

DISCUSSION

Lead a discussion using the general discussion questions and questions from one of the Crosscutting Concept categories. Intersperse pair talk with group discussion.

General Discussion

After all the observations have been completed, tell students to meet in small groups and discuss how the object changed over time, using their journal entries and measurements as data.

   a. "Meet with a group of four and look over your journal entries together."

   b. "Describe how the object changed over time. What changed, and when? Refer to your graphs, measurements, drawings, and notes."

   c. "Were there periods of slower change or more intense change? What might have caused that?"

   d. "What do you think caused the changes you observed in your object?"

Stability and Change

   a. "How did different parts of the object change?"

   b. "What parts of the object changed the most or least?"

   c. "Where did you find growth spurts on your graph? Where did you find periods of less change? What are some possible explanations for these periods of change and stasis?"

   d. "What changes may have occurred before our observations began? What changes do you think will take place next?"

e. "People tend to think that the state in which they see an object is how it has been and will always be. Are there any ways in which you could apply the lessons learned from observing this object over time to observations of other parts of the world?"

Structure and Function

a. "Plant structures perform different functions. Look at the structures you recorded and discuss what some of their functions might be."

b. "In your notebook, write some ideas about the possible function of structures as you have drawn them. Draw a box around your function notes so that they stand out from your other notes, and include a question mark if you're not sure."

c. "Is there a reason the structures develop in the order that they do? How might that help the organism survive?"

d. "Pick one structure and observe how it changed over time. Then discuss how its function might have changed over time, and the evidence and reasoning behind what you think."

Cause and Effect

a. "What mechanisms or forces could be behind some of the changes you observed?"

b. "Plant structures [or other organisms' structures] develop and are lost at different times. What environmental forces may affect the timing of the development of different structures?"

c. "Do you see any evidence of external forces or elements that have affected this object during the observation period?"

d. "What specific forces might have caused some of the changes you observed?"

Systems and System Models

a. "What outside factors influenced our subject when we observed it?"

b. "How might we label the parts of the subject of our study [e.g., orange skin, orange inside, orange stem, mold on orange]? How did those parts interact?"

c. "How might the external factors have influenced each of those parts and interactions?"

Matter and Energy

a. (If students have studied something decomposing) "What were some of the changes you observed over time?"

b. "Did any part of the object change form or appear to go missing?"

c. "Where might those parts be now? What could have happened to them? Did you see any evidence of this?"

d. (If students have studied something growing) "What were some of the changes you observed over time?"

e. "Did any part of the object change form, or grow?"

f. "Where did those parts come from? How did the plant seedling get so big?"

FOLLOW-UP ACTIVITIES ·

Observing an Area

Direct students to use a similar approach to study change at a larger scale and make regular observations in an area that is rapidly changing. This could be a section of a sandy beach that is close enough to the water to be affected by waves, or a meander in a sandy stream bank. (Entrenched river systems also change, but over a much longer period, and may take years to record.)

Watching Decomposition (Gross but So Cool)

In nature, we will encounter dead and decomposing animals. These are memorable phenomena for study and a way to learn about the way matter cycles, and they prompt interesting conversations. The FBI maintains several Forensic Anthropology Research Facilities, or "body farms," throughout the United States. These are used to study human decomposition under natural conditions. This research gives forensic experts vital clues to solving crimes.

Decomposing bodies go through several stages. Timing varies with environmental conditions.

Fresh. Rigor mortis (stiffening of muscles) and liver mortis (pooling of blood in lower parts of the body) set in; blowflies and flesh flies arrive.

Bloat. Gasses released by microbial decomposition fill the body cavities with air; the skin marbles; and fly larvae (maggots) hatch and begin to feed. Oh, and the smell…

Active decay. Maggots fill the body, rapidly consume tissues and organs, and pupate. Decomposing materials pool around the body, creating a "cadaver decomposition island" (CDI). The smell persists.

Advanced decay. Most of the soft body materials are gone. Soil may be stained from body fluids. The vegetation may be dead in the CDI.

Dry/remains. Only leathery skin, cartilage, and bone remain. Bare bones bleach in the sun. There is lush vegetation growth in the CDI.

If your students find a freshly dead animal, they can initiate their own investigation, visiting the carcass over successive days and months, noting changes and rates of change. Make sure that they take reasonable and appropriate precautions to avoid contact with carcasses and the potential for spreading disease.

INCORPORATING JOURNALING INTO LESSONS, FRAMEWORKS, AND ASSESSMENTS

JOURNALING OVER TIME

Setting up systems to support growth over time and embedding journaling within lessons increase student buy-in and enable us to use the practice to meet standards and learning goals.

When we create systems to support students in developing their skills, we make it more likely that they will gain the internal motivation to keep journaling. We can offer consistent feedback, new ideas about journaling, and opportunities for deliberate practice to kick students into a self-sustaining cycle of improvement and engagement. This chapter includes specific ways to use coaching and feedback to help students grow their journaling skills. We can also make sure our methods for grading and evaluating students' journals support this process of growth. This chapter also offers tools to help you structure evaluation and assessment in a way that builds student motivation and confidence.

Nature journaling can be a breakthrough method to get kids excited about learning. Making sure journaling is not a stand-alone activity but a practice woven into the fabric of your classroom, program, or organization will help you make the most of the activities in this book.

Long-term journaling projects that are incorporated into their lessons or units help students build their journaling skills and reach the goals of the curriculum. Students can take their learning further by discussing their observations, explaining patterns and processes, doing further research, and learning science concepts that help them better understand the processes they studied in their journals. When students see their own observations being used to drive a lesson, they can become more excited to get out the door and journal the next time. The activities in this book have value as stand-alone work, but they are not a full-service curriculum—they're learning experiences that you can build on and incorporate into your current teaching structure or curriculum. This chapter offers a guide to developing lesson plans around the activities, and points you to some further resources for integrating journaling into longer learning experiences and existing curricula.

Journaling within other lessons and curriculum also enables us to better meet standards or support educational frameworks. As educators who are also naturalists, we consider the development of the Next Generation Science Standards (NGSS), which are based on how scientists actually learn and practice their skills, one of the most exciting new possibilities in education. Because our methods for teaching journaling map so well onto the NGSS, we've gone into some depth here about that framework.

DEVELOPING SKILLS: IDEAS, PRACTICE, AND FEEDBACK

Students will develop journaling skills more quickly if given new tools, approaches for deliberate practice, and timely feedback.

Students will not start out as master journalers. That's to be expected, and it's OK! Give students repeated opportunities to use journals and learn, and they will build their muscles of focus, observation, and thinking on the page.

Students can keep improving their journaling skills over time. Together, new ideas and approaches, deliberate and reflective practice, and timely feedback or coaching create the support students need to grow and improve.

New Ideas and Approaches

Deliberate Practice   Appropriate, Timely Feedback

NEW IDEAS AND APPROACHES

When learning a new skill, we can pick up a lot through experimentation, but someone with more expertise can offer new ideas and approaches we won't come to through trial and error. This is true of any skill, whether it's playing music, cooking, repairing cars, or journaling. Offering students a specific strategy or "trick of the trade" for journaling will help them get better faster. In the Words, Pictures, and Numbers chapters of activities, we have also offered exercises to help build specific skills—for example, step-by-step processes for drawing in the field, a trick for sketching animals in motion, and ways to develop writing chops. Every new idea is a tool you could incorporate into lessons, repeating activities with a different skill focus each time.

Many of the journaling activities in this book can also be used to give students a new journaling skill or approach. After an activity like *Zoom In, Zoom Out* or *Comparison*, help students recognize the frame as a skill and approach that they can apply in future journaling entries.

You can offer strategies to individual students who need support with a specific skill, or to your class as a whole. You might, for example, give some new ideas on visual layouts to a student who is struggling to vary the structure of their journal page, or offer basic drawing approaches to an entire class that is new to journaling.

Your students will not be able to learn everything at once. Keep giving them new approaches and the time to become proficient, then add more as time goes on.

DELIBERATE PRACTICE

Generally, the more we do something, the better we get at it, but practice is more effective if it is deliberate.[33] In deliberate practice, we notice what works and what doesn't, adjust accordingly, and intentionally pick skills to work on over time. Doing the nature journaling activities in this book with students is a start. To engage students in deliberate practice, we can also coach them to set clear goals and reflect on their journaling process.

Setting Clear Goals

It is overwhelming for beginning journalers to think about getting better at every skill at once. Any project or skill can be broken down into sets of smaller, attainable goals. Help students set realistic goals, picking one or two at a time to focus on over a series of journal entries. For example, a student might spend several journal outings trying to ask more varied questions, integrate more numbers into their entries, improve a specific drawing technique, or refine their page layout. Focusing in this way does not mean that the student does nothing else in their journal entry. It is an area where they put their attention, taking time after making the entry to assess how they are doing with the skill and where they can improve. Reinforce the idea of a growth mindset, encouraging students to identify skills they want to deepen, and to trust that their work and focus will pay off.

Reflecting and Thinking about Thinking

Students will get more out of their journaling experience if they reflect on the learning process.[34] Questions to stimulate reflection are part of the general discussion questions we offer at the end of each activity, and you can extend those as you wish, using the examples in the box "Questions to Guide Reflection." Answering such questions as "What helped you learn during that journaling experience?" or tracing their thinking on the page solidifies the memory of what students put down in their journal and helps them become better learners. Even 2 or 3 minutes at the end of a journaling experience is enough to help students process and think through their learning.

We help students learn to reflect by offering questions to guide their thinking, and modeling the type of reflective thinking they might do (e.g., "Interesting to notice that I only included numbers next to questions in my journal entry, and did not make any measurements in my drawing. What's up with that?" or "This time, I tried an approach where I intentionally labeled more parts of my drawing early on. This made me feel less worried about making a pretty picture"). This modeling supports students (particularly younger students and language learners) to reflect on their process of skill development.

Students can reflect alone or have a "journal buddy" with whom they periodically check in, discussing challenges and successes in their practice. In reflecting together in a community of practice, students normalize the process of self-assessment. Doing so also helps students develop the capacity to notice where they can improve without tearing down their ego. This can be used as a natural lead-in to students' receiving and incorporating feedback in their journaling.

APPROPRIATE, TIMELY FEEDBACK

Every mistake, failed attempt, or place of struggle is an opportunity to learn if we have a way to identify a path for improvement.[35] As parents and teachers, we can give feedback to help students see the next step in their development, and coach them to push through difficult parts of the process.

Many teachers provide suggestions on students' work when they grade it, but this feedback often comes too late for students to make use of it. The most effective feedback is timely feedback, given when students can incorporate suggestions while they are still relevant.[36]

To give timely feedback, make a practice of circulating among students while they are journaling, engaging them in conversation about their work. This is a golden opportunity to build on your

relationship with individuals and to recognize patterns in students' work they might not be aware of. Another approach is to create a ritual of setting students' journals out after an experience, making a few quick comments verbally or using sticky notes.

Feedback: What to Say

When we set up journaling activities, we tell students that their goals are to create accurate observations of the world and follow their curiosity. Then arrives the moment when students show us their work. What is the spontaneous response to seeing a lovely journal page? "Good job! What a pretty picture. That looks great!" That pitfall is very hard to avoid, even for experienced teachers.

Students already comfortable with their artistic abilities might be encouraged in their artistic efforts. But this kind of feedback undermines our stated expectations and shows students that we were insincere—it's just another adult trick.

Instead, we give feedback in a way that advances the goal of the learning by noticing what the student has done on the page and relating this to the expectations we set out at the beginning of the activity. If we have told students they are to observe and carefully record what they have seen, then we give positive reinforcement when we find accurately observed details in their work. For example: "I see you have shown hairs on the stem. Details like that become important to botanists when identifying and studying plants" or "You have put a scale next to your drawing. That will really help you remember the dimensions of this when you review your notes later" or "I see you've used lines to connect the

QUESTIONS TO GUIDE REFLECTION

- What skill are you focused on in your journaling right now? What do you think are your opportunities for improvement? How did you use that skill in this journal entry? What are your strengths? What are you struggling with?

- What was challenging about this journal entry?

- Why did you structure the page the way you did? How did the page structure affect your thinking while you were journaling?

- Is there anything you would change about your entry (observations you wished you had added, a different way of organizing information, a question you wished you had followed up on)? Why?

- Compare this journal entry to your last few. What kinds of questions or observations do you tend to focus on? Is that focus intentional? Do you think you should change it? Why or why not?

- What role did [words, pictures, numbers] play in this entry?

- Are numbers, pictures, and words working together to show your observations and thinking? How could you have integrated them more on your page?

- While you were working on the entry, when were you surprised by something you observed, or when did you change your mind about something? What caused the shift in thinking?

- What helped you learn while doing that journal entry?

- What topics are you asking questions about? What are you not asking questions about?

- Trace the evolution of your thinking in the entry: When did you use drawing? When did you use writing? When did you use numbers?

numbers to your drawing and to the written observations on your page. That really helps me follow your thinking."

This type of feedback pays off. Students who are recognized for their observations, thinking, and strategies are more likely to succeed as learners than those who are told they are smart[37] or good at a skill.[38]

When you are reviewing your students' field notes or sketches, your attention will probably be drawn to the most detailed and attractive pictures. There is an unconscious tendency to comment on student work in the order of your aesthetic preference, beginning with the one you find most satisfying and ending with the one you like the least. Students will quickly pick up on this. Instead, skip around among the drawings. Keep the focus on students' observations and the pictorial and written recording of them. We make sure to comment on what they have written and on questions they pose in addition to the observations we see in their drawings.

Feed-Forward: What to Say

If we think of feedback as identifying what the student has done, feed-*forward* includes suggestions of what the student can do to improve a journal entry or further a skill. Both are essential to students' growth in their journaling practice.

The suggestions you give as feed-forward must be challenging yet attainable in order to keep fueling a growth mindset in your

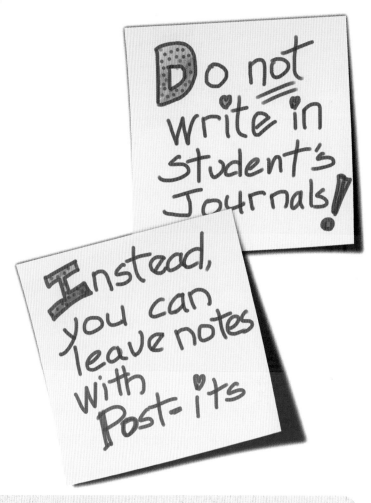

CONSTRUCTIVE FEEDBACK

Feedback can be the helping hand to guide students to take their next step. One way to make these supportive and constructive comments is to notice and observe what students have made and recorded and simply reflect it back to them—essentially, "I notice that you noticed this. Good noticing!" Another method is to highlight a journaling strategy that you want to reinforce, perhaps something you would like to see more students do or would like to see the student try again. Point out techniques they used, such as using enlargements, showing scale, or including labels, as successful strategies they should use again and that other students could include, too.

Don't Say

"That looks really good."

"What a pretty picture."

"You are a great artist."

"That looks so realistic."

"You are really good at shading."

Do Say

"The insect damage on that leaf you drew helps me pick out which flower you were looking at."

"The way you combine writing and drawing to describe this flower gives me a lot of information."

"I see you measured the distance between the branches and added a scale; those are two effective ways of using numbers."

"The way you have organized your page and added arrows and boxes shows your thinking on the page."

"Oh, you found a spider on top of the flower. Cool! What else do you notice?"

"That is an interesting question. What are some possible explanations?"

students. Giving a suggestion that is far beyond a student's ability will just lead to further frustration. Students who are struggling to write at all in their journal would likely balk if asked to write a paragraph of complete sentences. This is beyond their skill level at the moment, and that's OK. They might, though, be up to the challenge of listing three observations in bullet points. The trick is to notice where students are and offer suggestions that take them a step forward. Can you spot the next tool they need in their kit?

The more specific you can be, the better. "These details on the head show me how much you understand bird structure. On your next two drawings, pay particular attention to the bill length and width. Here's an easy trick: It helps to compare the size of the beak with the size of the head." If you do not know where to suggest improvement, ask, "What part of this page or drawing was the most difficult or challenging? What might you do differently next time?"

GRADING AND EVALUATION

Simple rubrics help students assess current skills and target areas in which they can improve.

WHAT'S IN THE GRADE?

Grading and feedback are ways to assess and communicate students' capabilities at one point in time and show them how they might be able to improve. But for many students, grades are taken as a reflection of their intelligence, capacity, or worth. The finality of a grade can make the grade feel more like a judgment than a signpost along the path of learning, reinforcing a fixed mindset. To help students take grading constructively, frame grades as a tool to reflect to students where they are on the continuum of their own learning, with the understanding that this is an assessment of a point in time, not a final statement. If you and your students can adopt a growth mindset, the purpose of a grade is to improve future performance. It is the finger pointing toward the next step, not the finger pointing to the flaw. The more that students understand the purpose of grades, the more useful grades will be.

GRADING JOURNALS: USING RUBRICS

When we give feedback in a timely manner, we help students think about their work and improve. This means that we must take the time to respond to journals more frequently than at the midpoint and end of the teaching term. It's better to regularly track, assess, and respond to student journals.

How do you evaluate student journals? It is easy to get distracted by a good drawing and lose track of the kinds of observation and thinking strategies we want to reinforce. Rubrics like the example on the next page can be a simple tool to help you highlight the important features students should include in their journal entries and show students where they have room to improve. Rubrics can also make it fairly straightforward to qualitatively and quantitatively assess and grade students' journals. We use the example rubric as a starting point and adapt it to the needs of each group of students by adding expectations appropriate to their context or by specifying further expectations for a certain activity.

We find that pairing rubrics and specific feedback on students' journaling skills encourages reflection and improvement. For example, we add comments to the rubric about places students could have improved the accuracy of their drawing, or notes on how they could have better integrated quantification and writing with their drawing. These specific suggestions reinforce the purpose of grading as a means of helping students improve their practice.

A rubric can also be a valuable tool to support students' reflection process. Give students a copy of the rubric you use so that they know what is expected of them; they can paste it into their journals. Occasionally set aside time for students to evaluate their own journals using the rubric, prompting them to add detailed notes about requirements or journaling skills that are proving challenging for them.

EVALUATION RUBRIC

The appendices include this rubric to copy or cut out (see p. 272). Use rubrics to help make your instructions clear and to hold students accountable to your expectations.

A straightforward rubric that is shared by teachers and students clarifies expectations and enables teachers to grade in an objective, fair, and supportive way. Before a graded journaling activity, list clear and reasonable expectations. During the exercise, students can use the list as a reminder of the tasks they should complete. As noted in the previous section, students can even use the rubric to evaluate their own work and to identify the next steps to improvement.

1. Students write their names and activity information at the top of the page.

2. As you tell students your expectations, have them mark the required item with a slash (/).

3. Add the total points possible and put this number at the bottom of the page.

Name **Dana Rho**
Date **11·20·2019**

Location **School Garden**
Project **Zoom In-Out**

Basics

- ☒ Date, place, and time
- ☒ Weather/temperature
- ☒ Words, pictures, and numbers used together
- ☐ Format features (titles, boxes, dividers, arrows, etc.)
- ☒ Observations and descriptions (I notice)
- ☐ Connections (It reminds me of)
- ☒ Questions from observations (I wonder)

Pictures

- ☐ Accurate drawings and diagrams show observations
- ☒ Enlargement (detail inset) of interesting parts
- ☒ Multiple views of same subject (top, side, etc.)
- ☐ Cross or longitudinal section views
- ☐ Color or notes about color
- ☐ Habitat sketches
- ☐ Maps

Words

- ☒ Paragraphs, single sentence, sentence fragments, bullet points, and labels used where appropriate
- ☐ Notes about methods or procedure
- ☐ Explanations (Could it be?)
- ☐ Distinguishes between observation and explanation
- ☐ Uses language of uncertainty

Numbers

- ☐ Counts or estimates features and individuals
- ☐ Times observations and speeds
- ☐ Measures lengths
- ☒ Shows scale (life-sized, magnification, etc.)
- ☒ Samples data (more than one measurement)

Additional expectations
☒ **Zoom out**

Comments **It is great to see you using the side and front views. What do you wonder?**

Total points received: **7**
Total points possible: **10**

---

4. Students may keep the rubric while they do the journaling activity as a checklist of things to do.

5. Once students have finished the activity, they may add a backslash (\) by each item they have included (to complete the X). Students may see that they forgot one of the required items (such as the date) and can add it in before they turn in the rubric. They then record the total points received.

6. Use the comments area to give encouragement, recognition, and targeted suggestions to help the student grow.

FROM ACTIVITIES TO LONGER LESSONS

Discussions and extended learning experiences build conceptual understanding based on the observations in students' journal entries.

We find that students are most engaged when journaling is just one part of a longer learning experience. Students participate more when they know that their own observations will contribute to the group's learning process. Students' journal entries can tie into longer lessons or units. This makes the practice of journaling feel real and important, not just like busywork.

FOCUSED ATTENTION

Choose phenomena carefully to tie journaling experiences into longer lessons. If we were to tell a group of students to "go journal about plants," they would likely produce a range of entries. One student might journal about a flower; another, a tree. Some students would focus on drawing plant parts; others might observe patterns of growth. This would not be a bad experience, but it would be hard to use these disparate observations if you are trying to teach about a specific topic. By contrast, if you direct students to compare two species of oak trees, all the students will share the same experience. This balances structure with student autonomy; each student makes observations of plants, decides what to record, and makes drawings and descriptions in their own journal. At the same time, the class generates a focused set of observations and questions about the structures of oak trees that can be used in discussions and further learning.

You can structure journaling activities around any topic or common theme. If you are teaching a unit on pollination, you could start with a comparison of two flower types and follow that up with a comparison of two pollinators. Then you could diagram and dissect flowers with the activities *Inside Out* and *Nature Blueprints*. Students could also make a map of a nearby meadow or vacant lot with a key showing different types of flowering plants, then tally the numbers of pollinators using different parts of the study area. Finally, students could create an infographic containing observations of a local plant and research about its pollinators. These journaling activities, interspersed with discussions and introduction of key concepts, form a deep learning experience focused on plants and life science.

The table here lists some examples of learning goals and the phenomena and activities that would help students reach those learning goals.

Examples of Learning Goals	Possible Phenomena	Activity
Learn how organisms' structures help them survive	Any two species of plants or animals	*Comparison*
Understand how rock type and composition affect erosion	Sedimentary and igneous rocks in a stream	Students notice differences and similarities between two subjects.
Learn how environmental factors can affect organisms' growth	Plants of the same species grown in different areas under different conditions	
Slow down and look at nature from different perspectives	A plant; geographic feature; small, slow-moving organism	*Zoom In, Zoom Out*
Learn about human impacts on water sources	An urban or rural pond, stream, lake, or other body of water	Students "zoom in" and "zoom out" to sketch a subject from close up, recording observations, then far away, focusing on surroundings and context.
Understand how environmental factors affect organisms' distribution	Distribution of spider webs in the schoolyard; distribution of plants on an unmowed ball field	*Mapping*
Learn about access to green space	Mature trees or public parks in two different neighborhoods	Students observe and record distribution of a subject or a part of nature across a specific area.

See appendix B on page 262 for a summary of each activity and the phenomena students could focus on. See that appendix for a list of learning goals, and suggestions of activities that will support students in meeting those learning goals.

After a journaling exercise, students will have had firsthand experience with a phenomenon. This gives everyone expertise and ideas to contribute in group discussions and future learning experiences.

DISCUSSIONS: EXPLANATIONS AND MEANING MAKING

"Why did we find thicker icicles on one side of the building?" "How do you think the lizard's different behaviors help it survive, and how do you think the most common behaviors shift throughout the day?" "What could have caused the pattern of the big patch of dandelions right around the sprinkler head?"

The discussion questions at the end of each activity help students to make explanations about what they journaled about. Making explanations deepens our understanding of natural processes and science concepts. When we make an explanation, we're starting to play with ideas and process information. We integrate what we know and what we've observed to try to figure out how or why something is happening. Actively trying to figure out how something works or why something happens often brings students to the edge of their understanding of natural processes and science concepts, and creates a "need to know"—an excitement to learn more.

If we were to ask students who have compared two different kinds of plants, "What were some of the differences between the two plants? How do you think those structures help the plants survive?" they might respond, "Well, we noticed one of the leaves was really broad and thin, but the other one was tiny and really thick and tough. We know plants need their leaves to get sunlight, so how would being really tiny and thick help the plant survive? Hmm, maybe if it's really thick and tough, insects can't eat it as easily. Or maybe even though the individual leaves are tiny, there are a lot of them, so they can still get enough sunlight?"

During this discussion, students will need to look back at their journal entry and review their observations. They'll also need to

use their current knowledge of science concepts (specifically, the survival needs of plants, and how plant structures work). This "productive struggle" is a key step in the learning process, leading to deeper understanding and retention of ideas than students would reach listening to a lecture.

OFFERING CONTENT: BUILDING CONCEPTUAL UNDERSTANDING

Students will make initial explanations for the patterns they observed in nature. These explanations reflect their current understanding of science concepts and natural processes. But to go deeper, students will need to come into contact with new ideas and concepts.

As students discuss their explanations, we try to notice: Where is the edge of their understanding? What are they struggling to figure out, and what will they not be able to explain with their current scientific knowledge? What will help them build on their explanation and understand what they are seeing? Our goal is to offer key science concepts that will help students explain and understand what they observed during the journal activity.

For example, after students compare plant leaves, we offer information from our own studies and reading about how plant structures aid their survival in a variety of environmental and weather conditions. Teaching students about transpiration (the process of water evaporating from plant leaves, which is slowed by having thicker leaves), adaptations for evergreen and deciduous trees, or how needle-shaped leaves can prevent freezing and water loss in winter would all shed light on students' initial explanations and give them more depth of understanding of what they noticed. This also contributes to an understanding of broader science knowledge, such as that organisms have different structures that help them survive in their environment, and that patterns of these structures can occur across ecosystems.

We might share content with students ourselves, in the classroom or in the field, or have students engage with a resource—a field guide, a local expert, or the research available through reliable online sources. Learning doesn't stop at hearing an answer from an expert: After encountering new content, students return to their meaning making, taking new ideas and using them to further their understanding and refine their explanations.

EXTENDED LEARNING AND FOLLOW-UP

Before students begin a journaling activity, we can offer opportunities for them to think about their prior knowledge of the topics they'll engage with during the activity—for example, what do they already know about dandelions, or plants in general? Then, after journaling, we provide extended opportunities for students to apply their new knowledge by communicating about what they learned, doing further research, or using that knowledge in a new context or discipline.

The example in the next section shows how journaling can be embedded within a longer learning experience that's been designed by the teacher to focus on the structure and function of seeds.

ADDITIONAL RESOURCES

Creating Effective Outdoor Science Activities, by the BEETLES Project (beetlesproject.org/creating-effective-outdoor-science/). This guide outlines a process for designing engaging, cohesive outdoor science lesson plans.

The Ambitious Science Teaching Project (ambitiousscienceteaching.org/). This website includes resources on designing engaging, cohesive science lesson plans.

EXAMPLE: JOURNALING IN A LESSON ON SEEDS

1. Context Setting, and Accessing Prior Knowledge

In pairs, students think about what they know about seeds and describe seeds they have seen before, then share out in the group.

2. Focused Journaling Activity

Students collect different types of seeds and complete the *Collection or Field Guide* activity, recording observations and questions about four different seeds, using words, pictures, and numbers.

3. Discussion and Beginnings of Meaning Making

The instructor guides students through the "Discussion" portion of the activity, focusing on the Structure and Function questions, such as "How are the seeds different from each other?" and "What are some possible explanations for those differences?" Students refer to their journal entries as they discuss their observations of seeds and start to come up with some explanations for the variation they observed.

4. Introduction to New Concepts

The instructor explains that the main function of seeds is to disperse themselves, and describes some different seed dispersal mechanisms—by wind, by birds and mammals, by rolling, and so on.

5. Deepening of Explanations and Integration of New Knowledge

Students look back at their journals, focusing on the structures of the seeds they drew and further discussing explanations for how different types of seed structures might function for dispersal. They use the knowledge the

instructor offered about seed dispersal mechanisms as evidence in their explanations, making specific connections between how seeds might be dispersed and the structures they drew and observed.

Students conduct research about seed types, comparing their sketches to what is in books, and use additional evidence to refine their explanations (in discussion or in writing) about how their seeds are dispersed.

6. Application of Observations and Knowledge

Students think about different environments and ecosystems, and they make predictions about the structures that seeds from those areas might have, using their journal entries and knowledge from their research as evidence for their predictions.

7. Reflection

Students look back at their journal entries and reflect, thinking about what they learned about seeds and what helped them learn. They then create an infographic about seed dispersal mechanisms.

Although students are not physically journaling during every moment of this extended learning experience, they refer back to the journal entry again and again. It is the backbone of the learning experience. Students' observations, questions, and documentation of the range of structures of seeds become the evidence for their early explanations of how seeds work. Students continue to look at their observations and questions as they think about what seeds might look like in different environments and when they reflect on their learning at the end of the lesson.

JOURNALING IN UNITS

Most set curricula break learning down into focused units. This isn't the only way to approach teaching, but it can be a helpful tool for planning out your time with students. Journaling experiences can fit into the existing curriculum you're using, or you can plan sequences of the activities in this book, with opportunities for follow-up. The sequence in the table on the next page is meant as an example of how journaling activities can build students' skills and understandings over time—it is not the one "right" way to teach these topics. This example focuses on life science for elementary school students, but journaling can also be used within units on other topics and in middle school, high school, and beyond. Students' journaling could focus on local geographic features and erosion patterns within a unit on earth science, or local land management approaches within a unit on human impacts on the environment. Middle and high school students can engage more deeply with modeling and math, using these science practices in their journals to study ecosystem dynamics, figure out how natural systems work at a molecular level, and study other more complex topics.

A SUPPORT, NOT A REPLACEMENT, FOR EXISTING CURRICULUM

Many teachers, educators, and organizations use curriculum packages, implement program-wide teaching, or are required to meet specific standards. Journaling can support these goals whether you use FOSS kits, a series of units designed by your school district, or some other prescribed set of lessons. Integrate the journaling activities in this book into existing units, finding opportunities for student observations of nature to become a focal point of a lesson. These observations and initial ideas become concrete memories, referred back to and elaborated on as students engage with the material and sources in the rest of the curriculum.

SHORT PROGRAMS

Almost all the journaling activities in this book can be used successfully in shorter, informal education contexts, such as outdoor science programs, nature centers, zoos, aquariums, or community organizations where learning experiences can range in length from an hour to a week of instruction.

In 1–2 hour programs, it's best to begin by giving students time to move around and explore before doing a journaling activity. Students who have just arrived in a new place are understandably interested in checking things out. After students have had some time to move around, even 15 minutes of focused journaling can frame the rest of a lesson or experience. In weeklong or multiday experiences, students could do a couple of activities per day focused on different parts of nature and in support of different learning goals.

SKILL-BUILDING ACTIVITIES

When we're working with students over an extended period of time, we intersperse conceptual journaling activities with those that focus on a specific skill (such as sketching or using numbers). The introduction of new journaling approaches followed by opportunities to practice them in context builds students' skills over time.

Grade 3–5 Elementary Classroom or Homeschool Unit:
How Organisms' Structures, Behaviors, and Interactions with the Environment Help Them Survive

					Learning Goals
Week 1	*I Notice, I Wonder, It Reminds Me Of*	*My Secret Plant Scavenger Hunt* or *To Each Its Own*			Learn observation skills and how to create a data-rich journal page
Week 2	*Mysteries and Explanations*	*Timeline*—focused on a plant	Students discuss structures of plants from seed to bud.	Students learn about pollination and plant reproduction.	Learn about making explanations; learn about plant life cycles; practice journaling
Week 3	*Inside Out*	*Collection or Field Guide*—focused on seeds	Students make explanations of seed structure and function.	Instructor talks about seed dispersal types; students conduct classroom research about seeds.	Learn diagramming skills, how organisms' structures help them survive, and basic plant biology
Week 4	*Collection or Field Guide*—focused on plants in vacant lot	*Landscape Cross Section*—focused on plants next to a pond	Students make explanations for plant distribution patterns.	Students learn how environmental factors affect plant distribution.	Practice visual layout skills; learn how environmental factors affect organisms' distribution
Week 5	*Writing to Think, Writing to Observe*	*Species Account*—focused on an animal in the schoolyard	*Event Comic* with narrative extension—focused on animal behavior	Students conduct field guide and video research of common animal behaviors.	Practice writing and story-telling; learn about animal behaviors
Week 6	*Mapping*—focused on distribution of spider webs	Analyze, explain, and discuss behavior data and spider distribution	*Timed Observations*—focused on squirrels	Students read about how animals interact with the environment to get their survival needs met.	Analyze data; consider mechanism behind animal behaviors and distribution across habitats
Week 7	*Team Observations*—focused on schoolyard habitats	*Infographic*—focused on habitats students observed	Students engage in an extended communication project about how organisms get survival needs met, what affects their distribution, patterns of organisms' interactions with the environment, and other learning.		Practice communication, storytelling, and teamwork; apply concepts; reflect

THE NEXT GENERATION SCIENCE STANDARDS

What would education standards look like if they were designed with real scientists and engineers to build the skills to understand and do science? The NGSS. Journals can support student engagement in these creative and useful standards.

There are many valuable and valid educational goals and frameworks that might guide what we do with our students. If you're intrigued by the NGSS approach to learning that we've shared in the "Curiosity Scaffolds or Question Generators" section of the book (p. 89), or you're already using the NGSS to teach science, you might find it helpful to delve a bit deeper into the structure of the standards and how they relate to journaling.

The NGSS have been adopted widely throughout the US, and many other states have adopted standards based on the NGSS. As we became familiar with the NGSS, we found them to be exciting, deep standards that emphasize getting students directly engaged in making observations, figuring out mysteries, and thinking scientifically—the very things we guide students to do in their nature journals. Here we're offering a deeper dive into the connections between the NGSS and journaling because of the depth and complexity of the NGSS, and to highlight the numerous opportunities for using journals as a tool to meet them.

The NGSS comprises three dimensions—the Science and Engineering Practices, the Crosscutting Concepts, and the Disciplinary Core Ideas—and there are ample opportunities for journaling to support students in meeting them.

Science and Engineering Practices are what scientists do. Scientists ask questions, design and carry out experiments, discuss ideas, read sources of information, and create models of phenomena in order to understand the world around them.

Crosscutting Concepts are how scientists think. Scientists use some overarching concepts and ideas about how the world works to ask questions and make explanations. The Crosscutting Concepts are big ideas that are found across all disciplines of science. For example, the idea that the structure of an object affects its function (and vice versa) is useful in physics, chemistry, and biology. These concepts have been used as the main scaffold for developing student discussion in our activities, helping students relate their learning to larger ideas about the world around them.

The Disciplinary Core Ideas are what science knows—for example, how water affects landscapes, how matter is transferred through an ecosystem, and how patterns of species change across different biomes. They are important scientific concepts for students to learn.

Many educators who first encounter the NGSS focus on the Disciplinary Core Ideas (e.g., "Matter moves among plants, animals, decomposers, and the environment") because they are the portion that is most similar to older standards. A key difference between the NGSS and earlier standards is that students need to

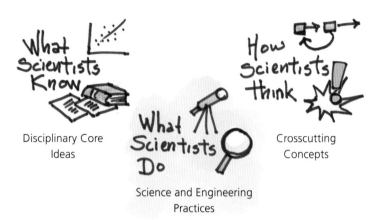

Disciplinary Core Ideas

What Scientists Do — Science and Engineering Practices

How Scientists Think — Crosscutting Concepts

do more than just memorize a fact (e.g., "Weather has predictable patterns"). In older standards, "process skills," such as conducting investigations or asking questions, were separated out from the ideas about science that students were supposed to learn. In the NGSS, students must engage in Science and Engineering Practices in order to learn concepts and understand science, and nature journaling can play a key role in fulfilling these standards.

In order for science lessons to truly meet the NGSS, all three dimensions must be present. Because the standards vary across grade levels, a full curriculum would need to lay out specific, thoughtfully sequenced lessons to teach the practices, concepts, and core ideas of science. That's not what we're attempting to do in this book; our goals are to help educators of all kinds—teaching students at all grade levels—get comfortable with teaching the integrated, holistic work that is nature journaling. We think journaling can be an effective and simple way to achieve the core goals of the NGSS—namely, engaging students directly with discovering knowledge themselves—and as naturalists, educators, and nature journalers, we find this development in mainstream education really exciting.

JOURNALING AND THE NGSS

Students can engage in many of the Science and Engineering Practices as they journal or as they discuss and build on their observations afterward. Journaling is an opportunity for students to make firsthand observations of the science concepts described in the Disciplinary Core Ideas.

Just as taking field notes is one part of a scientists' work, journaling is one step in the learning process for students, whether within the NGSS framework or in any other educational system. To meet these rigorous standards, students will need to continue their learning by discussing and analyzing their observations and ideas.

You can help students become better scientists when you situate journaling activities within wider learning experiences, offering opportunities for students to follow up and extend the learning done in their journals.

Here we will address how various journaling activities, approaches, and exercises can engage students with Science and Engineering Practices and help them apply Crosscutting Concepts in order to build understanding of Disciplinary Core Ideas.

SCIENCE AND ENGINEERING PRACTICES

Setting students up with a regular journaling practice and leading many of the activities in this book offer students numerous opportunities to engage in Science and Engineering Practices.

The next sections list examples of how students could engage with each practice through journaling. (See the NGSS website for complete descriptions of each Science and Engineering Practice.)

Our journaling activities always include opportunities for observation, discussion, and reflection. Reflection is key to helping students develop competency with these science practices. Even if students have used models in the past, if they have not had the chance to discuss models as tools for thinking, they might not be prepared to use models in new situations.

Observation isn't listed as one of the Science and Engineering Practices. That's because the team of scientists, educators, and researchers that developed the NGSS concluded that observation was *so fundamental to science* that it is a key part of every single science practice. Time spent teaching your students how to observe is time spent preparing them to succeed in science and with the NGSS.

Asking Questions and Defining Problems

A major part of science is asking questions—questions about phenomena, questions to drive an investigation, questions about data, or questions of peers and colleagues.

Students can engage in this science practice when they

- Ask questions during the activity *I Notice, I Wonder, It Reminds Me Of.* This routine gets students used to the act of asking questions, and cultivates curiosity.

- Practice asking questions based on their observations. This helps students start to build an intentional approach to questioning.

- Make a habit of recording questions in their journals. This helps students remember and build on past questions.

- Deepen their questioning skills by learning how to use scaffolds (such as the Crosscutting Concepts, the International Baccalaureate Key Concepts, and "Who, What, Where, When, How, and Why"—see page 92 for more).

- Think about how they might answer or investigate the questions they ask. This prepares students to sort questions based on whether they can be explored through science, and to decide on a method of inquiry.

- Follow up on a question during the course of a journal entry by generating multiple plausible explanations, attempting to answer the question through more observations, or asking more questions. This helps students use the journal to begin explaining phenomena, leading to engagement with other science practices.

The activity *Questioning Questions* is particularly helpful for supporting students' development with this practice. It is designed to expose students to different frameworks (including the Crosscutting Concepts) for generating questions.

Developing and Using Models

Using models to explain phenomena is an important practice in science and engineering. Models might include diagrams, drawings, physical replicas, mathematical representations, analogies, and computer simulations.

Students can engage in this science practice when they

- (In early grades) Make a labeled diagram, representing careful observations in a drawing. This fulfills the practice of modeling for early grades, and is a foundation for creating more complex models later.

- Use visual communication skills in any journal entry. This helps students get better at showing their thinking through diagrams, which is a foundational skill of modeling.

- Make precise observations, and distinguish them from explanations. Precise observations make for better models. The more substantial the observations, the more evidence students have with which to create an accurate model. As students learn about phenomena at hand through their own observations, they collect information they can use if they model it in the future.

- Describe components of a system in a journal. Recording the parts of a system can be a first step toward making a model of how the parts interact.

- Refine their models through discussion with peers. This is an essential part of the practice of modeling.

- Use the Crosscutting Concepts or other frameworks to ask questions about a phenomenon. Questions can lead to opportunities for generating explanations through the use of a model.

The activities *Phenomenon Model* and *String Safari* are especially suited to this practice. *Phenomenon Model* takes students through a process of constructing a model to explain a phenomenon, beginning with observations, then labeling unseen and seen forces, then using visual diagrams and descriptions to attempt to explain what is happening. This is a translatable approach. *String Safari* includes an introduction to system modeling.

Students can also take their work from any journaling activity during which they observed an effect, process, or phenomenon (such as that of a decomposing leaf in *Change over Time*) and use these observations as the foundation for a model to explain the phenomenon.

Planning and Carrying Out Investigations

Scientists and engineers plan and carry out investigations in the field or laboratory, working collaboratively and individually to gather data that will be used to support or refute explanations and to answer questions. These investigations must employ systematic methods for data collection.

Students can engage in this practice when they

- Sort questions in their journal and come up with initial approaches for answering them.

- Gather qualitative data by drawing and writing observations in their journal. Recording rigorous and specific qualitative observations is a skill that students can use in future investigations.

- Record quantitative data through counting, measuring, timing, or estimating. Recording rigorous and specific quantitative observations is a skill students can use in future investigations.

- Use their journal to record data over the course of a longer investigation.

- Write a detailed account of the methods of data collection.

- Learn field survey techniques and methods for gathering data, such as a sampling protocol. This gives students ideas for how they can gather information in future investigations.

The activities *Mapping, Cross Section, Hidden Figures, Timed Observations,* and *Biodiversity Inventory* are particularly helpful in supporting students' development with this practice. They all focus on using a field survey technique to gather and process information, and offer methods for data collection that students can use in future investigations.

Analyzing and Interpreting Data

Investigations and other scientific activities produce data. To use this data in explanations, scientists must analyze raw data to find patterns and trends.

Students can engage in this practice when they

- (In early grades) Do regular journaling activities, using writing and drawing to capture information and communicate ideas. This foundational approach to describing the world will

increase in sophistication later, and fulfills the expectations for this practice in early grades.

- Discuss interesting patterns or findings with peers after creating a journal entry. This practice leads students to use data and begin to analyze it to make sense of phenomena.

- Learn simple ways of visualizing data in the field (e.g., tally histograms, stem-and-leaf plots). These approaches to visualizing data make patterns apparent quickly.

- Use mapping or distribution data to develop explanations and make connections between spatial and temporal occurrences.

- In a different academic setting, learn new ways of representing data, then transfer it to their journaling approach.

- Analyze data when they're back in the classroom, using computers or other tools to process observations made in the field.

- Reflect on how errors in data collection might impact the validity of the information. This awareness is an essential part of scientists' thinking and this science practice.

- Revise an explanation based on new data or information.

The activities *Hidden Figures, Timed Observations, Change over Time, Mapping, Cross Section,* and *Comparison* are particularly helpful in supporting students' development with this practice.

Using Mathematics and Computational Thinking

Numbers are one language through which scientists describe observations and engage in study.

Students can engage in this practice when they

- Learn simple methods for organizing data as they collect it, such as stem-and-leaf plots or bar graphs.

- Measure, count, time, estimate, and otherwise use numbers to describe their surroundings.

- Make decisions about when to use quantitative rather than qualitative data to describe a specific part of a phenomenon.

- Use calculations to manipulate data in order to answer a question.

The activities *Hidden Figures, Timed Observations,* and *Biodiversity Inventory* are useful opportunities for students to develop this practice.

Constructing Explanations and Designing Solutions

One of the major goals of science is to come up with the best explanation based on all the available evidence. Scientists make explanations of cause-and-effect relationships to figure out how the natural world works.

Students can engage in this practice when they

- Make detailed observations of any part of the natural world. Students must construct explanations for observed phenomena or patterns. The richer the observations in students' journals, the more nuanced their explanations will be.

- Use the sentence starter "Could it be…" to come up with multiple possible explanations. This tentative approach helps students gain experience in explaining phenomena and cause-and-effect relationships.

- Engage in the discussion questions following a journaling activity, using their observations in order to construct explanations of patterns or phenomena.

- Refine an explanation through discussion with a peer, after more observations, or after consulting an outside source.

- Use journals to create a model to explain a phenomenon, effect, or other observed occurrence.

- Use Crosscutting Concepts or "big ideas" to frame, support, refute, or refine an explanation. Explanations should hold based on the tenets of science, and students should use their current understanding of science concepts to guide their explanations.

The Crosscutting Concept and general discussion questions after every activity in this book will all engage students with this practice. The activity *Phenomenon Model* offers experience with making visual explanations as they are used in modeling.

Engaging in Argument from Evidence

Argumentation is the process by which explanations and solutions are reached. Scientists refine their reasoning, engage in productive disagreement, critique evidence, and attempt to find the best possible explanation through argumentation.

Students can engage in this practice when they

- Discuss ideas, observations, and nature mysteries with peers. This culture of discussion and disagreement is the foundation on which to develop scientific argumentation skills.

- Use their firsthand observations as evidence in an argument.

- Learn to distinguish between observation and explanation.

- Make a habit of noticing when they change their mind in a journaling entry. This makes the act of changing their mind, which is key to argumentation, less intimidating.

- Trace their thinking throughout a journaling entry. This makes students more aware of their thinking process, which informs their argumentation and communication skills.

- Evaluate the quality of the sources of information they consult in response to a journal entry. A large part of argumentation is assessing the quality of sources. Making this a habit

in journaling supports students in developing the ability to assess the quality of a source as a skill.

- Make decisions about what information to record and what to leave out. Part of argumentation involves discerning what evidence is relevant to a question and what is not.

- Engage in regular discussions about the different approaches to recording information or structuring a journal entry, assessing the merits and drawbacks of each.

- Change their minds based on evidence.

The activities *Timeline* and *Photo, Pencil, and Found-Object Collage* are opportunities for students to engage in this science practice. The observations they generate after most other journaling activities could become evidence in students' argumentation experiences.

Obtaining, Evaluating, and Communicating Information

Communication of ideas and critical assessment of ideas (others' and one's own) are key parts of science. This includes discussing, writing about, or otherwise sharing one's own ideas, or engaging with other sources of information to understand concepts or critique information. This practice is particularly connected to journaling, and is also one that represents a significant part of what field scientists do every day.

Students can engage in this practice when they

- Complete any journal entry. Using writing, drawing, and numbers to communicate ideas in their journal is developing science literacy, and hones thinking and communication skills. Journaling helps build the foundation of students' ability to communicate their ideas in writing, speaking, and other types of media.

- Create visual graphics such as diagrams or graphs in order to communicate specific information. This helps students learn how to communicate visually, and also prepares them to look at and learn from diagrams found in other sources of information.

- Engage with the journal entries of classmates, noticing different approaches to recording observations and thinking. This gives students practice accessing information, and may offer ideas for refining their approach to visual communication.

- Discuss their ideas after creating a journal entry. These thinking, listening, and speaking skills are critical aspects of the practice.

- Engage with texts and sources of information to gather more evidence or answer a question. Literacy with technical subjects and gathering information are important parts of this practice.

All activities in this book include this practice to some extent. *To Each Its Own, My Secret Plant, Infographic,* and *Photo, Pencil, and Found-Object Collage* put specific emphasis on making decisions around what information to include in a journal entry.

CROSSCUTTING CONCEPTS

Summaries of the Crosscutting Concepts appear on page 90 in this book's introduction.

Students can apply Crosscutting Concepts to their nature exploration, journal entries, and thinking afterward to focus observations, spark questions, and guide explanations. The next sections list some ways that students might apply these Crosscutting Concepts while journaling.

The Crosscutting Concepts that are particularly useful for studying natural phenomena at the level we describe in this book are Patterns, Cause and Effect, Stability and Change, Structure and Function, Matter and Energy, and Systems and System Models. The concept Scale, Proportion, and Quantity can also apply to natural phenomena but is harder for students to pick up and start applying right away in their journals.

Patterns

In nature, patterns are everywhere—in where flowers grow on a bush and where they don't, in the angles of crashing ocean waves, in the shapes of seeds, and in the structure of bird songs. We can use patterns to organize and classify different parts of the natural world. The presence of a pattern also often indicates an underlying mechanism or process at work. This tends to lead toward cause-and-effect questions. In nature, students can look for patterns, ask questions to describe the pattern, think about underlying causes, and relate the pattern to what they have seen before.

Students can apply this Crosscutting Concept when they

- Observe a pattern in nature.

- Through an activity like *Comparison* or *Mapping,* document observations through writing, drawing, and numbers. This will reveal patterns.

- Create simple graphs and visualizations of data to reveal patterns.

- Look for patterns in quantitative data.

- Attempt to explain the cause of a pattern, using the "big idea" that patterns are clues to underlying phenomena.

Cause and Effect

Cause and effect and pattern are buddies. When you investigate one, it tends to lead to the other. It's hard to study them separately from each other. We can think of nature as a world full of mysteries. These mysteries might be a pile of feathers under a

tree, a preponderance of holes in a group of leaves, the pattern of sediment along a stream, or the distribution of organisms in a tide pool. We can think of these mysteries as "effects," or evidence of processes at work. A major part of science is investigating and explaining the mechanisms, or causes, behind what we observe. Students can observe their surroundings, then wonder about possible causes of what they noticed. Students can also use their knowledge or current observations to predict future effects or outcomes.

Students can apply this Crosscutting Concept when they

- Learn to think of what they can observe in nature as "effects" of processes and mechanisms, and attempt to explain the causes.

- Through journaling, develop an understanding of a phenomenon or process, then predict how it might behave under different conditions.

- Find a "nature mystery" and attempt to explain it.

- Attempt to distinguish between causation and correlation in their thinking and explanations.

- Use the observations generated from a journal entry as evidence when they attempt to explain a phenomenon.

- Look at a pattern they observed or a pattern in data, and attempt to explain it.

Scale, Proportion, and Quantity

Nature exists from the very small to the very large—from the atom to the cell, from organisms to the planets. When we observe the world at different scales, we notice different properties and functions of a phenomenon. We can use numbers to describe some of these properties, and this concept can help us remember to think about how to quantify accurately, given the scale at which we are observing. Students tend to struggle with understanding vast sizes and distances, such as with space objects, and it's worth making comparisons so that students can develop their understandings.

Proportions are another mathematical tool for using numbers to observe. Proportions also help us think about structure. For example, we don't see birch trees the size of redwoods because their branches would come crashing down. The structure of a birch tree would not be able to support the mass of the branch. Students can observe phenomena from different scales, using quantification and proportions to take their thinking deeper.

Students can apply this Crosscutting Concept when they

- (Especially in early grades) Collect quantitative information in their journals. This helps students begin thinking about numbers and scale when they interact with phenomena.

- Think about how an object's size might affect its properties as they attempt to explain how a structure functions.

- Think about how the size of a population of organisms relates to the population's stability.

- Make decisions about what to quantify and how to measure it.

Systems and System Models

We can look at a crab and observe its outside, but are we seeing everything that is the crab? Certainly not. The crab is made up of its internal structures, of the oxygen it breathes, of the food it eats. If we think about all the parts of the crab and how they interact with and influence one another, we will come to a deeper understanding of how the crab works as a whole. The same is true for other parts of nature, such as an ecosystem, cloud, or river. Systems thinking can help us explore and explain phenomena that are too large, small, or complex to observe directly. Students can look at any phenomenon or natural area through a systems lens, identifying the parts and their interactions, and asking questions about how this creates the system as a whole.

Students can apply this Crosscutting Concept when they

- Identify and make detailed observations of a group of organisms, a part of nature, or a phenomenon. Identifying the parts of a system is the first step in creating a system model.

- Think about the seen and unseen forces that affect a system or phenomenon.

- Create a system model through diagramming a phenomenon, labeling its parts, and thinking about interactions among those parts.

- Integrate a system model with other approaches for recording information so as to explain a phenomenon.

Energy and Matter

In the natural world, matter changes form constantly—plants take in carbon dioxide and build their bodies from it; deer eat leaves and digest them; the matter changes form again as it becomes a part of deer or their exhalation or their waste. The wastes of these organisms are reduced by bacteria and fungi to nitrates in the soil and carbon in the air. The carbon dioxide eventually finds its way to a tree again. All of these statements are oversimplifications of complex molecular processes. Tracking how matter cycles into, out of, and within systems takes our thinking about phenomena and processes a layer deeper, to a scale where we are able to more accurately understand them. Students can look at parts of nature and ask questions about how matter is changing form. They can also look for evidence of matter changing form, studying how these changes appear in context. Students can also trace cycles of matter and flows of energy within a system.

Students can apply this Crosscutting Concept when they

- Make an ecosystem model, detailing interactions among organisms. This type of model can be modified to trace matter cycling and energy flow.

- Gather evidence of how a part of nature, such as a flower or a decomposing leaf, changes over time.

Structure and Function

An object's shape and substructure determine many of its properties and functions (and vice versa). This is true for landform features, living things, and all natural objects as well as the human-built world. One of the main uses of this Crosscutting Concept in nature study is looking at organisms' structures and thinking about how they function. Students can study anything from bird wings to seeds to antennae to urchin spines, closely observing structures and wondering about how they function in the context of the organism's environment.

Students can apply this Crosscutting Concept when they

- Make careful observations of a structure of an organism or natural feature. These careful observations can become what students use to explain how the structure functions.

- Compare the structures of one or more organisms, looking for similarities and differences, then attempting to explain the differences in function.

- Learn how to make a structural diagram, identifying and drawing the repeating parts of an organism or natural structure.

- Watch an organism's behavior, focusing on observing its structures in action and in the context of the organism's environment.

Stability and Change

Change in nature can occur slowly and steadily, or rapidly. Understanding what factors maintain stability or catalyze change helps us understand a phenomenon and make predictions about future changes. We can develop this kind of understanding of a phenomenon by studying periods of change or looking at evidence of change. Some changes occur at rates we can observe. Others take place on a time scale beyond what we can comprehend. Students might track how ice on a pond melts and refreezes over the course of a week or how a leaf changes color throughout the fall. This kind of observation can lead to all sorts of questions about the factors maintaining stability and promoting change, and to a better understanding of the phenomenon as a result.

Students can apply this Crosscutting Concept when they

- Observe a phenomenon over time, recording data about rates of change, changes in size, or changes in other types of properties.

- Make a "timeline" for flowers, fungi, decomposing leaves, or some other phenomenon, noting how structures and properties change during development or decomposition.

- Return to a place, observing and recording changes.

- Look for evidence of change in a natural area.

- When creating a system or phenomenon model, think about factors that cause change or maintain stability.

- After a significant event that impacts the local landscape (e.g., a storm), observe the effects of change.

DISCIPLINARY CORE IDEAS

If students study different parts of nature by engaging in Science and Engineering Practices and applying Crosscutting Concepts, they can build understanding of many Disciplinary Core Ideas.

Through journaling, students can develop their understanding of a number of Disciplinary Core Ideas; which ones come into play depend on the phenomenon they study, any Crosscutting Concepts that guide their thinking, and the learning experience the journaling activity is a part of. Most activities could set students up to think about a range of Disciplinary Core Ideas, depending on the phenomena of study, so we will not describe a complete list of connections.

At younger grades, students could use nature journaling to engage with almost all life science topics, many earth science topics, and some space science topics. For some subjects in the middle school grades and many high school subjects, journaling and engaging directly with natural phenomena can still provide foundational understanding of Disciplinary Core Ideas, particularly if students focus on modeling. At this more sophisticated level, students will also need to engage with supplementary data sets or technologies to reach some of the deeper understandings required by the Disciplinary Core Ideas.

Physical science topics and some of the more abstract parts of life and earth sciences are best focused on in a classroom, with tools and technology available to help students engage with phenomena that they cannot directly observe in nature.

Learning through the NGSS is not about engaging in one activity or one process. Separating out practices and Crosscutting Concepts can be difficult, because students might start out asking questions, then shift to constructing explanations, and then collect quantitative data, all in a very short span of time. Regardless, the type of learning students do in their journals will align with the type of learning called for by the NGSS.

THE COMMON CORE STATE STANDARDS

The Common Core standards are used in many school districts throughout the country They promote critical thinking and creative problem-solving skills.

Collaboration, communication, thinking, and engaging with information are at the center of the Common Core. Students are required to learn basic skills in math and language arts. They are also required to focus on learning different processes and coming up with multiple approaches to solving problems in math, putting emphasis on the methods in addition to the answer. In English and language arts, students are to learn how to communicate their ideas in writing and in discussion, using evidence, information from other sources, and their own observations and experiences to develop arguments.

With scaffolding and support, students can meet these rigorous standards. The next sections describe some of the ways our approach to journaling overlaps with the Common Core standards for math and for English language arts/literacy and technical subjects.

MATH

Lower grades: Journaling and nature observation are an opportunity to learn important math skills in context. At lower grades, students can count, measure, time, estimate, and otherwise quantify observations in their journal. These connect to the Counting & Cardinality and Measurement & Data domains of the standards. Quick calculations of distances, area covered, or other values in the field are opportunities to use the approaches in Number & Operations in Base Ten, Measurement & Data, and Geometry.

Learning basic approaches to visualizing data will add another layer to students' use of math in their journals, offering them an opportunity to fulfill another aspect of the Measurement & Data standards by finding patterns in information.

Middle and high school: At slightly higher levels, students may collect and manipulate data samples, perform calculations, or,

in high school, use logarithms or algebra to solve and represent more complex aspects of data.

As students' knowledge in math, calculations, statistics, and other subjects grows, they will be able to improve and apply these skills in their journals, using algebra, logarithms, and other approaches for data analysis to deepen their learning about natural phenomena. Learning these math skills should go hand in hand with students' development in understanding science concepts, which will require them to have more sophisticated approaches to math and computation. Depending on the calculation and context, journaling might help students fulfill requirements in Ratios & Proportional Relationships, Expressions & Equations, or Functions.

Statistical thinking at the late middle and high school levels is particularly relevant to journaling. It can be a valuable approach to analyzing data in large sets that students gather over the course of extended observations and investigations, and can also fulfill Statistics & Probability standards.

At all levels, journaling is an opportunity to visualize data and numbers and find ways of showing meaning numerically through different graphs or plots. These activities not only fulfill math standards but also will enhance students' learning of the subject at hand. Beyond specific math standards met through journaling, the act of regularly using math in the authentic and exciting context of the outdoors to describe phenomena, analyze data, and solve problems empowers students to use math in other settings.

ENGLISH LANGUAGE ARTS/LITERACY AND TECHNICAL SUBJECTS

The communication students do in their journal is directly related to many aspects of the Common Core standards in English language arts and technical subjects. As students describe

INTERSECTIONS BETWEEN THE COMMON CORE AND THE NGSS

The NGSS Science and Engineering Practices include such skills as thinking, listening, reading, writing, communicating, and engaging with other sources of information. These are also skills required by the Common Core State Standards, and this overlap was intentional on the part of the developers of the NGSS. When students are writing in their journals, using drawings to communicate visually, and talking about their ideas, they are fulfilling the Common Core and meeting the NGSS. This intentional intersection

between the two sets of standards means that time spent on science (if taught to the NGSS) and journaling is not a distraction from language arts standards but an aid in meeting those standards. Particularly, the Science and Engineering Practice of Argumentation, which includes critiquing the ideas of others, speaking that is grounded in evidence, and discussion as a mode of learning, also appears in the Common Core for math and language arts.

observations through writing and drawing, and think about how best to record their ideas, they build foundational literacy skills that support their writing.

Speaking and listening skills are also a critical part of the Common Core. Discussion as a part of nature study gives students practice listening to one another's ideas and evidence, critiquing reasoning, and building on ideas. These activities meet many of the Writing and Speaking & Listening portions of the CCSS. The Common Core also focus on developing literacy with technical subjects. When students turn to a field guide to identify a plant or a scientific article to read about methods for studying a natural subject, or talk to someone with expertise, they are developing literacy authentically and in context.

The general approach to journaling in this book connects to the following portions of the CCSS for English language arts: Reading: Informational Text; Writing; Speaking & Listening; and Science & Technical Subjects.

TEACHING SCIENCE AND INQUIRY: A DEEPER DIVE

Scientific methods (not "the" scientific method) provide a road map to answering questions in the field with humility and rigor.

Journaling is a useful tool for observation and thinking, but it's just the beginning of where the mind can go. As students spend more time journaling and exploring nature, they can deepen their approaches to inquiry and investigation, learning to move from observations, questions, and ideas toward further, self-directed research. This process can be exciting and dynamic.

This section offers a deeper dive into the inquiry approaches "Let's go see," "Look it up," and "Could it be, maybe" we shared in the chapter Inquiry, Investigation, and Scientific Thinking. Here, we also include information on how to consider sample size, assess the quality of the sources we consult in the research process, and how to intentionally propose and test hypotheses. This range of techniques can add to students' inquiry tool kit, enabling them to come at the process from a place of humility and to circle closer to the truth.

This kind of scientific inquiry can build on observations students make in any journaling activity or nature exploration session, but activities in the chapters Observation and Natural History; Inquiry, Investigation, and Scientific Thinking; and Numbers: Quantitative and Mathematical Thinking are particularly useful jumping-off points.

ANSWERS FROM OBSERVATIONS

Observations on the Spot: "Let's Go See"

"Do all the leaves on this branch have the same number of spines?" "How many ladybugs are on this stump?" "How many times will the owl call in the next 10 minutes?" Some questions can be answered immediately through observation. It can be fun to find an answer to a question right on the spot. When we make these in-the-moment observations, we must remember that we are getting an answer for that place, in that moment. If we observe the number of spots on ladybugs, we cannot assume that all ladybugs will display the same pattern.

A quick observation during one moment in time gives us a fairly small sample size. Small samples are more likely to yield extreme results (because they are more easily skewed by outliers) and to miss the range of variation in the larger population. Stating the sample size (e.g., nine out of ten ladybugs, ninety out of a hundred) helps us maintain honesty with our data and awareness of its limits.

If we want to increase our confidence in our answer or get a more accurate picture of what's happening, we can increase our sample size. The larger the sample size, the stronger the data. Many journaling activities, such as *Biodiversity Inventory, Mapping, Timed Observations, Change over Time,* and *Species Account,* are approaches for collecting data adapted from techniques commonly used by scientists in the field. They are a basic tool kit of protocols for data collection that students can repeat over time to gather specific kinds of data.

It's also important to consider sampling bias in these on-the-spot observations. Were all the plants that we measured in the shade, next to the creek, or near the road? If so, there may be systematic bias in the sample. On a hot day, students might spend more time making observations in the shade than in the sun. Fewer animal tracks may be counted in an area that is dense with poison ivy or blackberry brambles, not because there are fewer tracks there but because no person wants to go in there and count them. Collecting random samples solves this problem, but is more easily said than done. Watch out for bias as you develop sampling methods, and specify how and where the samples were taken. You can avoid some bias by simply asking, "Am I collecting data in a way that might skew or bias my results?" and then taking steps to correct the problem.

Observation over Time: "Wait and See"

There are some questions we cannot answer through immediate observation but that we might answer over time through continued encounters with the natural world as we "wait and see."

"What are common weather patterns in this area?" "When will the dragonflies hatch?" "Do the ravens fly inland at dusk every night?" If you spend a lot of time outside and pay close attention, there is a lot you can learn. If you sit quietly for long enough, you might see a caterpillar munching a distinctive pattern on a leaf, or a falcon dive off a skyscraper to catch a pigeon. If you wonder how deer behave in the rain, and spend a lot of time outside watching deer, someday you might find an answer to your question. Other nature mysteries might require weeks, months, or years of observation before reaching resolution.

Kevin Beals, science educator and river guide, recounted this story: "I spend a lot of time in rivers and streams. My friends and I kept finding sticks that had material missing from them—it looked like a beaver had chewed and sliced part of the stick away, or at least that was the best explanation we could come up with at the time. But as far as we knew, there weren't beavers in the areas where we found the sticks! We tried to think of other animals that might have caused this to occur. This puzzled us for years. Years later, I found a stick that was wedged between two boulders, below that season's high water mark. Looking at it, I saw that it had the same pattern as the others! The stick, caught between the rocks, had rubbed against the stone as the river rushed past it, eroding the wood. It was a delight to find the answer to the mystery, but it was also fun to not know the answer."

People who are in frequent contact with the land and water, such as farmers, hunters, boat captains, and trackers, often build a wealth of knowledge through this wait-and-see approach. Through repeated encounters with the landscape and organisms, they incrementally push the edge of their understanding further. Engaging in this kind of sustained observation can be a rich and rewarding process, and it is powerful way to gain new insight over time.

We can also approach wait-and-see questions with more structured investigations, designing methods for systematic observations of phenomena over time. Methods might include designating a time to record observations at a site once a day or week, or taking repeated measurements over a period of time. The activities *Change over Time* and *Timed Observations* are useful structures for this kind of inquiry. When students take repeated measurements, coach them to plan how they will ensure that their measurements are consistent each time they take them, to avoid skewing the data. This might include using markers or flags to show where measurements should occur, using the same measurement instruments each time, using consistent protocols for how measurements will be taken, and taking measurements at consistent times.

Inference in action: In this journal entry, the student asks a question that cannot be directly observed (how the tree fell). He then finds evidence against three alternative hypotheses. This is how it's done!

Include anything in your journal that is significant or relevant. This student noted, "I rolled in this pile of poo." Even after this, he maintains scientifically tentative language ("Might belong to a bear").

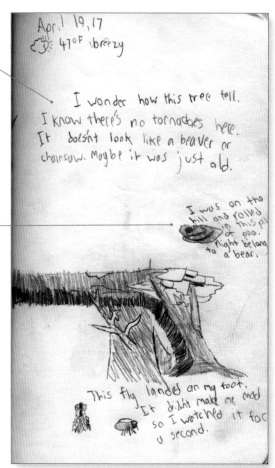

Seth, age 11

EXPLANATIONS, EVIDENCE, AND INFERENCE: "COULD IT BE"

When we have a "Why" question about a process or phenomenon in nature, it is difficult to come to a clear answer. "Why did the leaves at the end of this branch wilt?" The answer is likely to be a combination of causes and processes, many of which we cannot see or observe. We may not be able to answer such questions as "Who left behind these animal tracks?" "How high was the river when it was at flood stage?" "Why is this patch of ground drier than everything nearby?" or "Who let the dogs out?" through direct observation because we cannot directly observe the process, only the remaining evidence. These types of questions and most "Why" questions are rich fodder for inquiry and investigation. Students can use "Could it be," coming up with possible explanations in response to these questions. "Could it be that these leaves at the end of the branch are brown because the branch is broken here?" "Maybe this patch of ground gets more consistent sun during the hot times of day, and this has caused it to be drier."

Tentative Language

Scientists use the term *language of uncertainty* to refer to phrases such as "Could it be," "I think that," "Possibly," "Perhaps," "The evidence seems to show," and "Maybe." These phrases distinguish between what we know, what we do not know, and what we are still figuring out. They help the speaker and anyone listening maintain honesty about their level of certainty and

understanding. They also help students come up with more explanations because they do not get locked into their first explanation. Learning to use these sentence starters can help students develop critical thinking skills and scientific habits of mind. Coach students to use sentence frames such as "Could it be" and "Perhaps" as they make explanations and to come up with multiple plausible explanations for each question.

Evidence and Explanations

Nature is a place full of mysteries. Coaching students to infer explanations when they cannot directly observe an answer to a question (basing their explanations on evidence) helps them do the same work as scientists in all disciplines. Although students tend to need reminders to include their evidence, "Can you think of possible explanations for that?" is a question that students tend to immediately know how to respond to without needing to have the term *explanation* defined.

A statement such as "I think these tunnels may have been caused by termites, because there is dusty stuff in the tunnels. It looks like the termite poop I read about in a book" includes the explanation and evidence (the observations it was based on). It is a deeper and fuller statement than "That was caused by termites." Evidence connects an explanation to the observations that support it and makes thinking visible for others to critique or build on. Making explanations is fun and engaging, and students can add it in to their cycles of noticing and wondering. This can become a part of the fabric of your group's approach to nature study.

Digging In with Why

A more structured approach to exploring "Could it be" or "Why" questions is to form and test hypotheses and to infer which explanation is most likely to be correct.

To begin, generate multiple possible explanations (hypotheses), deduce predictions from each, and test the predictions by trying to find evidence that they are wrong. Each prediction opens a separate investigation, with its own methods for inquiry. To test a prediction, we might need to do a comparative survey (observing the same phenomenon under different conditions, or observing two different phenomena under the same conditions); a systematic observation (setting up a structured way to observe the predictions we have made); or a review of fossils, charts that show relatedness among organisms, or other records. A test may support (not prove) or show evidence against (not disprove) a hypothesis. Let's break down this process.

Forming Multiple Hypotheses

When we cannot directly observe the answer to a question, we may be able to come up with possible explanations based on what we can see and observe. Start your inference process by listing as many plausible explanations for a phenomenon as you can. "Could it be that the spots on the leaf were made by an insect feeding or a plant virus?" "Could it be that the round holes in the sand were caused by bubbles in sea foam popping, microcurrents in the water, or small animals?"

People tend to stop looking for alternative explanations once they think of one that seems reasonable. Help your students creatively generate more possibilities. Also remember that the real mechanism behind a phenomenon may be something you have not yet thought of. Keep an empty spot at the bottom of your hypothesis list or add a question mark to stand in for the inevitable "or something else."

Example: A student notices that there are few birds out on a winter day and wonders why. She develops three hypotheses: (1) that birds are less active when it is cold; (2) that the birds have migrated out of the area; (3) that the cold weather has killed many birds.

Making Testable Predictions

The next step in the process is to deduce predictions from a hypothesis. Ask, "If this were true, what would I expect to see?" If a prediction is something you can observe, it is testable. Each prediction comes embedded in a cloud of assumptions. If one of the assumptions is false, the prediction may also be invalid. Be explicit about your assumptions. (You know what they say…)

Example: The student chooses one of her hypotheses about the birds in winter, and develops predictions around it. If it is true that the cold weather has killed many birds, she would expect to see that

1. Bird populations drop during and after a cold snap.

2. There are more dead birds in roosting sites.

3. Smaller birds would be more affected than larger birds.

Next she considers the assumptions around prediction 1. She assumes the following: that the number of birds she observes indicates the number of birds in the area; that cold weather kills birds; that changes in the numbers of birds seen is due to bird deaths, not birds becoming less active in the cold or migrating.

Conducting Tests

Once you and your students have a clear testable prediction, go out and observe or measure it. Instead of looking for evidence that will support a hypothesis, try to falsify it. This is another good place to think about sample size and the potential for bias to creep into observations. More observations are better than fewer, and a defined system (ideally random) for picking samples is better than cherry-picking subjects that "look good."

Example: The student counts the species and numbers of birds around her school each morning and compares them to low nighttime temperatures. She finds no clear pattern in the data.

Evaluating Results

If a student reports that they've seen what they predicted, does that mean their hypothesis is true? Perhaps, but not necessarily. Some of their assumptions may have been wrong, making the test invalid. Even if their assumptions were correct, the hypothesis tested may not be what is driving the phenomenon; it may simply be correlated with the real driver.

In the world of science, we do not use the language "prove" or "disprove." That is the language of mathematics, where one can define the rules of a system, then do a proof. In science, we try to infer the rules by looking at the behavior of the system. We circle the truth, getting slowly closer, but knowing we may never arrive precisely on it. A good way to express this uncertainty is to employ a sliding scale from less confident to more confident. With each piece of new information, we can update our confidence level instead of stating a simple "right" or "wrong."

Example: The observed bird counts did not confirm the prediction. Does this mean that the hypothesis has been disproven? Perhaps. Again, there may be assumptions that invalidate the test. Our bird-watching student could conclude that if her assumptions are correct, it is less likely that the cold winter weather is killing the birds.

Provisional Acceptance

If the observations match predictions, they support the hypothesis. If many observations support the predictions, the evidence for that explanation is strong. If a hypothesis stands up to all our tests, it still may not be true. We have found a very useful explanation, however. We can use it to make more predictions. Instead of saying that the hypothesis is true, we may grant it *provisional acceptance*—acceptance, because this is the best hypothesis given the evidence; provisional, because we stay open to the possibility of observations that contradict the explanation, and are willing to discard it in the face of contradictory evidence or thinking.

Provisional acceptance is a powerful idea, but it is difficult to get used to. It is uncomfortable to accept an idea for the time being while acknowledging that it might not be correct. We want to have neat answers to our questions, but we live in a complex world. Causes and effects are often deeply intertwined, and provisional acceptance is as good as it gets.

Students can use this process of proposing possible explanations and testing hypotheses to refine their understandings of the natural world. Even if an explanation that a student lands on is likely to be wrong, it can offer valuable insight. In science, an idea that is not supported by the evidence is not a failure; it is a success. It is as important to find out what is not happening as it is to figure out what is. We can connect this to having a growth mindset, encouraging students to focus on the goal of gaining skills and knowledge over needing to be right.

CONSULTING SOURCES

Sources of information—field guides, scientific papers, and knowledgeable people, for example—can deepen our nature study with students. After students have conducted observations and discussions, their next step is to gather more information to start answering the questions initially raised. Many facets of nature have been studied before, and using the existing body of knowledge can take our students' understandings further. When faced with a question, many of us immediately type it into a search engine. Coach students to turn to resources mindfully, waiting to consult a source until they have a specific question and have exhausted their own knowledge or powers of observation.

For example, during nature study, students might

- Look at field guides to see range maps, behavior patterns, diet needs, and other information to refine explanations or deepen inquiry.

- Look at iNaturalist or other similar platforms to see where there have been recent sightings of organisms, or to discuss ideas with naturalists who have observed the same species.

- (For older students) See what studies are published about a similar phenomenon; look at the methods used in studies of similar phenomena or questions.

Field guides and natural history texts are helpful sources for identification, information about the life history of organisms, or descriptions of biological or earth systems and processes. Field guides with images and simple text are especially useful resources for younger students.

Scientific studies and papers can provide similar kinds of information, and also offer context for how the information was collected by including rigorous descriptions of methods, summaries of similar studies, and thoughtful analysis. College students and some high school students will be able to engage with these sources directly.

Talking to an expert allows students to ask specific questions and engage in dialogue about their ideas. Local scientists, naturalists, and native elders are rich sources of knowledge. Even if a teacher has an only slightly deeper understanding of a phenomenon, they can give students a bit of information that will take their thinking further, or direct students to a relevant resource at a key moment. But do not take expert opinion as the final answer. Always be willing to ask for the evidence behind an assertion. In science, we accept an explanation in proportion to the strength of the evidence supporting it.

Assessing the Quality of a Source

All sources are not created equal. Students must learn to assess the quality of their sources of information, particularly those found on the internet.

The internet can offer valuable scientific information, but the internet is also filled with misleading information, falsities, and faulty arguments. In an internet search, a list of articles pops up right away, but how reliable is the information in them? What evidence is it based on? How reliable are the studies that the arguments build on? Moreover, how has the algorithm of the software altered what is shown to you based on your past searches?

Older students should attempt to find and read the original studies being cited. Newspaper articles written about studies dilute and simplify the information and even get it wrong. If another person writes an article based on this simplified version of the original study, it becomes a game of Telephone, and the information is distorted. It is also removed from its original context, which would inform us of how the data was gathered in the first place. Reading an original study enables us to assess the methods for collecting data, see how data was analyzed, and look at the sources and previous studies cited by the author. This gives information that helps us decide how reliable the source is. Younger students can engage with such sources as nature videos, science books, field guides, science textbooks, or information they have learned from adults. These age-appropriate sources of information can support younger students in answering questions and making explanations; however, younger students may also refer to unreliable sources—movies, TV shows, fiction books, and the like.

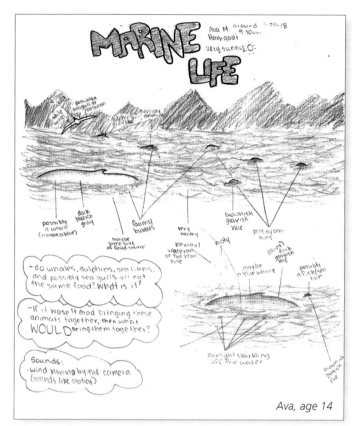

Ava, age 14

Take nothing for granted. If you are unsure of an identification, couch your ID with "Possibly," "Maybe," "Perhaps," or "Could it be."

An ideal time for students to engage with other sources of information is after they have done a journaling activity and discussed their ideas, or have done some initial research and thinking themselves. Students' firsthand journaling observations set them up with context and a frame of reference, enabling them to be more engaged, curious, and interested in the new information they encounter, and more likely to remember what they read or learn. For example, the activity *Infographic* begins with field observation, then prompts students to research the subject and add more information to the page from what they learned; this activity could be a follow-up to other journaling activities or learning experiences.

Citing Sources

When scientists refer to a fact, they will cite where it came from in the existing scientific literature. The statement, "Leaves are darker green on the top because there is more chlorophyll there" is interesting, but it says nothing about how the speaker obtained this information. Was it told to them, read in a textbook, found online, or observed from personal experience while dissecting a leaf under a microscope?

As educators, we should endeavor to maintain the connection between what we know and how we know it. Be clear with yourself and your students about what you know and how you know it; what you believe to be so, but have no data with which to back it up; and what you do not know. It's OK not to have the answer. You do not need to impress anyone by being the universal expert.

IT'S GOOD TO BE WRONG

Although we tend to feel that we have come to most of our beliefs through reason and observation, the likelihood is that we accepted many of them without real scrutiny or formed them before we had considered all the available evidence. To take in new knowledge, we often need to acknowledge that we're wrong about something. If we give ourselves and our students the permission to be wrong, we can fearlessly confront mysteries, and change our mind in the presence of evidence.

Making mistakes and being wrong are seldom portrayed as good things. In the media, those who change their minds can be portrayed as weak, unintelligent "flip-floppers." There is real pressure to demonstrate smarts by knowing facts. As educators, we can contradict this pressure by acknowledging changing our minds as a sign of strength, valuing flexible thinking and respect for evidence over being right or consistent.

Changing our minds in the presence of evidence is not easy to do. The ideas we currently hold have a rigorous defense system. Our brains selectively focus on information that reinforces those current ideas. When presented with evidence that goes against a preexisting explanation, many people tend to ignore or discredit the new information. We may even double down and believe an idea more when we're presented with a strong case against it (the backfire effect).

With careful attention, we can start to notice when the evidence contradicts our ideas, and learn to change our minds. Surprise is a tool for helping us change our minds and reframe being wrong. Psychologist Daniel Gilbert writes, "Surprise tells us that we were expecting something other than what we got, even when we didn't know we were expecting anything at all."[39] In other words, surprise is a sign letting us know we were wrong. Science teacher Charlie Toft encourages his students to keep a "surprise journal," in which they record every time they were surprised, including what they expected and what actually transpired. By noticing what we were surprised about and integrating this new information into how we think, we can change our minds. Toft says, "In the class culture, acknowledgment that you are mistaken has become dubbed a 'moment of surprise.'"[40] We can encourage our students to look for moments of surprise as they engage in nature observation and journaling. If students keep a record in their journals of how their thinking shifts and what causes them to change their minds, they will become more flexible thinkers. We can also look for opportunities to change our own minds and seek to deeply understand ideas different from our own.

QUICK STUDIES AND LONG INVESTIGATIONS

As we share inquiry skills and guide students to answer questions in nature using their journals as the foundation of their investigations, we also want to teach students how scientists explore the natural world, including acknowledging the extent and rigor of scientific studies. If students engage only in short investigations, answering questions in the moment with a few observations, they may not learn that scientists undertake careful studies over many years with large sample sizes and multiple observations. Teaching students about how scientists do their work and carefully design studies with rigorous approaches to data collection is a key part of building this science literacy. As students journal and learn about different facets of nature, they can read about methods employed in scientific studies to see how scientists have studied the same phenomena. This will deepen their understanding of scientific approaches and methods.

Yet students' short-term observations in the field do offer valuable information and can contribute to their understanding of how the natural world works. We don't want to invalidate students' ideas and observations by saying that they don't count. By teaching scientific inquiry based on their own investigations that begin with their observations and questions, we are helping students see themselves as part of the scientific community, learning the tools that real scientists use every day.

With a practice of nature journaling, we can encourage students to seek out the answers to their own questions and to think about the limits of the answers they come to. This practice builds the muscle of considering the reliability of information from any source, enabling students to be more thoughtful consumers of knowledge. It may even set them up to be thoughtful creators of knowledge in the future.

In fact, students may respect you more in the long term if they trust that you will be honest about the gaps in your knowledge, and they will be able to learn more because they will not always turn to you as soon as they have a question. When you speak to students, cite your sources and be honest about your level of confidence in your ideas. This practice demystifies the knowledge of the teacher, shows students how you learned what you know, and, in the process, models strategies for learning information.

You will understand part of any phenomenon you encounter. Start there, then ask, "What's going on here?" or "Why is this like that?" Do not stop there. Look beyond what you know to the edges of mystery. Find what is new to you or what seems odd and start digging. Let your students see you authentically wonder and explore.

252

Ask your students to cite their sources verbally and in their writing. For example, if they share a fact about deer during a discussion, they might say, "Male deer grow new antlers every year; I know this from a nature documentary I watched." You can also encourage students to remind one another to cite their sources and explain their thinking, asking one another, "How do you know that?" or "What is your source of information?" This practice can become a part of the group's culture in terms of how students interact, and helps students make a habit of keeping track of where their knowledge comes from.

Acknowledging how we know what we know is a practice of accuracy and also one of humility. Strive to trace the development of your own knowledge, and remember and credit the experiences, people, and sources that shifted your thinking.

ADDITIONAL RESOURCE

The BEETLES Project classroom activities *Evaluating Sources* and *Evaluating Evidence*, and field activity *Exploratory Investigation* (all found at beetlesproject.org)

FINAL THOUGHTS

Nature journaling is practical, powerful, creative, and fun. At its core, it is simply about being present and using words, pictures, and numbers to record observations, questions, and thoughts. Students can use these transferable skills to engage in other disciplines and subjects. Watch what happens as you expose children to journaling and give them opportunities to see how it works for themselves. As they make discoveries on their own and gain experience nature journaling, they'll buy in and build a learning tool kit that can last a lifetime.

You have everything you need right now to get started—not just with your own journaling process but with teaching it as well. Find your place to begin, and jump right in! There is no one right way to do it. It may be a little messy at first, but that's OK. Take risks and be willing to make mistakes in front of your children or students. The more you engage in the practice of journaling and share journaling with your students, the better it will get. This is a learning process that will unfold over time. Instead of getting lost in critiquing a single page or lamenting a challenging teaching moment, take the big view and play the long game. This is about learning how to pay attention, to wonder, to think. It is about guiding the young people in your life to meet the world with curiosity and awe. It is about finding joy and surprise in the daily delights of learning. The world—rich with beauty, wonder, and mystery—is waiting. Grab your journals and go outside.

APPENDICES

APPENDIX A: TOOLS AND MATERIALS

Nature journaling supplies are inexpensive, durable, and easy to manage. Putting the right materials in your students' hands will motivate them and make the logistics of journaling easier in the field.

You don't need special equipment to journal with students. Fancy paper and drawing pencils can make students feel that they need to Create Art instead of record observations and do science. Keep things simple, especially at the start. Once you and your students become more comfortable with journaling, slowly include more materials.

JOURNALS

The most important tool is a good journal. Ideally, each student should have a solid notebook that they can use throughout the year or program that can serve as a field writing surface and will stand up to field use. A well-made book will lend greater dignity to the process of taking notes. As students personalize and use their journal, they invest in both the book itself and the process of journaling. This journal also enables students, teachers, and parents to easily review or assess the student's development as a scientist, artist, and thinker.

Look for journals with these features:

- Sewn-in binding

- Stiff cardboard cover (many are too floppy)

- Medium size (composition book or equivalent)

- Blank, grid (quad ruled), or dot matrix paper (not lined paper)

- Heavyweight paper that prevents words and pictures showing through one side of the paper to the other

Homemade Journals

Students love using journals that they have made; they will often take better care of them and are less likely to lose them. You can make your own journals as a class project. There are many wonderful teacher-made tutorials online that give you creative ways to make a journal, along with step-by-step instructions to do it yourself. Make sure that the style of journal you use will hold up to field use. Journals held together by rubber bands tend to be short lived. Flimsy, messy, or awkward journals suggest to students that the journal and its contents are not important. Yet if the journal starts to feel precious, like something the students don't want to "mess up" by making a quick sketch, it becomes less useful as a scientific tool.

The following are two approaches to making solidly constructed journals:

Center-staple journals. If you have access to a center stapler, you can make small notebooks with ease. Stack one sheet of heavy 8.5-by-11-inch card stock (the stiffer the better) and five to ten sheets of white or graph paper. Use the center stapler to bind the sheets with two staples down the middle of the book, with the long portion of the staple on the card stock side of the pile. Fold along the line of the staples to make a 5.5-by-8.5-inch journal.

Bookbinding. You can find many simple bookbinding tutorials online. Look for options with a stiff cover, ease of opening, a reasonable number of pages, and no special tools required.

Low-Cost Nature Journals

You can also purchase journals. These journals tend to be sturdy and have the gravitas of being a "real book." Try to estimate how much journaling each student will do in a year to avoid having too much or not enough paper. If the cover is a light color, students can personalize their journals, adding art to the cover. Avoid using soft pencil or water-soluble pens, as they will smear and smudge.

The following are a few options available at the time of this printing:

Composition books. You may already use lined composition notebooks in your classroom. You can purchase the same kind of books with graph paper (or blank pages) for field journaling. We recommend graph paper. The horizontal lines help students write, and the grid supports drawing and open-ended note taking (placing titles, labels, call-out boxes, etc.). When students first get their journals, give them a little time to explore using the pages in creative ways: drawing vertical boxes for lists, using double or triple spaces for titles and subheadings, and so on. Look for composition books with a stiff cardboard cover for field use without a clipboard.

Bare Books (barebooks.com). These little hardbound journals are sturdy and ideal for field use. The standard Bare Book has twenty-eight pages (fourteen sheets). The Bare Book Plus has sixty pages (thirty sheets). The covers are white, allowing students to decorate their journals—just don't have them draw on the cover with water-soluble markers.

Sketch for Schools (sketchforschools.com). These spiral-bound customizable journals are made with quality paper and are fifty pages. We recommend the intermediate-weight textured paper. You can add your school logo or other custom page to the front (though this adds cost). You can also send the company any pages that you want to include, such as the cut-and-paste tool kit on page 274.

Artist's Loft™ Art For Everyone™ Sketchbook. This hard-bound sketchbook (110 pages) is available at some art supply stores and is usually sold at low cost.

Self-Printed Journals for Shorter Programs

Shorter outdoor programs in regional, state, and national parks or outdoor science schools often print their own journals for students to use. These journals are practical because they have an appropriate number of pages for the length of the program. These journals need not contain anything but blank pages, or pages with some simple structural elements (e.g., boxes or frames) to support students' visual layouts. A couple of additional pages that outline the schedule of the program, or show a site map, can be helpful for students. Avoid including pages that are "time killers"—word searches, crosswords, or fill-in-the-blank worksheets. Including these kinds of pages might seem like a helpful student management strategy, but they distract students from being in the moment and detract from the feeling of the journal being "their own."

For recommendations of types of pages to include in this style of journal, see the BEETLES Project resource "Model Field Journal Pages" (beetlesproject.org/resources/field-journal-pages/).

It is also fast, cheap, and simple to give students single sheets of paper and a hard surface for activities in shorter programs. Cut 9-by-11-inch sheets from cardboard boxes and attach binder clips to the top, and you have made cheap yet durable clipboards. This setup is ideal in circumstances where students will journal only once during a short program, or where materials must be as low cost as possible.

Customized Journals

A few modifications add greater functionality to any journal. When students get to modify their journals, they personalize the book and feel that their own book is special.

Name and contact information. Students should neatly write their name in permanent ink on the front cover to make passing out journals easier. On the inside cover, they should write: If lost, please contact (school or parent phone number).

Quantification tools, curiosity scaffolds, drawing tricks, prompts. Students can paste the naturalist quantification tool kit (appendix F) and the nature journal essentials (appendix E) pages into the last pages of their journal. These references help students engage with phenomena in the field, reminding them of the types of observations and thinking they can include in their journal.

Supply pocket. Use clear, wide packing tape and a piece of card stock or part of a file folder to create a flat storage pocket in the back of the journal.

Cover art. Encourage students to cover their journal with personalized art and quotes that are relevant to nature journaling. This does not need to be done in one sitting; students can add to the cover as they find new images or quotes that they like.

Wind band. Stretch an extra-large rubber band around the used or unused pages (whichever section is smaller). This makes the journal easier to open to the next blank page and prevents the pages from blowing around when sketching in a strong wind.

DRAWING AND WRITING TOOLS

No. 2 pencil. These are already in your classroom and do not feel like a fancy tool that could make students feel pressured to produce Art.

Pencil sharpener. Carry a handheld pencil sharpener in a Ziploc bag (for catching pencil shavings).

Non-photo blue pencil. These pencils are useful to block in basic shapes at the start of a drawing. Be sure to get Prismacolor's "Col-Erase 20028 NP Blue" pencil. A box of twelve pencils costs around $11.00.

Non-photo blue pencils are especially helpful for students who have a hard time drawing lightly. Changing pencils helps them create a boundary between the initial loose, light basic shape and deliberate drawing.

Ball-point pen. These are inexpensive and waterproof, and do not smear or smudge. Not being able to erase helps students move away from perfectionism and constantly erasing their lines; they just have to accept what is there and move on. These can also be useful for recording written notes in wetter conditions.

Colored pencils. A small set of around twelve pencils is all that is needed for field excursions. Before sketching in the field, make sure students take some time indoors to play with the pencils and learn how to overlap colors with a series of light layers. Colored pencils will smear and blur pencil lines, but play nicely over pen lines. If students' pencils are loose and prone to getting lost, try securing them with a rubber band before the next outing.

Pencil case. These can be handy for helping keep drawing tools organized.

NATURE STUDY TOOLS

A simple kit of nature study tools can greatly expand students' nature journaling approach. Give students a watch, and they will start timing everything they can think of. Offer a hand lens, and the micro world opens to them. Each new tool brings a different approach to observation.

The first time you introduce any new tool to your students, it will be a distraction. Expect this. It doesn't mean they can't use that tool, just that they need some time to explore how it works, and that means playing around with it. Try introducing a new tool in the classroom and giving students time to explore it when they are not in the field. Offer basic instructions for using the tool as well as ground rules for how to do so responsibly. This is especially important for tools such as binoculars or hand lenses, which can be easily damaged if handled improperly and can have greatly improved function if students know some basic techniques for using them.

The following are some essential items:

Cups and nets. Simple clear plastic cups (with lids!) and aquarium nets are great tools for empowering students to explore and catch critters. They are cheap, light, and easy to carry. Find cups at any drugstore or in bulk at a restaurant supply store, and aquarium nets (larger than the 2-inch model is best) at pet supply stores.

Hand lens or magnifying glass. A powerful magnifying loupe gets you close to the details of objects you can hold in your hand. It is worth investing in quality hand lenses; this is one product for which it is true that you get what you pay for. The least expensive plastic lenses scratch easily, do not give much magnification, or greatly distort the image. If possible, invest in quality glass lenses and think ahead about taking care of them. Look for models with a hole in the handle so that you can tie a string through the end and students can wear them around their necks. Store a class set in the compartments of a plastic bead box so that the cords don't tangle.

Magnifying box. A clear plastic box is indispensable for close looks at live insects. All the better if you can get one with a magnifying lens built into one side. Do not leave magnifying boxes in direct sunlight (with or without critters inside). Release any animals in the same place that you found them when you are done observing.

Ziploc bags. These have lots of uses, from catching pencil shavings to collecting leaves to bringing back interesting specimens from the field.

Small ruler. A small, hard millimeter ruler is useful for careful measurements. Measurements in millimeters are more easily converted and used in calculations than inches.

Retractable standard/metric measuring tape. A measuring tape is compact, lightweight, and useful for measuring distances larger than a ruler can (e.g., while measuring the distance between a set of tracks). Look for lightweight measuring tapes in hardware stores or sewing supply stores.

Watch. Students can use watches to document the time of observations, to time how long diving birds stay underwater, and to count how many times a bird sings in a minute or how many ants in a column pass one point in a minute. Stopwatch and countdown functions are very handy.

Tape. Clear plastic tape is great for adding flat found objects to journal pages. These could include pressed flowers (let them dry to avoid mold) or paper from an abandoned wasps' nest. Depending on your students' responsibility level, make a decision about whether to give students their own tape or distribute it yourself.

Binoculars. Use any leftover money in your budget to buy a few pairs of binoculars every year. Consider the Pentax Papilio 8.5-by-21. They are lightweight, durable, and great for kids. They magnify any distant object (as all binoculars do), but can also focus on objects as close as a foot-and-a-half away.

Thermometer. A small, portable thermometer can help students record more accurate thermal data.

Bags for carrying investigation kits. If students will be carrying their own journaling kits, large drawstring bags are an economical way to go. Avoid brightly colored cotton bags, as the colors may run if they get wet (and they will).

JOURNALING KITS FOR THE CLASS

Should you hold all the journaling supplies and pass them out when you want to use them, or let students carry and be responsible for their own materials? The answer depends on your students and the structure of your program.

The easier and more routine it is to get your gear together, the more you and your students will journal. A journaling kit is a simple set of supplies that is easy to grab and take into the field. These kits can be as simple as a notebook and a pencil. A more elaborate system might also include a shoulder bag with investigation tools (e.g., a ruler, hand lens, or loupe).

Long-Term Educational Settings

In long-term educational settings, such as a classroom or multi-week experience, develop a protocol for storing, maintaining, and transporting journaling kits. This could include an accessible place for the kits, set procedures for how to get them into the field and accessible to students, and student roles for maintaining the materials in them. For example, students could have rotating jobs: carrying journals, carrying investigation kits, sharpening pencils between outings, and doing a "sweep" outside to pick up forgotten items before the class heads back in. Students could also be individually responsible for the maintenance and transport of their own journaling kit.

Although it might take time initially to establish these routines, any teacher can tell you that the time spent developing a class routine will be worth the time saved later. If all you have to say is

"Let's get ready for a journaling outing!" and students know what to do without any further explanation, you and your students will journal more.

Some classes create nature journaling kits that contain all the students' favorite supplies and that hang on hooks near the door. Each student has their own kit. Students grab their journaling kits whenever the class goes outdoors to journal and explore together. Students can also bring their journaling kit out to recess or on a weekend family adventure to sketch and observe if they are responsible enough to return it.

Shorter Programs

For shorter-term experiences, such as those in informal or outdoor science settings, it is usually best for instructors and chaperones to carry journaling materials for students. It can be difficult for students to get into a routine of carrying their journal because they are in a new environment with many other protocols to learn. Because these students may also not be familiar with journaling, they will not have "bonded" with their journals and will easily leave them behind on the log where they had lunch. Initially, use protocols to collect materials at the end of a journaling experience. If you wind up using journals a lot, there is enough time for students to get invested in them, and if students seem responsible, you can let them carry the materials themselves. A backpack or journal bag for each student makes this easier.

In these shorter programs, program leaders should consider implementing program-level protocols to support the use of journaling. Educators in these settings often have many logistics to attend to, and little time to prep before teaching or to establish routines with students. The more the program can address the logistics of journals, the more likely instructors will be to use them.

This might include creating protocols for how journals are carried throughout the program and distributed to instructors (a backpack filled with journals and materials for a chaperone to carry works wonders), systems for collecting them at the end of the

day, and ways to make sure materials are readily accessible. Educators should be able to adjust this protocol if they want to, but having a system for journal use makes for one less thing to think about (and one less barrier to using journals).

Lost Journals and Materials

"But I lost my pencil!" Missing pencils and journals are the bane of every nature journaling leader. Having clear systems for managing journaling kits can help guard against this. Students who have been journaling for a longer time and are really into it are less likely to leave a journal behind, but it can still happen. In case a student loses a journal or forgets it on a field trip, carry a few extra journals, sheets of paper, and sharpened pencils so that a student can still complete the assignment.

If your program uses relatively light paper journals, you can punch a hole in one corner and tie a piece of string through the hole, enabling students to carry the journals around their necks. You can also attach a pencil to the string with masking tape.

Be careful if you are journaling around water. You do not want the journal to fall into a creek or tide pool. Keep the journals safely on shore to record observations from memory after the students have had a chance to explore, or use index cards binder-clipped to pieces of cardboard along with golf pencils to record data in the wet area. Students can glue these cards into their journals or transfer the data onto a clean page. And the golf pencils will float if dropped!

It can be extremely upsetting when a student loses a journal that contains a substantial amount of their work. This distress is magnified if the student's grade for the journal was not yet determined and they must repeat work. If a student knows they are prone to losing things and is worried they might lose a journal, offer the opportunity to take photographs of pages that are important to them. This is easy and quick to do with a smartphone or digital camera. The ideas remain intact, even if the journal does not. This should not, however, become an excuse for being careless with journals.

APPENDIX B: ACTIVITY SUMMARIES, LEARNING GOALS, AND POSSIBLE PHENOMENA

A single journaling activity can be repeated using different phenomena; Zoom In, Zoom Out, for example, is a very different experience when focused on a tree as opposed to a flock of pelicans resting on a rock. Similarly, the same phenomenon can be investigated through different activities. Many of the activities in this book could be used in an investigation of spiders. Although the same

Activity	Possible Learning Goals	Possible Phenomena
I NOTICE, I WONDER, IT REMINDS ME OF (INIWIRMO) Students use a three-part system to enhance observation, curiosity, and creative thinking.	• Learning a system that can be used to jump-start any observation • Focusing observation and improving memory • Helping students ask more questions • Helping students find relationships between observations and other things they have seen or experienced	An observable phenomenon; must be something that you, the teacher, also find interesting
MY SECRET PLANT Students choose a secret plant, observe it, and document it in their journal using pictures, words, and numbers. Then a partner tries to find their partner's plant using their journal entry.	• Learning how to create journal entries that include words, pictures, and numbers to communicate ideas and observations • Building visual literacy and communication skills • Building confidence and beginning journaling skills	Any area with enough plants that students can each study one and be physically spread out enough so that they are not right next to each other
TO EACH ITS OWN Students draw and describe one item from a set of similar objects (such as leaves or shells), then play a matching game in which they pair the objects to the notes their classmates made.	• Building confidence and beginning journaling skills • Building visual literacy and communication skills • Learning how to create journal entries that include words, pictures, and numbers to communicate ideas and observations	Leaves, shells, rocks, fossils, acorns, seed pods, pebbles—any objects smaller than your hand that are not too complex to draw
COMPARISON Students observe two similar natural objects or two areas subject to different conditions, recording similarities and differences between the two entities using words, pictures, and numbers.	• Generating a focused set of observations about a phenomenon's structures, patterns, or behaviors • Building toward conceptual understanding about structure and function of organisms, causes and effects of phenomena, or ecosystem functioning and dynamics	Soils; flowers of the same species; flowers of two similar species; plants of the same species in two areas with differing environmental conditions; rock types; rocks of the same type under different weathering conditions; grass under full sun vs. in the shade; leaves with and without insect damage; different types of insect damage; two different puddles; south- and north-facing slopes; bird activity in the morning vs. in the afternoon; structures (wings, beaks, feet, feathers, scales, antennae, etc.) of two different species

phenomenon would be used, each activity would help students think about spiders in a different way. Choose activities that will focus observations related to your learning goals. The more you know the phenomena in your area, the easier your planning will be.

Activity	Possible Learning Goals	Possible Phenomena
ZOOM IN, ZOOM OUT Students make a diagram of a natural phenomenon at life size, then "zoom in" to make a close-up of a feature and "zoom out" to make a diagram of the phenomenon in the context of its surroundings.	• Building visual literacy and communication skills • Learning how changing perspective affects the observations we make • Generating a set of observations about structures of an organism or phenomenon • Building understanding of and proposing explanations about structure and function, scale and proportion, and/or plant forms and growth	Any plant; landscape features; a herd of animals or a flock of birds (zoom in on beak, then focus on a single bird and its features, then flock structure or behavior); animal tracks (single track, track group, pattern in landscape); an ant colony (single ant structure, interactions between individuals on trail, and map of colony and trail network)
COLLECTION OR FIELD GUIDE Students record observations of things that belong in a category, such as seeds, leaf types, insects on a bush, or things that are gray. Students use words, pictures, and numbers to describe each object they observe, noticing differences and similarities.	• Noticing differences and similarities between natural phenomena and building a focused set of observations about things that belong in a specific category • Preparing students to think about possible causes of variation or likeness among the subjects they observe • Gaining a deeper understanding of a certain part of nature	Spider webs, seeds, leaf types, bark of different tree species, things that are red, insects on a bush, lichen, things that are dry, types of icicles, snow-form features, evidence of wind, things changed by water, things that are soft (or rough, spiky, etc.), clouds, things with strong odor (sort by category)
TIMELINE Students observe and make diagrams of an organism at different stages, such as flowers from bud to fruit, mushrooms at different stages, or leaves of the same species at different stages of decomposition.	• Generating a set of observations about life cycles of different organisms, and structures of plants • Building understanding of structure and function of organisms at different stages in the life cycle • Thinking about change over time, and causes and effects of changes	Flowers from bud to seed, mushrooms at different stages of growth, decomposing leaves or sticks of the same species, icicle growth, growth of seedling to mature plant, leaves turning color, any observable life cycle (frog, insect, etc.), livestock with observable age classes
STRING SAFARI Students use a circle of string to mark out a small study area on the ground, then draw and write about what they find inside. In an optional extension, students think of the area within the string as a system and make a model showing the interactions of the parts of the system, the inputs, and the outputs.	• Getting to know a habitat or environment through close and careful study • Generating a focused set of observations about the organisms, soil, and features in an area • Building a sense of place • Beginning to learn how to make system models, and to gain practice in doing so	Any habitat or environment where students can comfortably sit and study the ground (e.g., a local park, a vacant lot, forest floors, or open fields)

Activity	Possible Learning Goals	Possible Phenomena
ANIMAL ENCOUNTERS After seeing an animal, students verbally review observations of it, then translate their observations and memories to the page using words, pictures, and numbers.	• Building on a brief sighting of an animal and generating a focused set of common observations about it • Practicing how to make and remember precise observations • Building understanding of animal structures, behaviors, and life histories	Any brief sighting of an animal will do—a bird flying overhead, interacting insects, gulls squabbling, deer feeding, a snake slithering across the trail, salamanders walking...
SPECIES ACCOUNT Students focus on one species and record as many observations as they can using words, pictures, and numbers.	• Generating a focused set of observations about a species' structures, behaviors, distribution, or interactions with other species • Building understanding of ecosystem interactions, adaptations, and other science concepts	Any species that students can carefully observe—trees, bushes, plants, flowers, ducks, hawks, salamanders, spiders, insects, lizards, marine organisms...
FOREST KARAOKE Students listen to birdsong, then make diagrams to show pitch, loudness, and quality of the song.	• Learning an approach to diagramming sound • Building visual literacy and communication skills • Generating a set of observations about the patterns of different bird songs and bird behaviors • Building understanding of animal behaviors and strategies for survival	Any area where birds are singing
SOUNDSCAPE MAPS Students find a place to sit quietly, then make a graph to show natural, human, and machine sounds, using lines of different lengths, shapes, and colors to show variation in the quality of sound.	• Building a sense of place • Thinking about the interactions between living, nonliving, and human-generated things in an area • Thinking about how geography and history have influenced the distribution of things in a place • Thinking about the physics of how different kinds of sound travel	Any place students can spread out, sit, and listen—ideally, an area where students will be able to hear sounds from birds and other animals, as well as sounds from human activity (cars, tools, etc.)
MYSTERIES AND EXPLANATIONS Students explore nature while digging deep into curiosity. They propose and refine questions and pose possible explanations for what they see, without the need to be right.	• Developing and refining questions based on observations • Constructing natural explanations that are based on evidence • Practicing using the language of uncertainty	A small natural object that can be held in the hand, such as a leaf, pinecone, acorn, or seed pod; an area where students can spread out and check out their own mysteries

Activity	Possible Learning Goals	Possible Phenomena
QUESTIONING QUESTIONS Students create a journal entry about an interesting phenomenon, then learn how to use curiosity scaffolds (Crosscutting Concepts, 5W's1 + H, International Baccalaureate Key Concepts) to intentionally ask varied questions and make focused observations.	• Developing skills in asking varied questions and intentionally focusing observations • Asking more questions in journal entries	Earthworms in puddles after a rain, spider webs on a tree, lichen on a fence, cracks in mud, pigeons on a schoolyard, windblown sand or snow, ice formations, patterns of melting snow, patterns created by wind, animal tracks, leaf damage, reflections
MAPPING Students make a map of the distribution and arrangement of interesting natural features.	• Building visual literacy, communication, and spatial reasoning skills • Generating a focused set of observations about where a phenomenon occurs • Building understanding of the factors that affect the distribution of whatever is being mapped (e.g., how environmental factors affect the distribution of plants) • Building conceptual understanding of ecosystem dynamics and interactions	Ant trails, spider web locations, plants near a water source, regrowth after a forest fire, gopher holes, puddles after the rain, worm castings, trash in a schoolyard after lunch, windblown debris, areas used by birds around an agricultural field, weed growth in a vacant lot, snowmelt or icicle formation on roofs, tracks in sand or snow, sediment deposited by water (stream, pond, etc.), creek or river channels, currents
LANDSCAPE CROSS SECTION Students walk a transect (a straight line across a part of a landscape) and record zones of different types of vegetation, or other organisms.	• Building visual literacy, communication, and spatial reasoning skills • Generating a focused set of observations about how a phenomenon is distributed over an area with a gradient of environmental factors • Building conceptual understanding of how environmental factors (e.g., elevation, amount of sun, water, or soil type) affect organisms' distribution, or cause variation in a landscape	Tidal zonation, forest-to-meadow transition, snowmelt to bare ground, stream channel, across a ridge from north- to south-facing slopes, transition from managed field (agricultural, lawn, or sports field) to unmaintained ground, pond (and surrounding vegetation), snow drifts after a windstorm, dunes to beach to surf
PHENOMENON MODEL Students write and draw observations of a phenomenon in process (e.g., a cloud formation or eddies and standing waves in a river), ask questions about what they see, formulate possible explanations for the causes of the phenomenon, then make diagrams to show these explanations, using labels and arrows to show seen and unseen forces.	• Generating a focused set of observations about a phenomenon in process • Learning skills for and approaches to modeling, making explanations, and figuring things out • Building toward a longer learning experience to deepen understanding of the phenomenon and related science concepts	Ice formations, snow on rooftops, cloud formations, twisted leaves, sundogs or rainbows, patterns of ripples on water, leaves decomposing

Activity	Possible Learning Goals	Possible Phenomena
TEAM OBSERVATION Students form groups and study a natural object, such as a tree, landscape feature, or species. Then they discuss and share their findings.	• Building rapport and teamwork in a group • Building scientific practices and identity • Building collaboration and communication skills • Generating a focused set of observations and questions about a species or natural phenomenon • Building understanding of a species or phenomenon through shared study	Any organism or phenomenon large enough for all your students to study at once—a large tree, several shrubs of the same species, everything living within a large circle drawn on the ground, an ant mound and network of trails
WRITING TO OBSERVE, WRITING TO THINK Students focus on a subject in nature, and practice using different writing approaches to capture their observations and thinking.	• Using different forms of writing to express ideas and information • Building writing and communication skills • Integrating writing and drawing	Any part of nature that can be used for the full observation period—leaves, trees, slow-moving or cooperative animals, or landscape features
EVENT COMIC After witnessing an event or process, such as an animal behavior or sudden weather occurrence, students make a comic book–style journal page to describe what happened.	• Learning skills to creatively structure journal pages and show movement, behavior, and processes • Building a common set of observations and a clear memory of an event in nature • Building toward a meaning-making conversation in which students use science concepts to explain the event they saw • Building conceptual understanding of animal behavior, or other natural phenomena	Any interaction between two or more organisms (e.g., courtship behavior; a predation event, successful or not); any interesting behavior of a single individual; foraging/feeding behavior; an observed phenomenon with distinct stages (hailstorm after thunder and lightning, rain that clears to rising mist)
EVENT MAP At their own pace, students travel along a trail and stop to record what is interesting to them, using drawing and writing.	• Exercising choice in practicing journaling skills • Experiencing autonomy in movement and focus • Using a structure for reflection and connection to place	Any area where your group can move along a trail or through a natural area; ideally, a place with a few different ecosystems or features that students can explore
POETRY OF PLACE AND MOMENT Students use "I notice," "I wonder," and "It reminds me of" to write poems about nature and their personal experience.	• Self-expression and reflection • Remembering the place and moment • Sharing of personal and cultural perspectives	Locations that inspire, such as places with a view, below an ancient tree, or near water—any place that students can spread out, sit, and write comfortably

Activity	Possible Learning Goals	Possible Phenomena
SIT SPOT Students find a private place to sit, reflect, and connect with nature.	• Self-expression and reflection • Building a personal relationship with nature • Experiencing autonomy and choice in nature	Any place students can spread out and sit by themselves—a hillside, garden with several tall plants, creekside
INSIDE OUT Students use diagrams to describe different views of an apple or a mushroom, learn how architectural plans use diagrams to efficiently capture data, then discuss how they can use diagramming in future journal entries.	• Learning how to use diagrams to document observations, and building other basic drawing skills • Building visual literacy and communication skills • Generating a set of observations focused on structures and patterns • Building toward conceptual understanding of patterns among organisms or structure and function	Apples, button mushrooms, or any object that can be held in the hand and cut into sections
NATURE BLUEPRINTS Students learn how to efficiently diagram a complex natural object with repeating parts, such as a pinecone.	• Developing drawing skills and learning how to use diagrams to document objects' complex shapes • Building visual literacy and communication skills • Generating a class set of observations focused on structures and patterns • Building toward conceptual understanding of patterns or of structure and function in natural objects	Pinecones, bunches of lupine, some other complex plant feature with repeating parts
INFOGRAPHIC Students record observations in the field and add to these based on their research.	• Integrating observations and research • Exploring graphical display of information • Building written and visual communication skills	Any interesting found object or phenomenon that can be researched for more information
PHOTO, PENCIL, AND FOUND-OBJECT COLLAGE Students make a diagram of a natural object such as a rock, leaf, or flower, then take one photograph, which they paste into their journal. Then they discuss the types of information captured through each medium.	• Building science practices and visual literacy and communication skills • Thinking about different approaches to capturing information in their journals • Integrating technology thoughtfully into journaling experiences	A natural object that a group of four students can examine, such as a rock, small plant, flower, or leaf

Activity	Possible Learning Goals	Possible Phenomena
HIDDEN FIGURES Students learn skills of quantification, including counting, measuring, timing, and estimating, practicing each approach and adding their data to a journal entry.	• Building basic skills in quantification that will be applicable in the future • Learning how to include numbers in journal entries	An area with different plants, features, or phenomena; things that move or animals that repeat behaviors
BIODIVERSITY INVENTORY Students go to an area and use quick diagrams and notes to make a record of each plant species (or other type of organism) they find there. In an optional extension, students determine the richness and evenness of the area, then discuss the implications for ecosystem health and resilience.	• Learning about the range of organisms in an environment • Understanding how diversity relates to the functioning of ecosystems, and other concepts related to ecosystem resilience and functioning • Discussing impacts of humans and other factors on ecosystems	An area with a diversity of distinguishable plant species; should be accessible, not overgrown with brambles, poison ivy, or angry badgers, and where students can freely explore
TIMED OBSERVATIONS Students use a protocol to record organisms' behaviors over a period of time, then make a simple graph to determine the most common behaviors.	• Generating a focused set of observations about organism behavior and life history • Building understanding of behaviors that help organisms survive • Learning investigation techniques and sampling protocols	Ducks in a pond (particularly interesting when you can see courtship behavior in addition to feeding, resting, and alert birds), blackbirds in a picnic area, pigeons, lizards on a rock (basking push-ups), animals in a colony (ground squirrels, seals, seabirds), herds of elk or deer
CHANGE OVER TIME Students make a diagram of a natural object such as an apple or seedling, making notes about its growth or decomposition by observing structures and measuring distances between key features.	• Generating a focused set of observations about how a phenomenon changes over time • Building understanding of processes such as growth and decomposition • Learning techniques for systematically collecting data	A decomposing apple core or orange, plant seedling, metamorphosing insect, opening buds, growing mushrooms (these change more quickly than you would expect), icicles, seasonal changes in a deciduous tree, a decomposing carcass, seasonal changes in birds on a pond, or some other natural object students can make repeated observations of over time

APPENDIX C: NGSS CONNECTIONS

This table outlines possible connections between journaling activities and the NGSS. Because each activity can focus on different phenomena and you can bring out different concepts in the discussion at the end of a journaling activity and in follow-up experiences afterward, nature journaling can address a range of standards. We have listed many standards for each activity to reflect these possibilities, but when you teach an activity with students, it will support only one or two Crosscutting Concepts and Disciplinary Core Ideas.

The Disciplinary Core Idea (DCI) codes refer to different areas of content within the NGSS (LS, for example, is an abbreviation for life science). For the full list of DCI codes and abbreviations, see the NGSS website: nextgenscience.org.

Activity	Science and Engineering Practices	Crosscutting Concepts	Disciplinary Core Ideas
I NOTICE I WONDER, IT REMINDS ME OF	Obtaining, Evaluating, and Communicating Information	Patterns, Cause and Effect, Structure and Function	
MY SECRET PLANT	Obtaining, Evaluating, and Communicating Information	Patterns, Cause and Effect, Structure and Function	LS1.A, LS1.B, LS2.A
TO EACH ITS OWN	Obtaining, Evaluating, and Communicating Information	Patterns, Cause and Effect, Structure and Function	LS1.A, LS3.B, LS4.B, LS4.C
COMPARISON	Asking Questions; Obtaining, Evaluating, and Communicating Information; Constructing Explanations	Patterns, Cause and Effect, Structure and Function, Systems and System Models	LS1.A, LS1.B, LS3.A, LS3.B, LS4.A
ZOOM IN, ZOOM OUT	Obtaining, Evaluating, and Communicating Information	Structure and Function; Cause and Effect; Scale, Proportion, and Quantity	LS1.A, LS2.A
COLLECTION OR FIELD GUIDE	Obtaining, Evaluating, and Communicating Information; Constructing Explanations	Patterns, Cause and Effect, Systems and System Models, Structure and Function	LS1.A, LS2.A, LS2.C, LS4.C, LS4.B, ESS2.A
TIMELINE	Obtaining, Evaluating, and Communicating Information; Constructing Explanations	Patterns, Cause and Effect, Stability and Change, Structure and Function	LS1.A, LS1.B, LS4.C
STRING SAFARI	Asking Questions; Constructing Explanations; Obtaining, Evaluating, and Communicating Information; (optional) Developing and Using Models	Patterns, Stability and Change, Energy and Matter, Systems and System Models	LS2.A, LS2.B, LS2.C
ANIMAL ENCOUNTERS	Asking Questions; Obtaining, Evaluating, and Communicating Information	Patterns, Cause and Effect, Energy and Matter, Structure and Function	LS1.A, LS1.B, LS2.A, LS2.B LS2.D, LS4.A, LS4.B
SPECIES ACCOUNT	Asking Questions; Constructing Explanations; Obtaining, Evaluating, and Communicating Information	Patterns, Cause and Effect, Systems and System Models, Energy and Matter, Structure and Function	LS1.A, LS1.B, LS2.A, LS2.B LS2.D, LS4.A, LS4.B

Activity	Science and Engineering Practices	Crosscutting Concepts	Disciplinary Core Ideas
FOREST KARAOKE	Obtaining, Evaluating, and Communicating Information; Constructing Explanations	Patterns, Cause and Effect	LS1.A, LS2.D, LS1.B
SOUNDSCAPE MAPS	Obtaining, Evaluating, and Communicating Information; Constructing Explanations	Patterns, Cause and Effect	LS1.A, ESS3.C, LS2.C, LS4.D
MYSTERIES AND EXPLANATIONS	Asking Questions, Constructing Explanations	Patterns, Cause and Effect	
QUESTIONING QUESTIONS	Asking Questions; Planning and Carrying Out Investigations; Obtaining, Evaluating, and Communicating Information		
MAPPING	Planning and Carrying Out Investigations; Constructing Explanations; Obtaining, Evaluating, and Communicating Information	Patterns, Cause and Effect, Stability and Change, Systems and System Models	LS2.C, LS4.C, LS4.D, ESS2.A, ESS2.B, ESS2.C, ESS2.D, ESS3.C
LANDSCAPE CROSS SECTION	Planning and Carrying Out Investigations; Constructing Explanations; Obtaining, Evaluating, and Communicating Information	Patterns, Cause and Effect, Stability and Change	LS1.A, LS1.C, LS2.A, LS2.B, ESS2.A, ESS2.B, ESS2.C, ESS2.D, ESS3.C
PHENOMENON MODEL	Developing and Using Models, Constructing Explanations	Patterns; Scale, Proportion, and Quantity; Systems and System Models; Energy and Matter; Stability and Change	
TEAM OBSERVATION	Planning and Carrying Out Investigations; Constructing Explanations; Obtaining, Evaluating, and Communicating Information	Patterns, Cause and Effect, Systems and System Models, Energy and Matter, Structure and Function	LS1.A, LS2.A, LS2.B, ESS2.A, LS2.C
WRITING TO OBSERVE, WRITING TO THINK	Constructing Explanations; Obtaining, Evaluating, and Communicating Information	Patterns, Cause and Effect	
EVENT COMIC	Constructing Explanations; Obtaining, Evaluating, and Communicating Information	Patterns, Cause and Effect, Structure and Function	LS1.A, LS2.A, LS2.B, LS2.D, LS4.B

Activity	Science and Engineering Practices	Crosscutting Concepts	Disciplinary Core Ideas
INSIDE OUT	Developing and Using Models; Obtaining, Evaluating, and Communicating Information	Patterns, Structure and Function	LS1.A, LS3.B
NATURE BLUEPRINTS	Obtaining, Evaluating, and Communicating Information	Patterns, Cause and Effect, Structure and Function	LS1.A, LS3.B
INFOGRAPHIC	Obtaining, Evaluating, and Communicating Information	Patterns, Cause and Effect, Structure and Function	
PHOTO, PENCIL, AND FOUND-OBJECT COLLAGE	Obtaining, Evaluating, and Communicating Information; Engaging in Argument from Evidence		
HIDDEN FIGURES	Analyzing and Interpreting Data, Using Mathematics and Computational Thinking		
BIODIVERSITY INVENTORY	Planning and Carrying Out Investigations; Constructing Explanations; Obtaining, Evaluating, and Communicating Information; (optional) Analyzing Data and Mathematics; (optional) Computational Thinking	Patterns, Stability and Change, Cause and Effect	LS2.C, LS4.C, LS4.D
TIMED OBSERVATIONS	Asking Questions, Planning and Carrying Out Investigations, Using Mathematics and Computational Thinking, Constructing Explanations	Cause and Effect, Stability and Change	LS1.C, LS1.D, LS2.D, LS4.C
CHANGE OVER TIME	Asking Questions; Obtaining, Evaluating, and Communicating Information; Constructing Explanations; Planning and Carrying Out Investigations	Stability and Change, Structure and Function, Systems and System Models, Cause and Effect	LS1.A, LS1.B, LS1.C, LS2.B, LS3.A

APPENDIX D: EVALUATION RUBRIC

Name

Date

Location

Project

Basics

- ❑ Date, place, and time
- ❑ Weather/temperature
- ❑ Words, pictures, and numbers used together
- ❑ Format features (titles, boxes, dividers, arrows, etc.)
- ❑ Observations and descriptions (I notice)
- ❑ Connections (It reminds me of)
- ❑ Questions from observations (I wonder)

Pictures

- ❑ Accurate drawings and diagrams show observations
- ❑ Enlargement (detail inset) of interesting parts
- ❑ Multiple views of same subject (top, side, etc.)
- ❑ Cross or longitudinal section views
- ❑ Color or notes about color
- ❑ Habitat sketches
- ❑ Maps

Additional expectations

Words

- ❑ Paragraphs, single sentence, sentence fragments, bullet points, and labels used where appropriate
- ❑ Notes about methods or procedure
- ❑ Explanations (Could it be?)
- ❑ Distinguishes between observation and explanation
- ❑ Uses language of uncertainty

Numbers

- ❑ Counts or estimates features and individuals
- ❑ Times observations and speeds
- ❑ Measures lengths
- ❑ Shows scale (life-sized, magnification, etc.)
- ❑ Samples data (more than one measurement)

Comments

Total points received:

Total points possible:

Name

Date

Location

Project

Basics

- ❑ Date, place, and time
- ❑ Weather/temperature
- ❑ Words, pictures, and numbers used together
- ❑ Format features (titles, boxes, dividers, arrows, etc.)
- ❑ Observations and descriptions (I notice)
- ❑ Connections (It reminds me of)
- ❑ Questions from observations (I wonder)

Pictures

- ❑ Accurate drawings and diagrams show observations
- ❑ Enlargement (detail inset) of interesting parts
- ❑ Multiple views of same subject (top, side, etc.)
- ❑ Cross or longitudinal section views
- ❑ Color or notes about color
- ❑ Habitat sketches
- ❑ Maps

Additional expectations

Words

- ❑ Paragraphs, single sentence, sentence fragments, bullet points, and labels used where appropriate
- ❑ Notes about methods or procedure
- ❑ Explanations (Could it be?)
- ❑ Distinguishes between observation and explanation
- ❑ Uses language of uncertainty

Numbers

- ❑ Counts or estimates features and individuals
- ❑ Times observations and speeds
- ❑ Measures lengths
- ❑ Shows scale (life-sized, magnification, etc.)
- ❑ Samples data (more than one measurement)

Comments

Total points received:

Total points possible:

APPENDIX E: CUT-AND-PASTE JOURNAL STRATEGIES

Fits a standard composition book.

I NOTICE... (OBSERVATIONS)

Notice all that you can. Describe how it looks, what it does, and where it lives. Pay attention to little details and the big picture. Say your observations out loud, then put them all on paper.

I WONDER... (QUESTIONS)

Get curious and find mysteries. Ask questions about your observations, and use this list to come up with more questions:

Who Identification: Who is it? Who was it? Who will it be? Whose song is that? Who made these tracks?

What Process: What happened? What is happening? What will happen next? What does it do? What causes this? What does that do?

Where Space and location: Where is it? Where was it? Where will it be? What is the territory of this animal? Where does it rest, feed, or complete its life cycle? Where does this kind of animal live?

When Time: When did it happen? When will it happen? How long will it stay underwater? When does the migration start/end? When did river otters arrive in this area? When do the birds begin to sing each morning?

How Structure, function, and process: How does it work? What are the parts? How do they work together? How does this flower?

Why Cause and effect, meaning, and purpose: Why did this happen? Why is it this way? What events led to this outcome?

IT REMINDS ME OF... (CONNECTIONS)

Make connections and comparisons between what you observe now, and things you have seen, learned, or felt before. Ask yourself: What does this remind me of? How is this like that? How is it different?

COULD IT BE...(EXPLORE AND EXPLAIN)

Try to figure out the mysteries. Make your own explanations: "Maybe the bird keeps going to that tree because it has a nest there..." and include evidence: "...and my evidence is that I can see it carrying insects to the tree." Use words, pictures, and numbers to show your ideas and explanations. Try to disprove your ideas with evidence and be willing to change your mind. It is OK to be wrong.

PICTURES

Make fast sketches and simple diagrams; draw the same object from different angles; create maps, cross sections, storyboard sequences, and close ups; zoom out for the big picture. Connect related items with frames, arrows, and divider lines. Use icons to add emphasis.

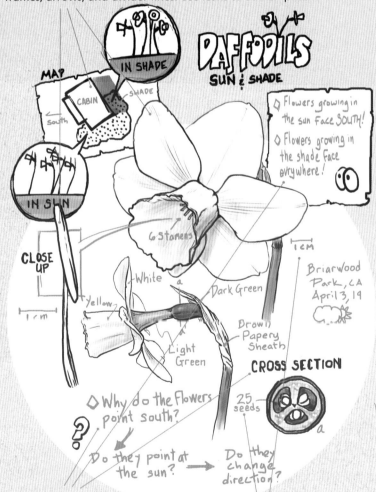

WORDS

Use labels, bullet points, lists, sentence fragments, and full paragraphs all on the same page. Change your fonts and writing style to create titles, headings, and to add emphasis. Record the date, location, weather, time, or any other relevant context.

NUMBERS

Count, measure, and time. If you can not count, estimate. Show scale on maps and drawings. Take more than one measurement to try to get more accurate data. Create tables and graphs to help you track and visualize your data.

APPENDIX F: CUT-AND-PASTE QUANTIFICATION TOOL KIT

This page fits a standard composition book. Do not reduce or enlarge it (to preserve ruler scale).

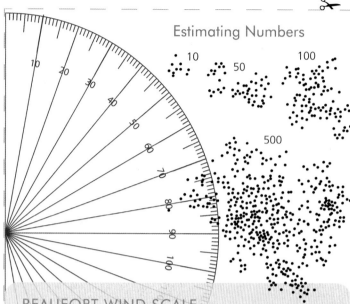

Estimating Numbers

10 · 50 · 100 · 500

Percent Cover

How cloudy is the sky? How much of the ground is covered in leaves? Use these circles to help you estimate.

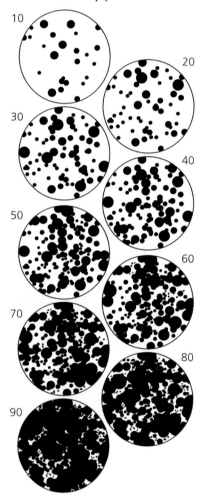

10 · 20 · 30 · 40 · 50 · 60 · 70 · 80 · 90

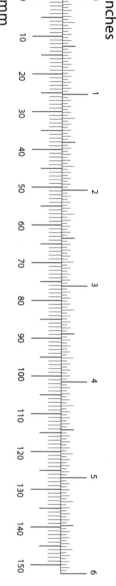

Inches / mm ruler (0–6 inches / 0–150 mm)

BEAUFORT WIND SCALE

0 Calm <1 mph (< 1 km/h) Smoke rises vertically. Flat, glassy water.

1 Light air 1–3 mph (1.1–5.5 km/h) Smoke drift indicates wind direction. Leaves and wind vanes are stationary. Ripples without crests.

2 Light breeze 4–7 mph (5.6–11 km/h) Wind felt on face. Leaves rustle. Wind vanes begin to move. Small wavelets. Crests have glassy appearance, not breaking.

3 Gentle breeze 8–12 mph (12–19 km/h) Leaves and small twigs constantly moving, leaves blown into air, light flags extended. Large wavelets. Scattered whitecaps.

4 Moderate breeze 13–17 mph (20–28 km/h) Fairly frequent whitecaps. Dust and loose paper raised. Small branches begin to move. Small waves with breaking crests.

5 Fresh breeze 18–24 mph (29–38 km/h) Branches of a moderate size move. Small trees in leaf begin to sway. Moderate waves of some length. Many whitecaps. Small amounts of spray. Crested wavelets form on inland lakes and large rivers.

6 Strong breeze 25–30 mph (39–49 km/h) Large branches in motion. Whistling heard in overhead wires. Umbrella use becomes difficult. Empty plastic bins tip over. Long waves begin to form. White foam crests are very frequent. Some airborne spray is present.

7 High wind, moderate gale, near gale 31–38 mph (50–61 km/h) Whole trees in motion. Effort needed to walk against the wind. Sea heaps up. Some foam from breaking waves is blown into streaks along wind direction. Moderate amounts of airborne spray.

8 Gale 39–46 mph (62–74 km/h) Some twigs broken from trees. Cars veer on road. Walking very difficult. Moderately high waves with breaking crests forming spindrift. Well-marked streaks of foam are blown along wind direction. Considerable airborne spray.

Biometrics (Use metric or standard)

height = _____   one step = _____

arm span = _____   10 meters/feet = _____ steps

fingers-elbow = _____   20 meters/feet = _____ steps

shoe length = _____   50 meters/feet = _____ steps

Degrees of Arc

1°   5°   10°   15°   25°

NOTES

1. Richard A. Menary, "Writing as Thinking," *Language Sciences* 29, no. 5 (2007): 621–32, doi:10.1016/j.langsci.2007.01.005; V. A. Howard, "Thinking on Paper: A Philosopher's Look at Writing," in *Varieties of Thinking: Essays from Harvard's Philosophy of Education Research Center*, ed. V. A. Howard, 84–92 (New York: Routledge, 1990).

2. Pam A. Mueller and David M. Oppenheimer, "The Pen Is Mightier Than the Keyboard: Advantages of Longhand over Laptop Note Taking," *Psychological Science*, 2014, 1–10, doi:10.1177/0956797614524581; Myra A. Fernandes, Jeffrey D. Wammes, and Melissa E. Meade, "The Surprisingly Powerful Influence of Drawing on Memory," *Current Directions in Psychological Science* 27, no. 5 (2018): 302–8.

3. Melissa E. Meade, Jeffrey D. Wammes, and Myra A. Fernandes, "Drawing as an Encoding Tool: Memorial Benefits in Younger and Older Adults," *Experimental Aging Research* 44, no. 5 (2018): 369–96, doi:10.1080/0361073X.2018.1521432.

4. Carol A. Carrier and Amy Titus, "The Effects of Notetaking: A Review of Studies," *Contemporary Educational Psychology* 4 (1979): 299–314; Verena Paepcke-Hjeltness and Teddy Lu, *Design for Visual Empowerment: Sketchnoting, Breaking the Rules* (2018), available at www.researchgate.net/publication/328346780_DESIGN_FOR_VISUAL_EMPOWERMENT_SKETCHNOTING_BREAKING_THE_RULES.

5. Misty Antonio, "Drawing to Support Writing Development in English Language Learners," *Language and Education* 27, no. 3 (2013): 261–77.

6. Zaretta Hammond, *Culturally Responsive Teaching and the Brain* (Thousand Oaks, CA. Corwin, 2015).

7. C. Twohig-Bennett and Andy Jones, "The Health Benefits of the Great Outdoors: A Systematic Review and Meta-Analysis of Greenspace Exposure and Health Outcomes," *Environmental Research* 166 (October 2018): 628–37, doi:10.1016/j.envres.2018.06.030; Gregory N. Bratman, J. Paul Hamilton, and Gretchen C. Daily, "The Impacts of Nature Experience on Human Cognitive Function and Mental Health," *Annals of the New York Academy of Sciences: The Year in Ecology and Conservation Biology* 1249 (2012): 118–36; Gregory N. Bratman, J. Paul Hamilton, Kevin S. Hahn, Gretchen C. Daily, and James J. Gross, "Nature Experience Reduces Rumination and Subgenual Prefrontal Cortex Activation," *Proceedings of the National Academy of Sciences of the United States of America* 112, no. 28 (July 14, 2015): 8567–72; first published June 29, 2015, doi:10.1073/pnas.1510459112.

8. Mathew P. White et al., "Spending at Least 120 Minutes a Week in Nature Is Associated with Good Health and Well-being," *Scientific Reports* 9, no. 7730 (2019), doi:10.1038/s41598-019-44097-3.

9. Pamela Pensini, Eva Horn, and Nerina J. Caltabiano, "An Exploration of the Relationships between Adults' Childhood and Current Nature Exposure and Their Mental Well-Being," *Children, Youth and Environments* 26, no. 1 (2016): 125–47; Stanley T. Asah, David N. Bengston, Lynne M. Westphal, and Catherine H. Gowan, "Mechanisms of Children's Exposure to Nature: Predicting Adulthood Environmental Citizenship and Commitment to Nature-Based Activities," *Environment and Behavior* 50, no. 7 (2018): 807–36.

10. P. K. Piff, Paul Dietze, Matthew Feinberg, Daniel M. Stancato, and Dacher Keltner, "Awe, the Small Self, and Prosocial Behavior," *Journal of Personality and Social Psychology* 108, no. 6 (June 2015): 883–99, doi:10.1037/pspi0000018.

11. Rachel Carson, *The Sense of Wonder* (New York: Harper & Row, 1965), 59.

12. As remembered by the poet Naomi Shihab Nye (personal communication).

13. Arne May, "Experience-Dependent Structural Plasticity in the Adult Human Brain," *Trends in Cognitive Sciences* 15, no. 10 (2011): 475–82, doi:10.1016/j.tics.2011.08.002.

14. Bogdan Draganskia and Arne May, "Training-Induced Structural Changes in the Adult Human Brain," *Behavioural Brain Research* 192, no. 1 (2008): 137–42, doi:10.1016/j.bbr.2008.02.015; Arne May, "Experience-Dependent Structural Plasticity"; Carol Dweck, *Mindset: The New Psychology of Success* (New York: Random House, 2006).

15. Eleanor A. Maguire, K. Woollett, and H. J. Spiers, "London Taxi Drivers and Bus Drivers: A Structural MRI and Neuropsychological Analysis," *Hippocampus* 16, no. 12 (2006):1091–101; E. A. Maguire, David. G. Gadian, Ingrid. S. Johnsrude, Catriona D. Good, John Ashburner, Richard S.J. Frackowiak, and Christopher D. Frith, "Navigation-Related Structural Change in the Hippocampi of Taxi Drivers," *Proceedings of the National Academy of Sciences of the United States of America* 97, no. 8 (2000): 4398–4403.

16. Carol Dweck, "What Is Mindset," Mindset, mindsetonline.com/whatisit/about/ (site discontinued).

17. Arne May, "Experience-Dependent Structural Plasticity"; Elizabeth Ligon Bjork and Robert A. Bjork, "Making Things Hard on Yourself, but in a Good Way: Creating Desirable Difficulties to Enhance Learning," in *Psychology and the Real World: Essays Illustrating Fundamental Contributions to Society*, ed. Morton A. Gernsbacher, R. W. Pew, L. M. Hough, and James R. Pomerantz, 56–64 (New York: Worth Publishers, 2011).

18. Harold Pashler, Mark McDaniel, Doug Rohrer, and Robert Bjork, "Learning Styles: Concepts and Evidence," *Psychological Science in the Public Interest* 9, no. 3 (2009): 105–19, doi:10.1111/j.1539-6053.2009.01038.

19. Lynn Waterhouse, "Inadequate Evidence for Multiple Intelligences, Mozart Effect, and Emotional Intelligence Theories," *Educational Psychologist* 41, no. 4 (2006): 247–55.

20. Stephen Porges, *The Polyvagal Theory: Neurophysiological Foundations of Emotions, Attachment, Communication, and Self-Regulation* (New York: Norton, 2011).

21. Catherine Broom, "Exploring the Relations between Childhood Experiences in Nature and Young Adults' Environmental Attitudes and Behaviours," *Australian Journal of Environmental Education* 33, no. 1 (2017), 34–47.

22. Sophie von Stumm, Behedikt Hell, and Tomas Chamorro-Prezumic, "The Hungry Mind: Intellectual Curiosity Is the Third Pillar of Academic Performance," *Perspectives on Psychological Science* 6, no. 6 (2011): 574–88; Adrian Raine, Chandra Reynolds, Peter H. Venables , and Sarnoff A. Mednick, "Stimulation Seeking and Intelligence: A Prospective Longitudinal Study," *Journal of Personality and Social Psychology* 82, no. 4 (2002), 663–74, doi: 10.1037/0022-3514.82.4.663.

23. Matthias J. Gruber, Bernard D. Gelman, and C. Ranganath, "States of Curiosity Modulate Hippocampus-Dependent Learning via the Dopaminergic Circuit," *Neuron* 84, no. 2 (2014): 486–96.

24. Gruber, Gelman, and Ranganath, "States of Curiosity."

25. J. Cummins, "Language Proficiency in Academic Contexts," *Bilingual Education and Bilingualism* 23 (2000).

26. Terry Tempest Williams, *Red: Passion and Patience in the Desert* (New York: Pantheon Books, 2001), 75.

27. Melissa E. Meade, Jeffrey D. Wammes, and Myra A. Fernandes, "Comparing the Influence of Doodling, Drawing, and Writing at Encoding on Memory," *Canadian Journal of Experimental Psychology/Revue canadienne de psychologie expérimentale* 73, no. 1 (2019): 28–36.

28. Jeffrey D. Wammes, Melissa E. Meade, and Myra A. Fernandes, "Creating a Recollection-Based Memory through Drawing," *Journal of Experimental Psychology: Learning, Memory, and Cognition* 44, no. 5 (2018): 734–51; Margaret Anne Defeyter, Riccardo Russo, and Pamela Louise McPartlin, "The Picture Superiority Effect in Recognition Memory: A Developmental Study Using the Response Signal Procedure," *Cognitive Development* 24, no. 3 (July–September 2009): 265–273; Jeffrey D. Wammes, Melissa E. Meade, and Myra A. Fernandes, "The Drawing Effect: Evidence for Reliable and Robust Memory Benefits in Free Recall," *Quarterly Journal of Experimental Psychology* 69, no. 9 (2016): 1752–76, doi:10.1080/17470218.2015.1094494; Meade, Wammes, and Fernandes, "Drawing as an Encoding Tool"; Fernandes, Wammes, and Meade, "Surprisingly Powerful Influence."

29. James M. Clark and Allan Paivio, "Dual Coding Theory and Education," *Educational Psychology Review* 3, no. 3 (September 1991): 149–210.

30. Diana I. Tamira, Emma M. Templeton, Adrian F. Ward, and Jamil Zakid, "Media Usage Diminishes Memory for Expe-riences," *Journal of Experimental Social Psychology* 76 (May 2018): 161–68; Julia S. Soares and Benjamin C. Storm, "Forget in a Flash: A Further Investigation of the Photo-Taking-Impairment Effect," *Journal of Applied Research in Memory and Cognition* 7, no. 1 (March 2018): 154–60.

31. Haley A. Vlach and Sharon M. Carver, "The Effects of Observation Coaching on Children's Graphic Representations," *Early Childhood Research and Practice* 10, no. 1 (2008): 1–26.

32. Jonathan Kingdon, *East African Mammals* Vol. 1 (Chicago, IL: The University of Chicago Press, 1974).

33. Karl Ericsson, Ralf T. Krampe, and Clemens Tesch-Romer, "The Role of Deliberate Practice in the Acquisition of Expert Performance," *Psychological Review* 100, no. 3 (July 1993): 363–406; Brooke N. Macnamara, David Z. Hambrick, and Frederick L. Oswald, "Deliberate Practice and Performance in Music, Games, Sports, Education, and Professions: A Meta-Analysis," *Psychological Science* 25, no. 8 (2014): 1608–18, doi:10.1177/0956797614535810.

34. Carlo Magno, "The Role of Metacognitive Skills in Developing Critical Thinking," *Metacognition and Learning* 5 (2010): 137–56.

35. Raffaella Borasi, "Capitalizing on Errors as 'Springboards for Inquiry': A Teaching Experiment," *Journal for Research in Mathematics Education* 25, no. 2 (1994): 166–208; Patrick J. Eggleton and Carla Moldavan, "The Value of Mistakes," *Mathematics Teaching in the Middle School* 7, no. 1 (2001): 42–47.

36. Bryan Goodwin and Kirsten Miller, "Good Feedback Is Targeted, Specific, Timely," *Educational Leadership* 70, no. 1 (2012): 82–93.

37. Claudia M. Mueller and Carol Dweck, "Praise for Intelligence Can Undermine Children's Motivation and Performance," *Journal of Personality and Social Psychology* 75, no. 1 (July 1998): 33–52; Shannon R. Zentall and Bradley Morris, "'Good Job, You're So Smart': The Effects of Inconsistency of Praise Type on Young Children's Motivation," *Journal of Experimental Child Psychology* 107, no. 2 (2010): 155–63.

38. Li Zhao, Gail D. Heyman, Lulu Chen, and Kang Lee, "Praising Young Children for Being Smart Promotes Cheating," *Psychological Science* 28, no. 12 (2017): 1868–70, doi:10.1177/0956797617721529.

39. Daniel Gilbert, *Stumbling on Happiness* (New York: Vintage Books, 2005), 8.

40. Charlie Toft, quoted in Julia Galef, "Surprise! The Most Important Skill in Science or Self-Improvement Is Noticing the Unexpected," *Slate*, January 2, 2015, slate.com/technology/2015/01/surprise-journal-notice-the-unexpected-to-fight-confirmation-bias-for-science-and-self-improvement.html.

ACKNOWLEDGEMENTS

THIS BOOK HAS EMERGED from nearly a decade of conversation and collaboration, and we have been deeply fortunate to have the support of many organizations, individuals, and communities along the way.

Early on in our work, Catherine Debs identified this as an education project she would like to support. She and John Debs organized a team of patrons—including themselves, Ned and Carol Spieker, and Rick and Donna Fluegel—who enabled us to commit to over a year of writing. Their funding also paid for the creation of the How to Teach Nature Journaling website and related videos, as well as our attendance at educational conferences and workshops. This targeted and timely patronage freed us up to create and share the resources you now see.

Emilie Lygren wishes to offer the deepest of appreciations to her family, friends, and mentors for their support throughout the course of this project. She also would like to thank the community of outdoor and environmental educators for their commitment to the field and their abundant joy.

Jack Laws wishes to acknowledge his parents, Robert and Beatrice Laws, who taught him patience, kindness, and wonder. Thanks and love to his brother, James Laws, explorer of the Sierra; his wife, Cybele Renault, loving teacher; and his children, Amelia and Carolyn, the adventure girls.

Together, we thank the whole Heyday team for understanding our message and for working to make it shine clearly. To our editor, Marthine Satris: we are so grateful for your thoughtful guidance and dedication. Your incisive edits and your feedback on the structure and tone of the book made it immeasurably better. We would like to thank our copy editor and proofreader, Michele Jones, for her thorough and scrupulous work. Diane Lee, thank you using your artistic and design sense to sharpen the look and feel of the book and make it easy to use. Thank you to Leigh McLellan for your work on the design and layout of the book: your precision and skill made it a professional and beautiful guide. We would like to send the utmost gratitude to Steve Wasserman, the executive director of Heyday, for his support and insight, and for facilitating a collaborative and inspiring experience for us as authors. We would also like to thank Emmerich Anklam, Emily Grossman, Ashley Ingram, Christopher Miya, Anna Pritt, Marlon Rigel, Gayle Wattawa, and founding director Malcolm Margolin for their dedication to our book and for making Heyday a fixture and a leader in the world of publishing.

We offer our sincerest gratitude and appreciation to the educators, scientists, thinkers, and community members who read the book at various stages of the writing process and who provided invaluable insights and feedback: Cherine Badawi, Kevin Beals, Rachel Economy, Darrow Feldstein, Kirsten Franklin, José González, Catherine Halversen, Art Middlekauff, Kevin Padian, and Sarah Rabkin.

We are also indebted to the writers, researchers, educators, thinkers, community members, and scientists whose work has influenced our thinking over the course of this project (and throughout our careers): Nicole Ardoin, Megan Bang, Jo Boaler, Adrienne Maree Brown, Rodger Bybee, Michael Canfield, Rachel Carson, George Washington Carver, Anne Caudle, Louise Chawla, Robin DiAngelo, Carol Dweck, Paulo Freire, Julia Galen, Beth Gillogly, Erick Greene, Zaretta Hammond, Hannah Hinchman, Cathy Johnson, Jenny Keller, Robin Wall Kimmerer, Alfie Kohn, Jay Lemke, Clare Walker Leslie, Richard Louv, Rue Mapp, Charlotte Mason, Kristin Meuser, Nalini Nadkarni, Todd Newberry, Ken Norris, Naomi Shihab Nye, Mary Oliver, Jonathan Osborne, Marley Peifer, Tim Pond, Kurt Rademacher, Allan Ridley, Carl Rogers, Fred Rogers, Bob Ross, Evert Schlinger, Arnold Shultz, Carroll Smith, Charles Henry Turner, Lev Vygotsky, Laurie Wigham, Terry Tempest Williams, and E. O. Wilson.

To the members of the Nature Journal Club, thank you for your commitment to finding curiosity, joy, and connection through the practice of journaling. You continue to be a source of new ideas, support, and inspiration.

To the Learning and Teaching Group at the Lawrence Hall of Science, and specifically the BEETLES Project, Craig Strang, Kevin Beals, Lynn Barakos, Jedda Foreman, Ramya Sankar, Xiomara Batin, José González, and Luana Rivera Palacio: thank you for being thought partners throughout this project. Your work has immensely informed our understanding of teaching and learning, and we are grateful for your exemplary leadership in the field of outdoor science and environmental education.

The San Francisco State Sierra Nevada Field Campus has been pivotal to our collaboration and to our individual careers as naturalists and educators. We would like to thank the directors, Jim Steele and J. R. Blair, for their continued efforts to champion dynamic and affordable field classes. We are also deeply appreciative of our fellow instructors, whom we are honored to consider peers, and of the students who visit the field campus: your enthusiasm and curiosity make it a true learning community.

Early in the writing process, we sought out example journal pages from experienced and exemplary scientists, writers, naturalists, and nature journalers. We wanted these journal pages to demonstrate the potential and value of this powerful learning tool, and we are thrilled and honored to feature their work. Thank you to Diana, Mark, and Hannah Berry for permission to use pages from William D. Berry's journals.

We also have the upmost gratitude for individuals and organizations who field-tested and provided feedback on our activities: Veronica, Lisa A, Lynn Akers, Richard Annan, Esther Arnusch, David Benterou, Terri Brenner, Kristine Brown, Jessica Burbank, Colby Burrow, Brenda Crosier, Katharine Dickanson, Dawn Duran, Kirsten Franklin, Heather Gabel, Betty Gatewood, Lloyd

Goldwasser, Sherry Green, Elizabeth Hlibichuk, Jack Howells, Elissa Ikeda, Laurel Johnson, Tessa Keath, Aneta Koehn, Jason Maas-Baldwin, LeKeshua Malone, Kasha Maslowski, Rachael McCaffree, Erin Miller, Bekki Page, Logan Rosenburg, Alli Taggart, Nancy Werner, Jean Yacono, and instructors at Exploring New Horizons Outdoor Schools, Great Smoky Mountains Institute at Tremont, and San Mateo Outdoor Education. Homeschooling parents and families throughout the Charlotte Mason network tested our lessons and advised us how to improve them. We are particularly grateful to the Charlotte Mason Institute for connecting us with so many families. Thank you to the Fireweed Academy and students in Akutan, Alaska for contributing their work.

Thank you to the students of Malcolm X Elementary School in Berkeley and to their garden teacher, Rivka Mason, for welcoming us to your school. We are also deeply appreciative of the parents, teachers, and educators who shared the sample journal pages and photographs of young people journaling that are found throughout the book—thank you for supporting our work. Specifically, we'd like to thank the following organizations for sharing photos and giving us permission to use them in our book: the BEETLES Project at the Lawrence Hall of Science, the Bird School Project, and Camp Tyler Outdoor School. We are also grateful to master photographer and naturalist Robert Hirsch for the use of his photographs in this book.

And, to the thousands of students who we have had the honor of meeting: thank you for your authenticity, humor, and curiosity. Learning with you has been one of the greatest joys of our lives.

ABOUT THE NATURALIST JOURNAL CONTRIBUTORS

William Berry

Avid Alaskan artist William D. Berry (Bill Berry, 1926–1979) gained recognition for his acute nature observation and detailed wildlife sketches. Some of his sketches and original work are published in the book *Alaskan Field Sketches*. He was also known for cartoon work and mural painting.

Paola Carrasco

Paola Carrasco studied biological sciences at the National University of Córdoba (UNC), Argentina. She is a professor at UNC and researcher at the National Council of Scientific and Technical Research in Argentina. She researches the systematics and evolution of snakes of the family Viperidae, with a focus on pit vipers from South America.

Laura Cunningham

Laura Cunningham is an artist-naturalist who paints landscapes and nature. Her thoroughly researched book *A State of Change: Forgotten Landscapes of California* shows hundreds of drawings, sketches, and oil paintings depicting the landscapes of California before European contact. She holds a graduate certificate in natural science, illustration, and science communication from UC Santa Cruz, and has painted illustrations for numerous museums, books, technical journals, and magazines.

Liz Cunningham

Liz Cunningham's mission is to be a voice for the life of the seas and the people who are working to save it, and to inspire and empower others to join the effort to save our seas and forge a more sustainable and just future. She is the author of the award-winning *Ocean Country: One Woman's Voyage from Peril to Hope in her Quest to Save the Seas*, with a foreword by Carl Safina. Her illustrations have been exhibited at a variety of venues including the Berkeley Art Center, Fort Mason Center, and the Oakland Museum. More information about her work is available at www.lizcunningham.net.

Chloé Fandel

Chloé is a Ph.D. student at the University of Arizona, working on modeling groundwater flow in karst aquifers. She also leads hands-on science field trips for middle and high school students at the University of Arizona Sky School. Before starting her Ph.D. work, she received a B.A. in geological sciences from Brown University, and worked as a field technician for consulting firms in France and for the US Forest Service. She enjoys scientific illustration and animation projects.

Fiona Gillogly

Fiona Gillogly, born in 2003, is a passionate birder, nature journaler, artist, musician, and actor who lives in the Sierra Nevada foothills in California. She loves to spend time in nature, looking for mysteries and asking lots of questions. For the past three summers, she has received scholarships from her local Audubon chapter to attend teen birding camps. To see more of Fiona's work, visit fionasongbird.com.

José González

José González is the founder of Latino Outdoors. He is an experienced educator—a K–12 public education teacher, environmental education advisor, outdoor education instructor and coordinator, and university adjunct faculty. He is also an illustrator and science communicator. His commentary on diversity and environmental/outdoor equity has been featured by *High Country News*, *Outside* magazine, *Bay Nature* magazine, *Earth Island Journal*, and *Latino USA*. He received his B.A. at UC Davis, and his M.S. at the University of Michigan School for Environment and Sustainability.

Ruth Heindel

Dr. Ruth Heindel is an environmental earth scientist who conducts field research on atmospheric dust deposition, nutrient cycling in soils, and the impact of winds on landscape evolution. Ruth first became fascinated by polar landscapes when she traveled to Churchill, Manitoba, as a high school student, and she has since traveled to and conducted research in Alaska, Svalbard, Greenland, Antarctica, and the Colorado Front Range. Wherever she travels, Ruth brings her notebook and colored pencils, and she loves getting to know a landscape through the focused observation that comes from sketching. Ruth grew up in Vermont, attended Brown University and Dartmouth College for her undergraduate and graduate degrees, and was a postdoctoral scholar at the University of Colorado Boulder. Ruth is currently the Dorothy & Thomas Jegla Assistant Professor of Environmental Studies at Kenyon College, where she teaches courses on earth systems, climate change, and landscape evolution.

Marcelo Jost

Marcelo Jost is a naturalist who sees nature as a never-ending source of wonders and challenges (depending on whether he understands what he is seeing or not). His nature journal is a place to graphically debate what he is seeing, to dialogue with nature, and to (he hopes!) produce wonder out of challenges.

Jonathan Kingdon

Dr. Jonathan Kingdon, who grew up in Tanzania, is a zoologist, science author, artist, and research associate with the University of Oxford. He is the author of over a dozen illustrated books and field guides depicting the diverse flora and fauna of Africa, including the acclaimed seven-volume series *East African Mammals*.

Eriko Kobayashi

Eriko Kobayashi was born in Mitaka city, in the suburbs of Tokyo. She started her career as a nature artist and illustrator in 1996. Recently, she has started nature journaling and including writing in her journals. She finds that drawing and writing together lead her to observe nature more and more deeply. In 2018, Eriko started the Japan Nature Journal Club and has gathered people of all ages to look at and learn from nature. She lives in Zama, Kanagawa.

Akshay Mahajan

Akshay Mahajan is a hardware engineer from Mumbai, India. He loves the experience of learning absolutely anything; today he finds this in nature journaling, swing dancing, Mandarin, social psychology, music, and, of course…engineering !

Nalini Nadkarni

Dr. Nalini Nadkarni is a professor of biology and a forest ecologist at the University of Utah. Her research is on the biota of rainforest canopies in Costa Rica and Washington State. She also engages the public in science, focusing on groups that do not or cannot access traditional science education, including faith-based groups, urban youth, and the incarcerated. Her work is supported by the National Science Foundation and the National Geographic Society.

Marley Alexander Peifer

Marley is a naturalist, educator, and artist. He longs for a re-integration of art with science and words with images, a synthesis that he develops in his journals. Nature journaling has been a fundamental practice for Marley ever since he discovered how it improved his observation and learning.

Sarah Rabkin

Sarah Rabkin (www.sarahrabkin.com) is the author and illustrator of *What I Learned at Bug Camp: Essays on Finding a Home in the World*. A former high school biology teacher and UC Santa Cruz faculty member in writing and environmental studies, she now works as a freelance editor and oral historian. Sarah leads retreats and workshops that foster awareness of the more-than-human world and the true nature within each of us.

Robert Stebbins

Robert Stebbins (1915–2013) was a researcher, herpetologist, and field guide author of many popular books, including the *Field Guide to Western Reptiles and Amphibians*. He was also a professor of zoology at UC Berkeley and a curator of herpetology at the Museum of Vertebrate Zoology.

Terry Tempest Williams

Terry Tempest Williams has been called a "citizen writer," a writer who speaks and speaks out eloquently on behalf of an ethical stance toward life. A naturalist and fierce advocate for freedom of speech, she has consistently shown us how environmental issues are social issues that ultimately become matters of justice. Williams, like her writing, cannot be categorized. She has testified before Congress on women's health issues, been a guest at the White House, has camped in the remote regions of Utah and Alaska wilderness and worked as "a barefoot artist" in Rwanda. She is the author of dozens of books, short stories, and articles, and her writing has appeared in the *New Yorker*, the *New York Times*, *Orion* magazine, and numerous anthologies worldwide.

ABOUT THE AUTHORS

JOHN MUIR (JACK) LAWS

John Muir (Jack) Laws is a scientist, educator, and author who helps people forge a deeper and more personal connection with nature through keeping illustrated nature journals and understanding science. Jack has kept nature journals since he was a child. Being dyslexic, Jack struggled in school. He found his place and delight in learning through spending time in nature and keeping notebooks of his observations, discoveries, and adventures. Trained as a wildlife biologist and scientific illustrator, he now observes the world with rigorous attention and awe. He looks for mysteries, plays with ideas, and seeks connections in all he sees. He has found that attention, observation, curiosity, and creative thinking are not gifts, but instead are skills that grow with training and deliberate practice. As an educator and author, Jack shares ways to make these skills a part of everyday life.

EMILIE LYGREN

Emilie Lygren is an educator, facilitator, and poet committed to building a more connected and just world. Throughout her career, Emilie has intentionally cultivated a broad range of teaching skills and expertise in curriculum development, science education, social emotional learning, discussion facilitation, learning theory, outdoor leadership, and culturally responsive teaching practices. Emilie has coauthored dozens of publications and curricula focused on outdoor science education with the BEETLES Project at the Lawrence Hall of Science, and has published poems in several literary journals. Connection is at the foundation of Emilie's work in science education, poetry, and outdoor learning. Through her writing and teaching, Emilie offers practices of authenticity, awareness, and curiosity, and she strives to bring people into deeper relationship with themselves, their communities, and the natural world.

AMY TAN

Amy Tan's novels are *The Joy Luck Club*, *The Kitchen God's Wife*, *The Hundred Secret Senses*, *The Bonesetter's Daughter*, *Saving Fish from Drowning*, and *The Valley of Amazement*. She is the author of two memoirs, *The Opposite of Fate* and *Where the Past Begins*, and two children's books, *The Moon Lady* and *Sagwa, The Chinese Siamese Cat*. Tan coproduced and cowrote the film adaptation of *The Joy Luck Club*; served as creative consultant on *Sagwa*, the PBS children's television series; and wrote the libretto for an opera based on her novel *The Bonesetter's Daughter*. Her passions include wild birds, drawing, and wildlife conservation.